# MOLECULAR AND SUPRAMOLECULAR PHOTOCHEMISTRY

*Series Editors*

**V. RAMAMURTHY**

*Professor*
*Department of Chemistry*
*Tulane University*
*New Orleans, Louisiana*

**KIRK S. SCHANZE**

*Professor*
*Department of Chemistry*
*University of Florida*
*Gainesville, Florida*

*ADDITIONAL VOLUMES IN PREPARATION*

# Organic Molecular Photochemistry

# Organic Molecular Photochemistry

edited by

## V. Ramamurthy
*Tulane University*
*New Orleans, Louisiana*

## Kirk S. Schanze
*University of Florida*
*Gainesville, Florida*

ISBN: 0-8247-6606-7

**Headquarters**
Marcel Dekker, Inc.
270 Madison Avenue, New York, NY 10016
tel: 212-696-9000; fax: 212-685-4540

**Eastern Hemisphere Distribution**
Marcel Dekker AG
Hutgasse 4, Postfach 812, CH-4001 Basel, Switzerland
tel: 41-61-261-8482; fax: 41-61-261-8896

**World Wide Web**
http://www.dekker.com

The publisher offers discounts on this book when ordered in bulk quantities. For more information, write to Special Sales/Professional Marketing at the headquarters address above.

# Preface

The third volume in the Molecular and Supramolecular Photochemistry series focuses on organic molecular photochemistry. This volume presents six chapters in the general area of photochemical and photophysical studies of organic molecules. Although the thrust of the individual chapters vary, each deals with the excited-state behavior of organic molecules in their molecular form.

Tremendous activity in photochemical studies over the past 40 years has resulted in a reasonable understanding of the photobehavior of a large number of organic molecules, in the discovery of innumerable reactions, and in establishing the basic mechanistic framework for a large number of photoreactions. Such a rapid development has resulted in photoscience becoming a tool in the hands of chemists who want to exploit light for useful ends. Because of this, there is a need to periodically consolidate and critically evaluate the information that becomes available through the tireless efforts of our colleagues. The Molecular and Supramolecular Photochemistry series aims to serve this role.

In Volume 3 of this series, *Organic Molecular Photochemistry*, eight active photochemists summarize and critically evaluate the literature in their area of expertise. Chapters 3 and 4 deal with a well-investigated and important reaction in photochemistry: geometric isomerization. Arai (Chap. 3) deals with the one-way cis-trans isomerization of aryl-substituted alkenes. Rao (Chap. 4) summarizes recent mechanistic developments in the area of geometric isomerization of olefins. These two chapters combined provide over 300 references that should be valuable to those seeking current knowledge of the excited-state geometric isomerization of olefins. It is unnecessary to remind the readers that one of the classic reviews on geometric isomerization, by Saltiel and coworkers, appeared in an earlier version of this series (Org. Photochem., 1973, 3, 1).

One of the important activities of chemists is to seek similarity among apparent dissimilarity. Fleming and Pincock have successfully done this in their chapter on photochemical cleavage reactions. Cleavage of the C—X bond prompted by light has been investigated by a number of workers (X being different in each case) and mechanisms for the cleavage process have been proposed. Further progress in the field depends on one's ability to connect all that is known on the photocleavage process. Fleming and Pincock have efficiently achieved this in Chapter 5. This chapter, along with one by Cristol and Bindel that appeared in the Organic Photochemistry series (1983, 6, 327), is indispensable to those interested in basic mechanistic studies of cleavage process, as well as phototriggers, photocleavages, photoaffinity labels, and photoacid generation.

Despite intense activity in discovering new reactions and establishing mechanisms of photoreactions, control of the chirality of photoproducts is still unknown. While strategies for achieving high (e.e., >95%) asymmetric induction in a number of thermal reactions have been developed, such is not the case for reactions initiated by light. Activity in this area can certainly be expected in the coming years. The chapter by Everitt and Inoue (Chap. 2) provides a thorough critical summary of the literature on asymmetric photoreactions in solution. This chapter, along with a previous review by Inoue (*Chem. Rev.*, 1992, 92, 741), should be a valuable information package in the hands of photochemists wishing knowledge of asymmetric photochemistry.

Although the field of photochemistry has reached its maturity, certain critical gaps in our understanding of the photobehavior of molecules in organized assemblies exist. Two chapters are devoted to presenting the current status of activities in this area. Ito (Chap. 1) provides a summary of the literature on solid-state photoreactions of two component crystals. The literature in this area is quite spread out and Ito has done a wonderful job of distilling it into a single chapter. Molecular photochemistry has been a valuable tool in characterizing the interior of organized assemblies in terms of the parameters that we use to understand isotropic solvents. Bhattacharyya (Chap. 6) provides a critical evaluation of the various probes and their photoproperties that can be used to understand the reaction cavities of organized assemblies such as zeolites, cyclodextrins, and micelles.

It is our hope that the chapters presented in this volume will serve not only as a valuable resource for experts and active workers, but also as supplementary reading material for graduate students. As editors we have enjoyed reading the work of the authors, who have done a wonderful job of presenting interesting and current material in a critical and consolidated manner. We hope that you will benefit from this book and support this series.

*V. Ramamurthy*
*Kirk S. Schanze*

# Contents

# Contributors

**Tatsuo Arai, Ph.D.**  Department of Chemistry, University of Tsukuba, Tsukuba, Ibaraki, Japan

**Kankan Bhattacharyya, Ph.D.**  Physical Chemistry Department, Indian Association for the Cultivation of Science, Calcutta, India

**Simon R. L. Everitt, Ph.D.**  Inoue Photochirogenesis Project, ERATO, Japan Science and Technology Corporation and Osaka University, Toyonaka, Japan

**Steven A. Fleming, Ph.D.**  Department of Chemistry and Biochemistry, Brigham Young University, Provo, Utah

**Yoshihisa Inoue, Ph.D.**  Inoue Photochirogenesis Project, ERATO, Japan Science and Technology Corporation and Osaka University, Toyonaka, Japan

**Yoshikatsu Ito, Ph.D.**  Department of Synthetic Chemistry and Biological Chemistry, Graduate School of Engineering, Kyoto University, Kyoto, Japan

**James A. Pincock, Ph.D.**  Department of Chemistry, Dalhousie University, Halifax, Nova Scotia, Canada

**V. Jayathirtha Rao, Ph.D.**  Organic Chemistry Division II, Indian Institute of Chemical Technology, Hyderabad, India

# 1

# Solid-State Organic Photochemistry of Mixed Molecular Crystals

**Yoshikatsu Ito**
Kyoto University, Kyoto, Japan

## I. INTRODUCTION

Organic photoreactions in the solid state are strictly controlled by the crystal structure, which in turn is the result of an interplay of a range of noncovalent, intermolecular interactions, e.g., van der Waals forces, hydrogen bonding, and donor–acceptor interactions. Thus, in principle, the solid-state photoreactivities and selectivities must be freely controllable by the high-level design of the crystal structure by exploiting these intermolecular forces. However, the balances between these forces are so subtle that one cannot predict, at present, the crystal structure from the molecular structure. In this connection, our efforts to devise various methods that can "forcibly" change chemo-, regio-, stereo-, and enantio-selectivities of particular solid-state photoreactions to desired directions through "crystalline supermolecule" formation is important [1].

This chapter surveys unimolecular and bimolecular photochemical reactions in mixed molecular crystals and solid mixtures. Various photoreactions occurring in mixed crystals (solid solution), hydrogen-bonded cocrystals, donor–acceptor crystals, crystalline organic salts, and solid mixtures are described. In contrast to one-component crystals [2], the organic photochemistry of such multi-

1

component crystals is still very young and hence has an abundant serendipity and possibility. In a recent review article [1], the author overviewed solid-state photoreactions in two-component crystals. It covered mixed crystals, hydrogen-bonded cocrystals, donor–acceptor crystals, inclusion crystals, asymmetric syntheses by using chiral hosts, crystalline organic salts, solid mixtures, solid-state sensitization and quenching, and solid-state asymmetric syntheses. Therefore, a substantial part of this chapter is based on this previous publication. However, photoreactions in inclusion crystals and of Lewis acid complexes and photophysical investigations of mixed crystals were not mentioned. For these subjects, the reader might be referred to Ref. 1. Finally, in order to save space, drawings of the X-ray crystal structure were not reproduced here. Those who want to know these details should consult the original papers.

## II.  WHAT IS THE MIXED MOLECULAR CRYSTAL?

The two-component crystal may be divided into three categories: 1.) mixed crystal (solid solution), 2.) crystalline molecular compounds, 3.) a simple mechanical mixture of component crystals. I propose the term "mixed molecular crystal" to represent both mixed crystal and crystalline molecular compounds and have used it as such in the title of this chapter.

Mixed crystal is a homogeneous crystal of two or more substances. Solid solution is a homogeneous solid mixture of substances. Both are virtually equal, although the latter can be amorphous. In general, mixed crystals are obtained from molecules of similar shapes and sizes. The forces between molecules are usually weak. Compositions can vary throughout a certain range and crystal structures are disordered.

When the intermolecular forces are relatively strong and directional, crystalline molecular compounds (crystalline molecular complexes) are formed. They have fixed stoichiometries and ordered structures. These two-component molecular crystals are also called *cocrystals* or maybe *adduct crystals*. Hydrogen-bonded cocrystals, donor–acceptor crystals (charge transfer crystals), and inclusion crystals (host–guest crystals) are examples of crystalline molecular complexes. Crystalline organic salt is a special case of hydrogen-bonded cocrystal or donor–acceptor crystal, i.e., proton (or electron) transfer from the acid (or donor) to base (or acceptor) occurred.

A solid mixture is any mixture of crystals of different substances, regardless of its solid-state structure. It may be a simple mechanical mixture of component crystals, a mixed crystal, a molecular complex, or an ionic salt. In this chapter, a simple mechanical mixture as well as all uncharacterized two-component crystals are grouped under solid mixtures.

Figure 1 describes three representative phase diagrams for a two-component system [3]. Diagram a has a eutectic point E and the region *l* is a homoge-

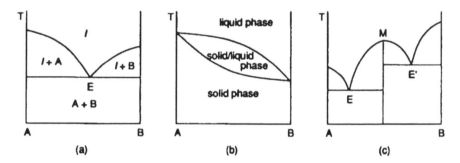

**Figure 1** Three typical temperature versus composition diagrams for a binary system (A and B): (a) diagram with a eutectic point E; (b) diagram for the case of unlimited solubility; (c) diagram showing formation of a molecular complex M. Diagrams a–c represent the simplest case, respectively, where a simple mixture of component crystals, a solid solution (mixed crystal), or a crystalline molecular complex is brought about.

neous liquid phase. Below the eutectic temperature, there is a two-phase region (A + B) of a solid, which consists of crystals of component A and crystals of component B. This two-phase region conforms with a simple mechanical mixture of component crystals mentioned above.

When molecules of the two components have unlimited solubility both in the liquid and in the solid state, diagram b is obtained. The solid phase corresponds to a solid solution or a mixed crystal. Diagram c represents a case whereby a crystalline molecular complex M of the composition $A_m B_n$ ($m = n = 1$) is formed and, besides, the complex M does not form solid solutions with its components A and B. This diagram is the sum of two diagrams similar to example a.

## III. PREPARATION AND CHARACTERIZATION OF TWO-COMPONENT CRYSTALS

Two-component crystals are usually prepared either 1.) by traditional cocrystallization or slow evaporation of solvent from a solution containing the components, 2.) by grinding the components together in a mortar and pestle [4] or in a ball grinder [5], or 3.) by melting together the components and subsequent solidification of the melt by cooling [6]. Recently, the second method has commonly been used. It has now been proved that molecular complexes prepared either by grinding (method 2) or by cocrystallization (method 1) exhibit not only similar chemical and photochemical reactivities but also similar spectroscopic and physical properties [4,5,7–12].

Solid materials thus prepared may exist as a mixed crystal (= a solid solution), a molecular compound (e.g., a hydrogen-bonded complex, a donor–acceptor complex, and an inclusion complex), an ionic salt, or a simple mechanical mixture of homocrystals of each component. Structural information of these solid samples is obtained by various physical means such as single-crystal X-ray diffraction, powder X-ray diffraction (PXD), solid-state $^{13}C$ nuclear magnetic resonance (NMR), IR and UV, electron microscopy, thermal analysis [differential scanning calorimetry (DSC) and thermogravimetric analysis], phase diagram determination, etc.

For example, each batch of crystals from cocrystallization can be examined under a polarizing microscope to determine whether the crystals have homogeneous morphologies or whether mixtures of crystals are present. Solution NMR results and elemental analyses can give the stoichiometry of components. Melting points of complexes are usually sharper than those of mixtures of components. A clearer distinction can be made from measurements of PXD, solid-state $^{13}C$ NMR, IR and UV, and DSC spectra. Because the spectral patterns for a simple mechanical mixture are composites of those for each component, observation of new peaks in these spectra indicates the formation of a molecular complex. When a molecular complex is available as a single crystal of sufficient size and quality, its crystal structure and molecular conformation can be unequivocally determined by single-crystal X-ray diffraction [4–11,13].

The PXD pattern of a solid solution is very similar to that of the host component. Differentiation of a solid solution from a simple mechanical mixture and a molecular complex may be best made on the basis of phase diagrams (Fig. 1), which are constructed from DSC measurements [3]. Probably, however, interpretations of these data are not always straightforward [14–16]. For weak complexes, X-ray quality single crystals are not available in many cases. Hence, the characterization of certain two-component crystals may potentially lead to dispute. Furthermore, a few institutions for organic chemistry are well equipped with the facilities mentioned above.

Irradiations of solid samples are carried out in various manners, depending on the researcher [1]. For example, in the author's group, solid samples are usually ground into powders and are irradiated externally in a specially designed apparatus (Fig. 2) [1].

## IV. DESIGN OF CRYSTALLINE MOLECULAR COMPOUNDS

Hydrogen bonding is the most important directional interaction responsible for supramolecular construction [17]. Appendix 1 illustrates several typical hydrogen bond patterns which are present in two-component molecular crystals. Appendix 2 exemplifies a variety of reported cocrystals selected mainly from the recent papers [18]. Not only strong hydrogen bonds (i.e., O—H···O, O—H···N,

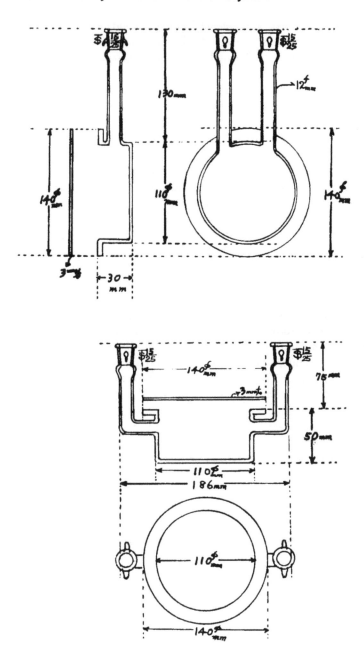

**Figure 2** Two types of vessels for solid-state photolysis.

N—H···O, and N—H···N) but also weaker hydrogen bonds and other weak forces [i.e., C—H···O, C—H···N, O—H···$\pi$, N—H···$\pi$, and C—H···$\pi$ interactions], halogen atom interactions (Cl,Br,I)···(Cl,Br,I), (Cl,Br,I)···(N,O,S), and C—H···Cl; chalcogen atom interactions (S,Se)···(S,Se), S···N, S···Cl, and S—H···S) are utilized for cocrystal formation.

Appendix 1 also describes aryl···aryl and hydrophobic interactions. Aliphatic side chains longer than around five carbon atoms may lead to preferential close packing of the side chains. Aromatic rings may associate via either a herringbone (edge-on) mode or a stacking mode. The herringbone geometry may be considered to be one manifestation of the C—H···$\pi$ interaction. In the stacking mode, either face-to-face stacking or offset stacking is possible. A favorable geometry for the aryl···aryl interaction is determined by the electrostatic interaction, i.e., by the balance between $\pi$-$\sigma$ attractions and $\pi$-$\pi$ repulsions, although the magnitude of the overall interaction energy is controlled by the van der Waals interaction and probably by the charge-transfer interaction [19]. Planar $\pi$ donors and $\pi$ acceptors tend to form offset overlapped structures, e.g., cocrystals a–e in Appendix 2. Note, however, that the interaction between benzene and hexafluorobenzene in cocrystal a is electrostatic in nature, not a charge transfer type. Solid-state complex f is an example of O···C donor–acceptor interactions.

According to Kitaigorodskii's close-packing principle, molecules will pack in a manner that minimizes void space or, in other words, in a way to maximize van der Waals interactions [20]. Hence, effects of both molecular shape and size are important in crystal engineering. For example, since racemic crystals tend to pack into centrosymmetric space groups, this statistical preference was utilized to prepare cocrystal g in Appendix 2, which is a rare molecular compound called a quasiracemate. The formation of cocrystal g was made possible due to the isosteric nature of the isopropenyl and dimethylamino substituents, indicating that shapes rather than dipoles can be in fact the dominant factor (cf. Ref. 31).

Recently, many studies on the design of metal-containing crystalline supramolecular assemblies have been carried out [17f,17l,21]. A variety of coordination numbers and geometries associated with transition metals as well as various metal–ligand bonding reactions, including M—H···O, O—H···M, C—H···O≡C—M, and other hydrogen bond–like interactions, should be useful for construction of new functional materials.

As described below, investigations on the design and synthesis of photoreactive mixed molecular crystals are still very limited in number and scope. However, by very ingenious manipulation of the above-mentioned intermolecular interactions, construction of new crystals bearing desired reactivities should be possible for many types of bimolecular photoreaction. More investigations in this direction should be carried out.

## V. MIXED CRYSTALS

Mixed crystals are useful for controlling reactivities and competitive reactions [6,13–16,22–27,30–32,36a], for inducing changes in crystal symmetry [28,29], for sensitization, quenching, and photophysical studies [33,37], for the oriented incorporation of guests into crystalline hosts [34], and for others [35].

[2+2] Photocycloadditions between different two *trans*-cinnamic acids, different two *trans*-cinnamamides, or different two *trans*-stilbene derivatives in the mixed crystalline state were first reported as early as 1972 by the pioneers of this field [6,22,23]. Mixed crystals of two monomers A and B were prepared by cooling the melts and were irradiated to produce [2+2] homodimers ($A_2$ and $B_2$) and [2+2] heterodimer (AB). Their relative yields were interpreted to indicate that the mixed crystals are nearly random substitutional solid solutions and that there is no transfer of excitation energy between A and B molecules. For example, 1.) the substantial yields of the heterodimers AB imply that the two components A and B do not self-aggregate separately and 2.) the suppression of, say, the $B_2$ formation under selective excitation of, say, A by using filtered light suggest that there is no transfer of excitation energy from A to B. The structures of the dimers were either syn head-to-tail ($\alpha$-truxillic dimer) or syn head-to-head ($\beta$-truxinic dimer), depending on the crystal form ($\alpha$ form or $\beta$ form, respectively) of the component monomer crystal, i.e., 1.) a mixed crystal prepared from a $\gamma$-form acid and an $\alpha$-form (or $\beta$-form) acid gave two $\alpha$-truxillic (or $\beta$-truxinic) homodimers and one $\alpha$-truxillic (or $\beta$-truxinic) heterodimer (e.g., Scheme 1); 2.) a mixed crystal of two $\alpha$-form (or $\beta$-form) acids gave two $\alpha$-truxillic (or $\beta$-truxinic) homodimers and one $\alpha$-truxillic (or $\beta$-truxinic) heterodimer; and 3.) a mixed crystal of an $\alpha$-form acid with a $\beta$-form acid gave two $\alpha$-truxillic homodimers, two $\beta$-truxinic homodimers, one $\alpha$-truxillic heterodimer, and one $\beta$-truxinic heterodimer (Table 1). It is clear from finding 1 that the photostable $\gamma$ form was made to adopt a photoreactive crystal structure ($\alpha$ or $\beta$ form) as a result of mixing.

Similarly, a heterodimer was formed as a major product upon irradiation of mixed crystals which had been prepared by cocrystallization of Cl- and Me-substituted 2-benzyl-5-benzylidenecyclopentanones (Scheme 2) [24]. As a consequence of cocrystallization, the photoinert chloro compound (X = Br, Y = Cl) has been induced to adopt both the crystal structure and molecular conformation of the photoreactive methyl derivative (X = Br, Y = Me).

Isomorphous $P2_12_12_1$ crystals of 1-(2,6-dichlorophenyl)-4-phenyl-E,E-1,3-butadiene and its 4-thienyl analog formed mixed crystals (substitutional solid solutions) upon cooling from the melt or upon cocrystallization from ethanol. A large single mixed crystal containing 25% of the latter in the former compound was powdered and irradiated (mainly the thienyldiene was selectively excited by using an appropriate cutoff filter) to give a chiral mixed dimer in a 70% enanti-

**Scheme 1**

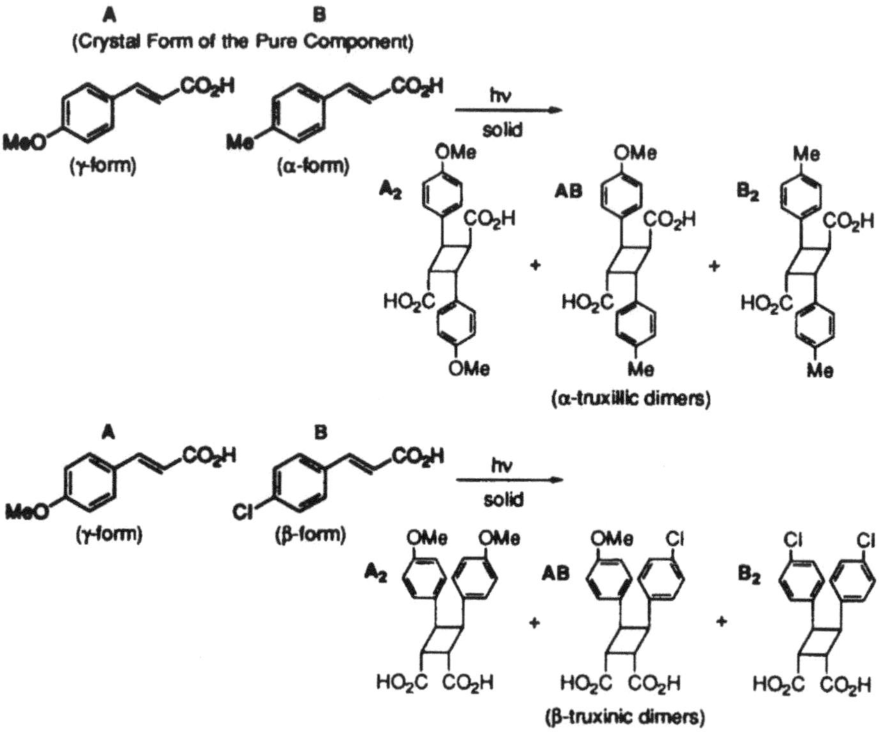

omer excess. This indicates the existence of a preferred direction of dimerization along the short unit-cell axis [25,26] (Scheme 3).

Several 1,4-disubstituted phenylenediacrylates such as **1–3, 4a, 4b, 5a, 5b**, racemic **5, 6a, 6b**, racemic **6**, and the 1:1 mixture of **2** and **9** crystallized into chiral structures and, except for the enantiomeric **4a** or **4b**, they photodimerized into either (SSSS)-cyclobutanes or (RRRR)-cyclobutanes with medium to quantitative enantiomeric (diastereomeric) yields (Scheme 4, Table 2) [27]. In addition to these dimers, the corresponding trimers and oligomers were also produced with high enantiomeric yields. Other modes of photodimerization reaction are schematized in Scheme 5.

The chiral homocrystals **1** and chiral mixed crystals 1:1 **2/9** were designed on the basis of the chiral crystal structure of **5** by using the so-called principle of isomorphous replacement (Scheme 6: Eqs. (3) and (1), respectively). The same approach was applied to the monomers **7** and **10** (Scheme 6: Eqs. (4) and (2), respectively), but they failed to crystallize into a chiral space group (Table 2).

**Table 1**  Other Mixed Crystals That Yielded Homodimers $A_2$ and $B_2$ and Heterodimer AB

| A | Ar~~COOH | | | | |
|---|---|---|---|---|---|
| Ar = | Ph (α) | Ph (α) | p-MePh (α) | o-MePh (γ) | 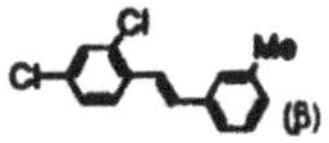 (β) |

| B | Ar~~COOH | | | | |
|---|---|---|---|---|---|
| Ar' = | p-MePh (α) | p-ClPh (β) | p-ClPh (β) | o-ClPh (β) | (β) |

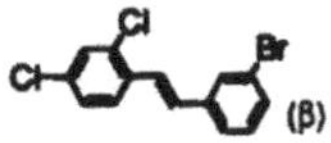

| Structure of A₂, B₂ and AB | α-struct | both α-struct and β-struct | both α-struct and β-struct | β-struct | β-struct |
|---|---|---|---|---|---|

| A | Ar~~CONH₂ (α form in all cases) | | | | | | | | |
|---|---|---|---|---|---|---|---|---|---|
| Ar = | Ph | Ph | Ph | p-MePh | p-MePh | p-MePh | p-ClPh | p-ClPh | Ph p-ClPh |

| B | Ar'~~CONH₂ (α form in all cases) | | | | | | | | |
|---|---|---|---|---|---|---|---|---|---|
| Ar' = | p-ClPh | p-MePh | 2-thienyl | p-ClPh | p-MeOPh | 2-thienyl | p-MeOPh | 2-thienyl | Me Me |

| Structure of A₂, B₂ and AB | α-structure in all cases | | | | | | | (No B₂ formed) |
|---|---|---|---|---|---|---|---|---|

Photostable Solid Mixtures:   A = 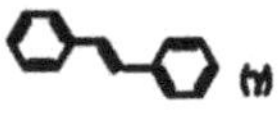 (γ)   B = 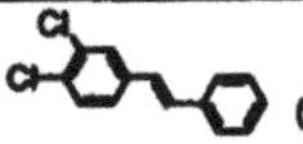 (γ)

**Scheme 2**

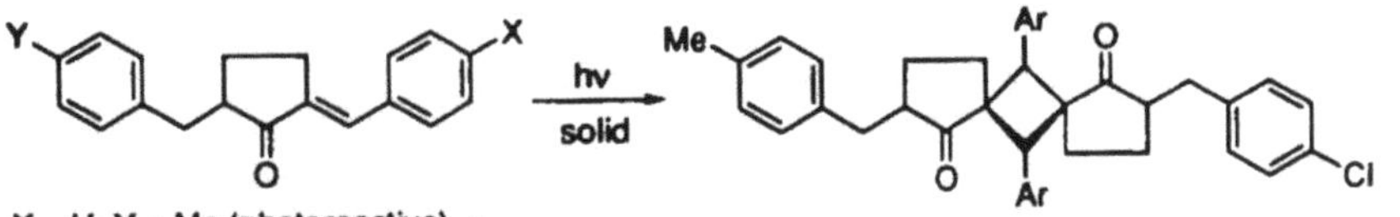

X = H, Y = Me (photoreactive)  ⎫
X = H, Y = Cl (photoreactive)  ⎬ mixed crystal
X = Br, Y = Me (photoreactive) ⎫
X = Br, Y = Cl (photoinert)    ⎬ mixed crystal

heterodimer

+  two homodimers

**Scheme 3**

total conversion to dimers = 29 %

9        :        1

70 % e.e.

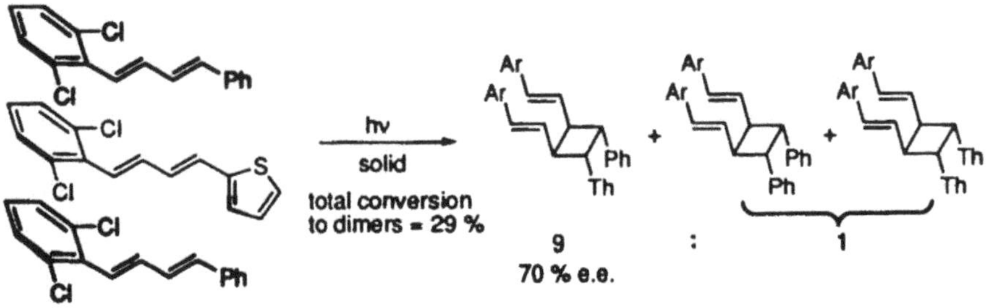

**Scheme 4**

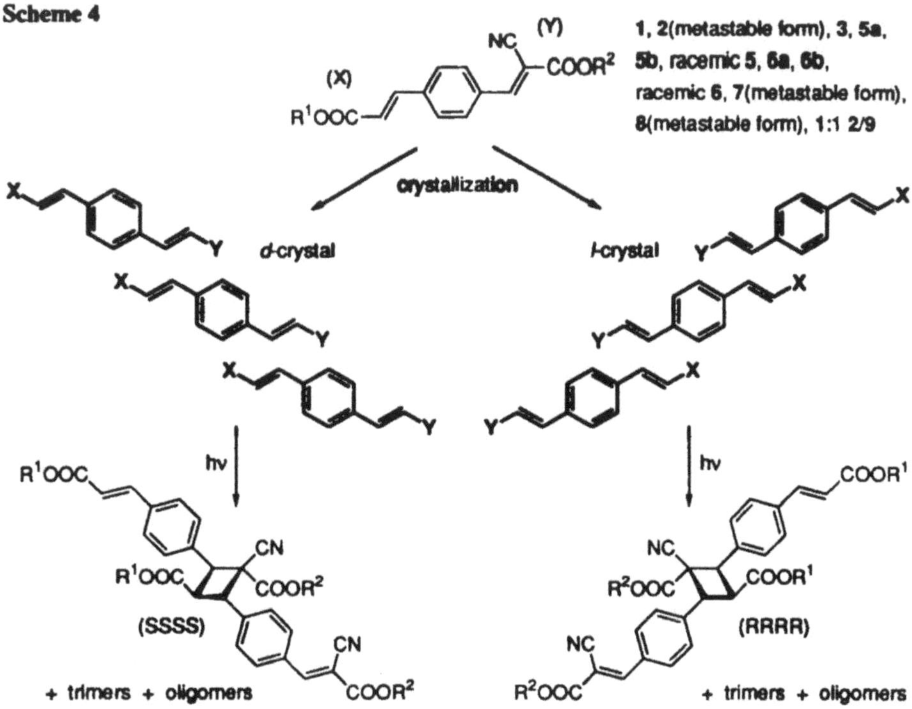

Because of the effect of the chiral *sec*-butyl group, **5a** or **5b** always yields one enantiomorphic chiral crystal. On the other hand, since each of **1–3**, racemic **5** or **6**, and 1:1 **2/9** gives a mixture of chiral crystals of opposite chiralities, the optical rotation of their photoproducts is unpredictable in magnitude and direction. However, in the presence of small amounts of the resolved photoproduct (dimer, trimer, or oligomer), the homochiral crystals whose chirality is opposite to that of the added photoproduct were obtained (chirality reversal), e.g., the (RRRR)-cyclobutane rings induced the precipitation of *d* crystals. Thus, absolute asymmetric syntheses were successfully achieved as summarized in Table 2. A chiral dimer (trimer, oligomer) replaces two (three, several) monomers on the surface of a growing monomer crystal with the same absolute configuration and thus inhibits the growth of monomer crystals of the same chirality.

Centrosymmetric crystals can be transformed into chiral or polar mixed crystals and this enables us to obtain novel means for absolute asymmetric synthesis. The principle is based on selective introduction of a guest molecule into a centrosymmetric host structure, thus reducing the symmetry of the mixed crystal. Crystallization of (E)-cinnamamide (space group P2₁/c) in the presence of (E)-

**Table 2**   Solid-State Photolyses of 1,4-Disubstituted Phenylenediacrylates[a]

| Monomer | $R^1$ | $R^2$ | Space group | Enantiomeric yield (diastereomeric yield) of photodimers, % |
|---|---|---|---|---|
| 1 | 3-Pen | Me | $P2_1$ | ~100 |
| 2 | 3-Pen | Et | Stable form, $P\bar{1}$ | — |
|   |       |    | Metastable form, $P1$ | 65 |
| 3 | 3-Pen | Pr | $P1$ | 80 |
| 4a or 4b | (R)-(−)- or (S)-(+)-s-Bu | Me | α form, $P2_1$ | Photostable |
|   |   |   | β form, $P2_1$ (pseudo-$P2_1$/a) | 0 |
| Racemic 4 | (R,S)-s-Bu | Me | $P2_1$/a | 0 |
| 5a or 5b | (R)-(−)- or (S)-(+)-s-Bu | Et | $P1$ | 100 |
| Racemic 5 | (R,S)-s-Bu | Et | $P1$ | 45 |
| 6a or 6b | (R)-(−)- or (S)-(+)-s-Bu | Pr | α form, $P2_1$ | 100 |
|   |   |   | β form, $P1$ | 100 |
| Racemic 6 | (R,S)-s-Bu | Pr | $P1$ | 50 |
| 7 | t-Bu | Et | Stable form, $P\bar{1}$ | Not determined |
|   |   |   | Metastable form | 0 |
| 8 | Pr | Et | Stable form, $P\bar{1}$ | Not determined |
|   |   |   | Metastable form | 0 |
| 9 | i-Pr | Et | $P\bar{1}$ | 0 |
| 10 | i-Pr | Pr | $P\bar{1}$ | 0 |
| 1:1 2/9 | 3-Pen/i-Pr 1:1 | Et | $P1$ | 95 |

[a] Powdered crystals were irradiated through Pyrex at low temperatures ($<5°C$). Conversions were ~90%, producing dimers, trimers, and oligomers. Dimer yields could be increased up to 60% by cutoff of the shorter wavelength light ($<350$ nm).

**Scheme 5**

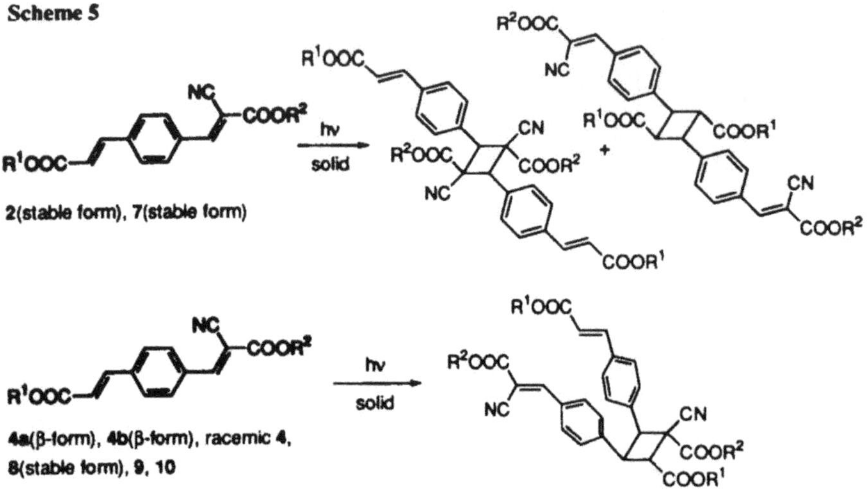

2(stable form), 7(stable form)

4a(β-form), 4b(β-form), racemic 4,
8(stable form), 9, 10

**Scheme 6**

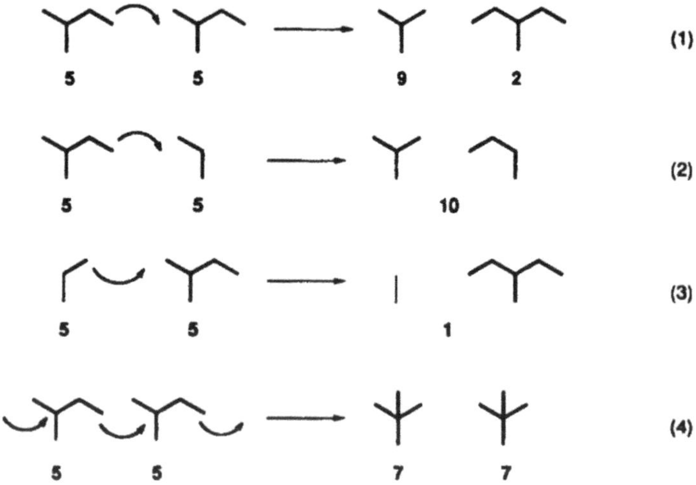

cinnamic acid results in a mixed crystal composed of two enantiomorphous halves (space group P2₁) each containing 0.5–1% acid. A O(hydroxyl)···O(amide) lone pair repulsion between guest and host at the chiral surfaces of the growing crystal is responsible for the stereoselective occlusion of the guest. Irradiation of each half separately yielded the optically active cyclobutane heterodimer with an enantiomeric yield in the range of 40–60% (Scheme 7) [28].

Similarly, the symmetry of the mixed crystal system composed of (E)-cinnamamide (host) and (E)-2-thienylacrylamide (guest, 8% occluded) is lower than that of the host crystal (P2₁/c). The selective occlusion of the guest arises from repulsive sulfur···π interactions. The +b end and the −b end of the mixed crystal are enantiomorphic (space group P1) and underwent topochemical [2+2] photodimerization in 40–69% enantiomeric excess (Scheme 8) [29].

Dichloro-substituted, methylenedioxy-substituted, or sulfur-containing planar aromatic molecules often crystallize in β structures. These effects were utilized to design [2+2] photoreactive molecular complexes [30]. Cocrystallization from EtOH of 6-chloro-3,4-methylenedioxycinnamic acid with 2,4-dichlorocinnamic acid and with 3,4-dichlorocinnamic acid (all three cinnamic acids are β-type [2+2] photoreactive crystals) gave the 2:1 and 1:1 complexes, respectively. Irradiation of these complexes yielded mixtures of three β-truxinic dimers in a statistical ratio (Scheme 9). This result, together with the crystal structures of the component acids, indicates that each sheet that is stabilized by an interstack

Scheme 7

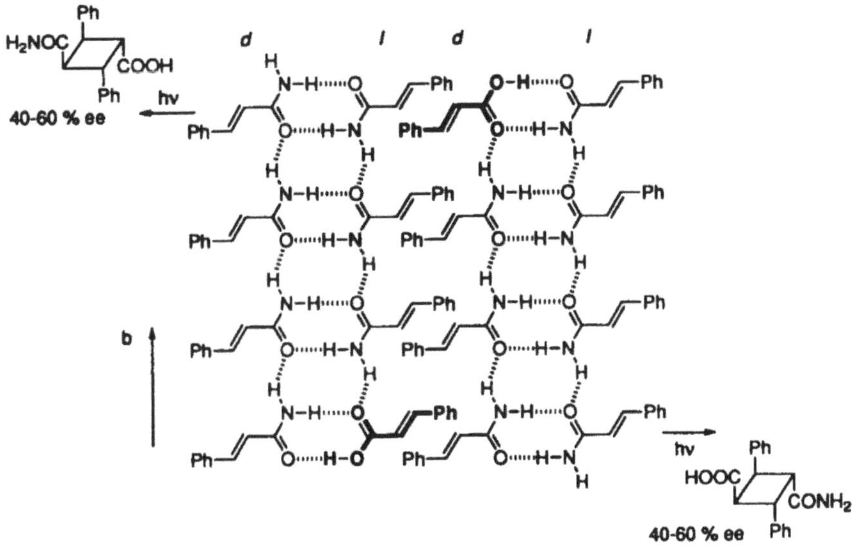

**Scheme 8**

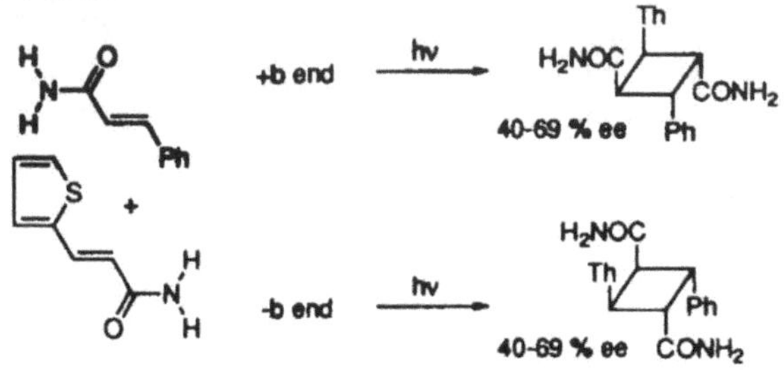

**Scheme 9**

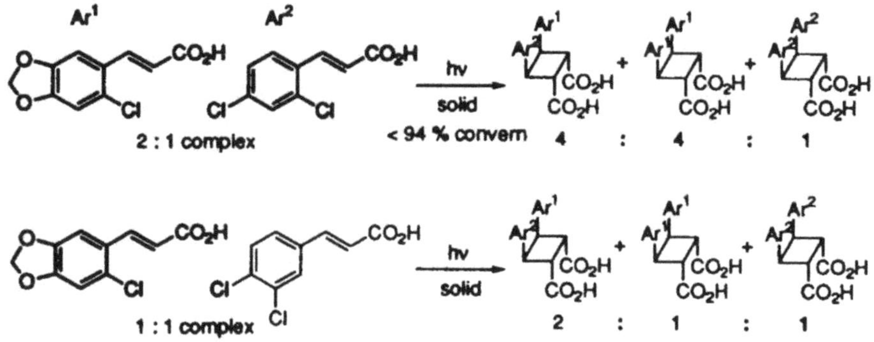

C—H···O, O—H···O, Cl···Cl, C—H···Cl, and O—H···Cl interaction is ordered with 2:1 or 1:1 stoichiometry, but the stacking along a 4-Å axis is random.

Two compounds of similar chemical structures and isomorphous crystal structures sometimes tend to form a molecular complex. Cocrystallization of 3-cyanocinnamic acid (photostable at room temperature, but β-type [2+2] photoreactive at 130 °C) and 4-cyanocinnamic acid (β-type [2+2] photoreactive) yielded crystals of a 1:1 complex, not a solid solution. This complex, however, was photostable [31]:

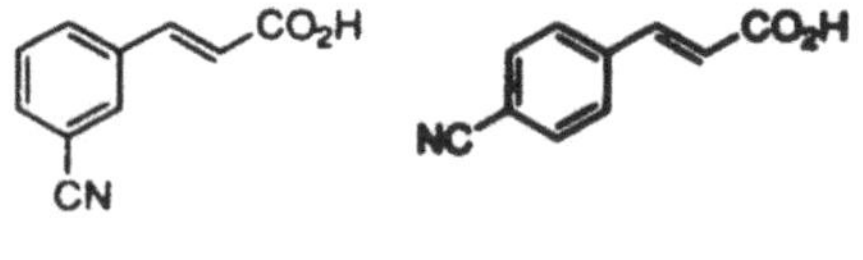

Cocrystallization or grinding together of 2,5-distyrylpyrazine (**11**) and ethyl 4-[2-(2-pyrazinyl)ethenyl]cinnamate (**12a**) forms a 1:2 molecular complex with

**Scheme 10**

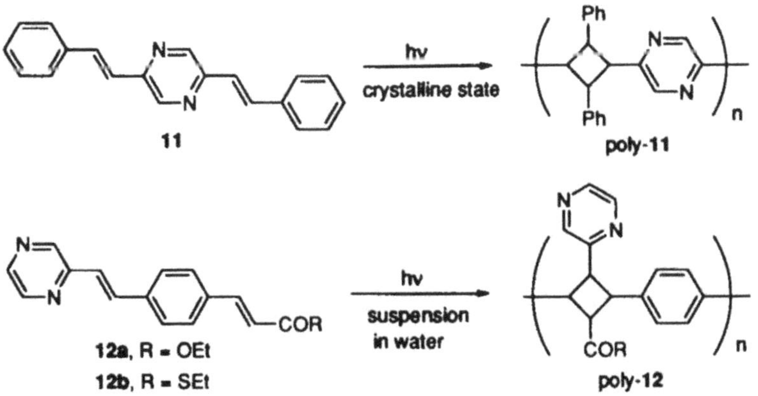

each component **11** and **12a** in a separate column. Molecules **11** and **12a** are interacting through a weak (Ph)C—H···O(carbonyl) hydrogen bonding. UV irradiation of this cocrystal afforded a perfectly ordered (alternately layered) crystalline polymer composite of poly-**11** and poly-**12a**. A copolymer was not produced [13].

On the other hand, cocrystallization or grinding together of **12a** and *S*-ethyl 4-[2-(2-pyrazinyl)ethenyl]thiocinnamate (**12b**) forms a 45:55 molar mixed crystal (a solid solution), which upon UV irradiation gave a crystalline random copolymer (Scheme 10) [14].

Ethyl and propyl α-cyano-4-[2-(4-pyridyl)ethenyl]cinnamates (**13a** and **13b**) also form a mixed crystal of various compositions. Irradiation of mixed crystals having 70:30 to 5:95 **13a/13b** afforded unsymmetric monocyclic dimers **14** and tricyclic [2.2]paracyclophane-type dimers **15** quantitatively, whereas each homocrystal gave the corresponding symmetric dimer **16** (Scheme 11) [15,16].

9-Cyanoanthracene has a β-type crystal structure but produces a head-to-tail photodimer. Unlike cinnamic acids and their derivatives, the stereochemistry of the photoproduct dimer from anthracenes cannot often be predicted on the basis of their crystal packing [78]. 9-Methoxyanthracene has a photostable γ-type crystal structure. 9-CNA host crystals doped with 9-MeOA form a solid solution, which upon irradiation yield a heterodimer (Φ = 0.12). A reaction mechanism via an exciplex is proposed. The exciplex fluorescence was observed [32]:

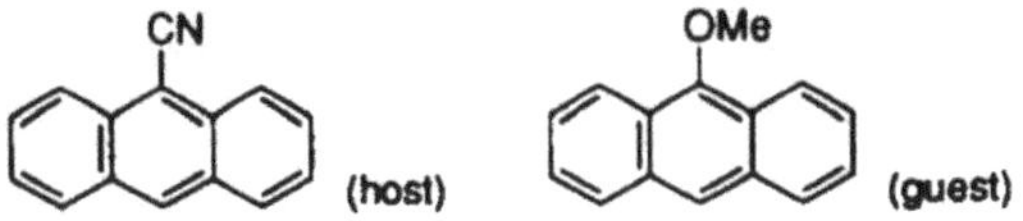

**Scheme 11**

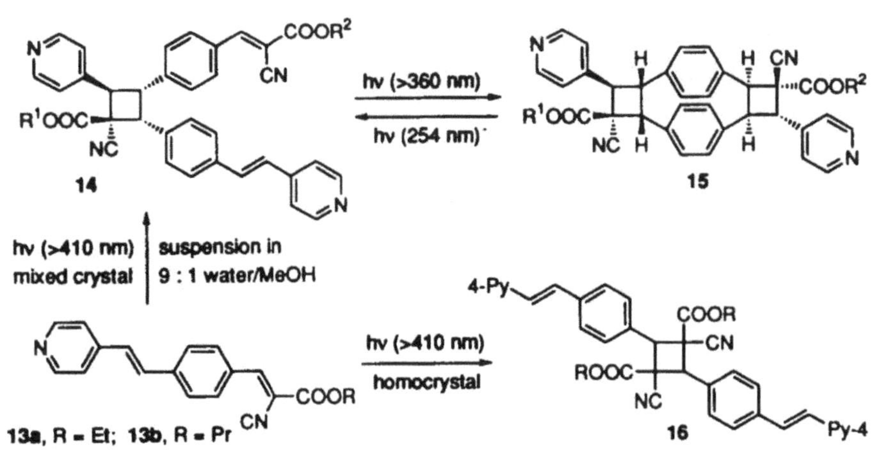

Photolysis of the solid mixture of dibenzobarrelene with various triplet sensitizers was carried out (Scheme 12) [33]. Unlike the direct irradiation whereby two products were formed, dibenzosemibullvalene, the triplet photoproduct, was exclusively obtained. It was estimated on the basis of maximum conversions that one xanthone molecule can photosensitize the formation of up to 24 dibenzosemi-

**Scheme 12**

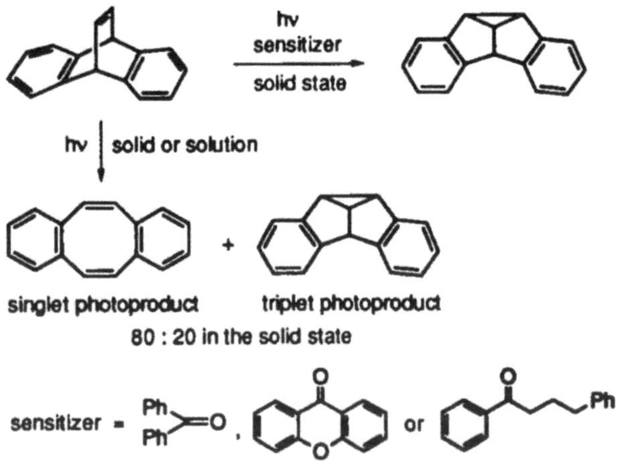

Scheme 13

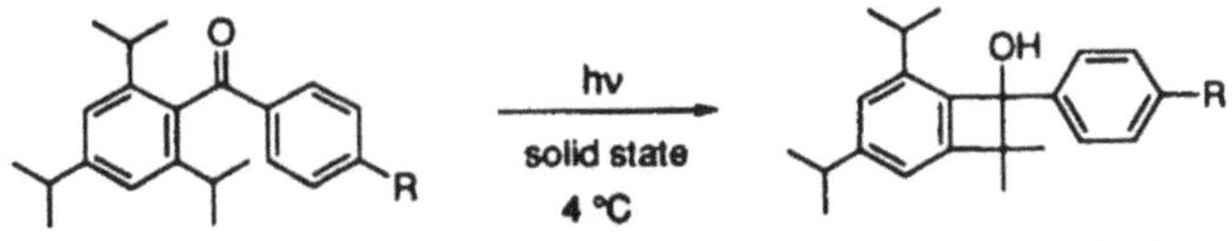

| R | conversion | yield |
|---|---|---|
| p-CO$_2$Me | no reaction | – |
| p-CO$_2$Et | 76 % | |
| p-CO$_2$Me + p-CO$_2$Et (1:1, ground together) | 30 % (R = p-CO$_2$Me) + 78 % (R = p-CO$_2$Et) | ~100 % in all cases |
| p-CO$_2$Me (irradiated at 84 °C) | 54 % (apparent E$_a$ ~ 20 kcal/mol) | |
| p-CO$_2$Me (thoroughly pulverized) | 4 % | |

bullvalene molecules, i.e., the conversion is 24% when 1% (mol/mol) xanthone in dibenzobarrelene was employed. The X-ray crystal structure of dibenzobarrelene indicates that there are 24 neighbors within a distance of 12 Å from a reference molecule. These findings are reasonable, since an upper limit for triplet–triplet energy transfer to occur in rigid media is 10–15 Å.

We have applied the method of mixed crystal formation together with a temperature effect and a thorough pulverizing effect in order to prove a narrow reaction cavity for the solid-state photocyclization of 2,4,6-triisopropyl-4′-me-thoxycarbonylbenzophenone (Scheme 13) [36a]. The crystals of this ketone were photostable, unless the crystal lattice was disturbed either by solid–solid mixing with 2,4,6-triisopropyl-4′-ethoxycarbonylbenzophenone, by raising the irradiation temperature, or by complete powdering.

Photophysical investigations of reaction in doped crystals [37] are not reviewed here. Their brief summaries are found in Ref. 1.

## VI. HYDROGEN-BONDED COCRYSTALS

We have found that indole and trans-cinnamic acid form a weak crystalline molecular complex [38]. This finding triggered extensive investigations of photoreactive hydrogen-bonded cocrystals by Ryukoku workers [39–41].

The 1:1 hydrogen-bonded cocrystals between aza aromatics NArH and aralkyl carboxylic acids R-CO$_2$H (17·a, 17·b, 17·bD, 17·c, 17·d, 18·a, 18·b,

18·c, and **18·e**) were irradiated. Except for the photoinert **18·e**, these cocrystals underwent decarboxylation to yield RH with high selectivity, as exemplified by acridine·1-naphthylacetic acid **17·b** and phenanthridine·3-indoleacetic acid **18·a** (Scheme 14) [39]. Solution photoreaction gave a mixture of radical coupling products HNArH-R, NAr-R, R-R, and HNArH-)₂ in addition to RH. The reaction mechanism (modified) is displayed in Scheme 15.

From the crystallographic analysis, the OH···N hydrogen bonding with H···N distances 1.42–1.77 Å and O···N distances 2.60–2.72 Å were found. The back H transfer from the heterocyclic NH group to R· (Scheme 15) was regarded to occur readily over short distances of 3.2–3.5 Å, which were approximated by C1···H1 distances (Scheme 14). Similarly, the inefficient coupling of the geminate radical pair (hetero coupling) was explained by a long separation between the radical centers (>5 Å), which were estimated from the C1···C2 or C1···C2′ distance. At this stage, I would like to point out that one photon cannot excite both members of the neighboring hydrogen bonding pair. The authors in Refs. 39b, 40a, 41, and 53a, however, did not care about this restriction and hence some of their mechanistic interpretations need to be reconsidered.

In contrast to the above cocrystals (Scheme 14), cocrystals **17·f** and **17·g** gave the corresponding hetero coupling products HNArH-R and NAr-R as major products (Scheme 16) [40]. This difference may be ascribed to their short C1···C2′ (and C1···C3′) distances (4.2–4.8 Å). For the cocrystal **17·g**, the stereocenter in R-(−)-2-phenylpropionic acid (**g**) was retained to a certain extent in the course of the decarboxylative condensation.

Thanks to the exploratory work of Schmidt et al. [42], the concept of absolute asymmetric induction starting from chiral crystals of achiral molecules is now very commonly known [1,43]. A 1:1 hydrogen-bonded cocrystal of acridine and diphenylacetic acid **17·h** is chiral (space group P2₁2₁2₁). Upon irradiation, the cocrystal underwent mainly hetero coupling to give chiral HNArH-R in a relatively low optical yield (37% ee, Scheme 17) [41].

Acridine and phenothiazine cocrystallized to give two kinds of hydrogen-bonded CT crystals. Both crystals showed some photoreactivity and appear to have given many photoproducts (Scheme 18) [44]. Although this crystalline complex is complicated in terms of stoichiometry, crystal structure, and photoreaction, a transient study by femtosecond diffuse reflectance spectroscopy was carried out, as had been done for durene·pyromellitic dianhydride cocrystal [45]. For the yellow cocrystal, a transient absorption spectrum with maxima around 600 and 520 nm was obtained, which decayed biexponentially with lifetimes of 2 and 50 ps. The two absorption maxima were ascribed to the acridine anion radical and the phenothiazine cation radical, respectively.

Irradiation of cocrystals between 4-methoxy-6-methyl-2-pyrone and maleimide gave an endo-5,6-[2+2] adduct (Scheme 19) [46]. Sensitized solution photolysis furnished a 1:2 adduct as a major product.

**Scheme 14**

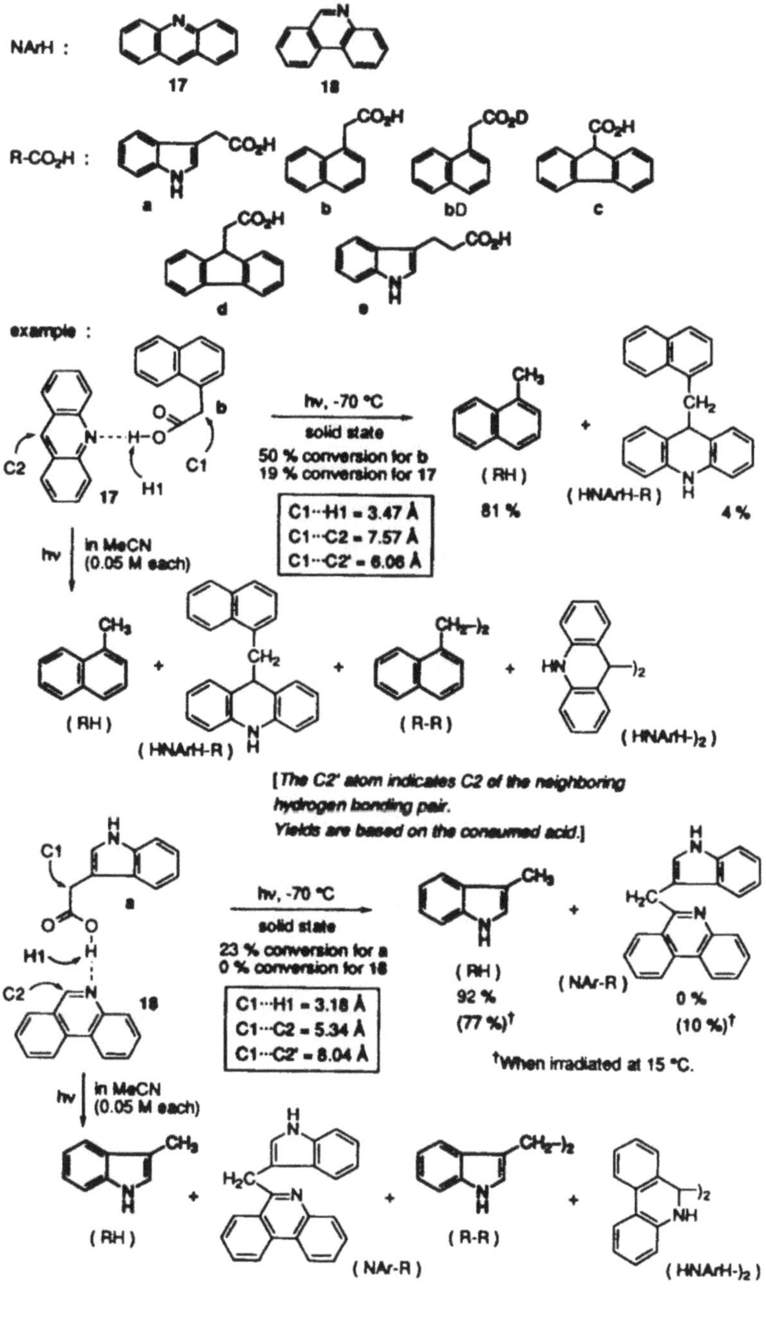

NArH :

17    18

R-CO₂H :

a    b    bD    c

d    e

example :

17    H1    b    C1

hv, -70 °C
solid state
50 % conversion for b
19 % conversion for 17

C1···H1 = 3.47 Å
C1···C2 = 7.57 Å
C1···C2' = 6.06 Å

( RH )    81 %

( HNArH-R )    4 %

hv | in MeCN
(0.05 M each)

( RH )    ( HNArH-R )    ( R-R )    ( HNArH-)₂ )

[*The C2' atom indicates C2 of the neighboring
hydrogen bonding pair.
Yields are based on the consumed acid.*]

18    H1    C1    C2    a

hv, -70 °C
solid state
23 % conversion for a
0 % conversion for 18

C1···H1 = 3.18 Å
C1···C2 = 5.34 Å
C1···C2' = 8.04 Å

( RH )    92 %    (77 %)†

( NAr-R )    0 %    (10 %)†

†When irradiated at 15 °C.

hv | in MeCN
(0.05 M each)

( RH )    ( NAr-R )    ( R-R )    ( HNArH-)₂ )

**Scheme 15**

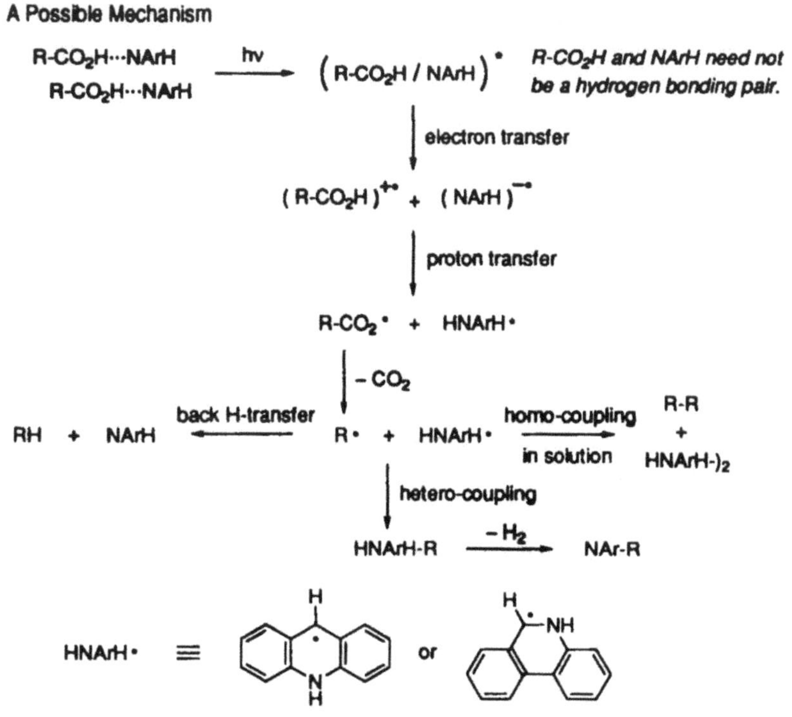

A Possible Mechanism

R-CO$_2$H···NArH
                    $\xrightarrow{\text{h}\nu}$    $\left(\text{R-CO}_2\text{H / NArH}\right)^{\bullet}$    R-CO$_2$H and NArH need not
R-CO$_2$H···NArH                                                     be a hydrogen bonding pair.

$\Big\downarrow$ electron transfer

( R-CO$_2$H )$^{+\bullet}$  +  ( NArH )$^{-\bullet}$

$\Big\downarrow$ proton transfer

R-CO$_2$$^{\bullet}$  +  HNArH$\bullet$

$\Big\downarrow$ $-$ CO$_2$

RH  +  NArH  $\xleftarrow{\text{back H-transfer}}$  R$\bullet$  +  HNArH$\bullet$  $\xrightarrow[\text{in solution}]{\text{homo-coupling}}$  $\begin{array}{c}\text{R-R}\\+\\\text{HNArH-)}_2\end{array}$

$\Big\downarrow$ hetero-coupling

HNArH-R  $\xrightarrow{-\text{H}_2}$  NAr-R

HNArH$\bullet$  $\equiv$  [structure]  or  [structure]

Hydrogen bonds as well as coordination bonds to metal ions are the most attractive intermolecular forces to design organic building blocks for functional materials (see Sec. IV). One drawback of organic molecular crystals is their thermal, mechanical, and/or chemical instability. Solid-state photodimerization (or photocrosslinking) between suitably preorganized unsaturated bonds may be utilized to improve stability of these materials. Two approaches in this direction have been reported [47,48]. One of them is hydrogen bond–enforced preorganization of cinnamic acid derivatives (Scheme 20) [47].

Another approach is cocrystallization of a substituted urea (host) and a diacetylene (guest) [48]. Substituted ureas are used to prepare layered diacetylene crystals. Two examples are shown in Scheme 21. Such a host–guest/cocrystal approach to supramolecular synthesis should be general. In this strategy, the host is used to control the structure and the guest provides the function (optical, electrical, chemical, or physical).

We have also utilized hydrogen bonding and ionic interactions for the pur-

**Scheme 16**

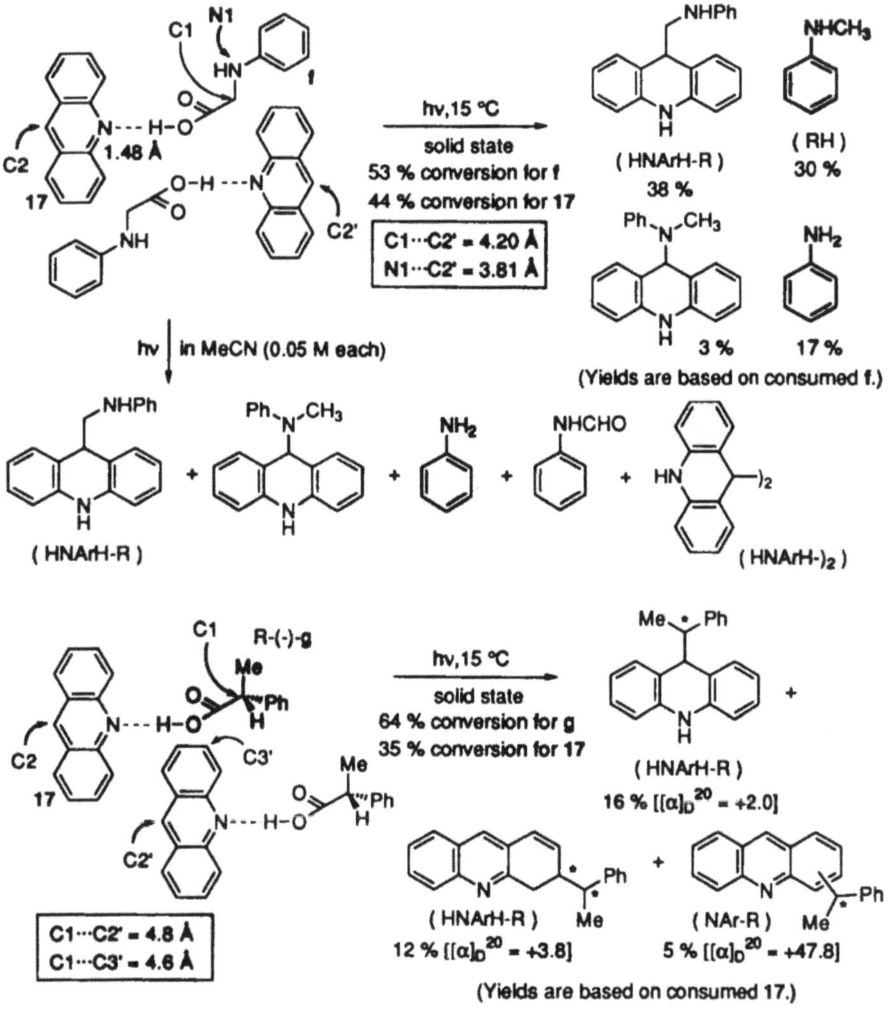

pose of forcibly changing the stereochemistry of photochemical [2+2] or [4+4] cycloadditions to a desired direction, as described in Sec. VIII.

## VII. DONOR–ACCEPTOR CRYSTALS

Electrical and magnetic properties of donor–acceptor crystals (charge transfer crystals) are interesting in view of their potential applications to organic conduc-

**Scheme 17**

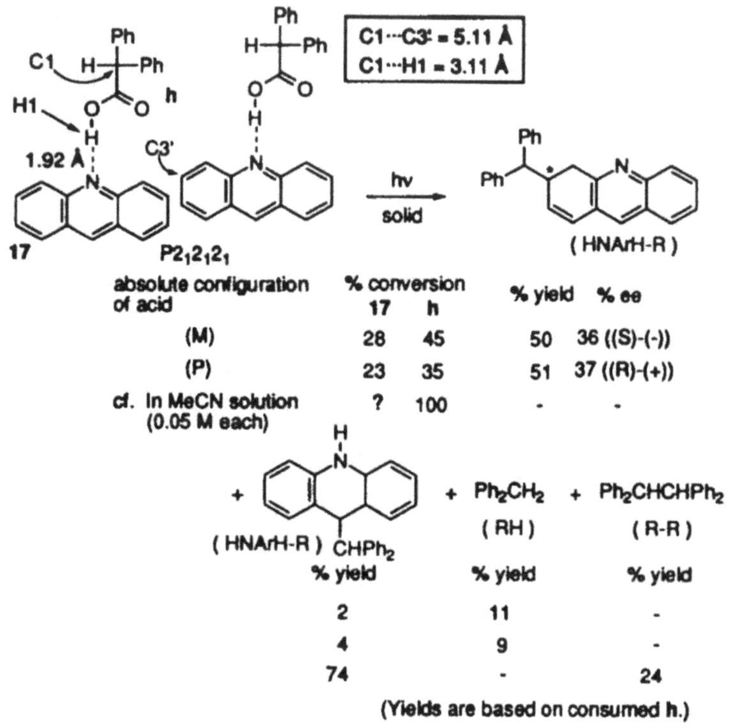
C1···C3' = 5.11 Å
C1···H1 = 3.11 Å
C1
H1
C3'
1.92 Å
17
$P2_12_12_1$
hν
solid
( HNArH-R )
absolute configuration of acid
% conversion
17 h
% yield % ee
(M) 28 45 50 36 ((S)-(-))
(P) 23 35 51 37 ((R)-(+))
cf. In MeCN solution (0.05 M each) ? 100 - -
( HNArH-R ) CHPh2
Ph2CH2
( RH )
Ph2CHCHPh2
( R-R )
% yield % yield % yield
2 11 -
4 9 -
74 - 24
(Yields are based on consumed h.)

tors, superconductors, photoconductors, ferromagnets, and nonlinear optical materials. Recently, their photoreactivities are attracting attention. However, at the time we undertook the study of donor–acceptor crystals involving nitro and cyano acceptors, only a few reports by Desiraju had dealt with the photoreactivity of donor–acceptor crystals [49].

Despite the parallel arrangement of double bonds with 3.8 Å distance, the 1 : 1 donor–acceptor complex **19a · 19d** was found to be photostable (Scheme 22) [49,50]. The photostability was ascribed to the presence of the low-lying CT state. The complexes **19a · 19e** and **19a · 19f** were also photostable. On the other hand, the 1 : 1 molecular complex of 3,5-dinitrocinnamic acid with 2,5-dimethoxycinnamic acid **19b · 19c** (the double bond center-to-center separation 3.54 Å) as well as **19b · 19d** and **19b · 19e** were photoreactive, producing the corresponding unsymmetric β-truxinic dimers in ~60% yields.

The face-to-face stacking interaction between phenyl and perfluorophenyl groups was used to orient olefinic molecules and to cocrystallize them [51]. The

**Scheme 18**

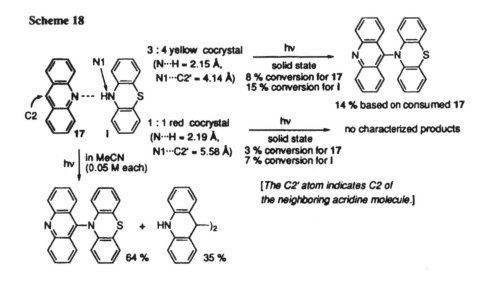

3 : 4 yellow cocrystal
(N···H = 2.15 Å,
N1···C2' = 4.14 Å)

hv
solid state

8 % conversion for 17
15 % conversion for I

14 % based on consumed 17

1 : 1 red cocrystal
(N···H = 2.19 Å,
N1···C2' = 5.58 Å)

hv
solid state

no characterized products

3 % conversion for 17
7 % conversion for I

[*The C2' atom indicates C2 of
the neighboring acridine molecule.*]

hv | in MeCN
(0.05 M each)

64 %      +      35 %

cocrystals shown in Scheme 23 were successfully prepared and they underwent
an expected [2+2] photodimerization or photopolymerization. Although I
grouped these cocrystals here, the nature of the phenyl–perfluorophenyl interaction is considered to be an electrostatic attraction rather than a charge transfer
interaction.

Charge transfer excitation of 1 : 1 CT crystals derived from bis [1,2,5] thia-
diazolotetracyanoquinodimethane and arylolefins gave [2+2] cycloadducts

**Scheme 19**

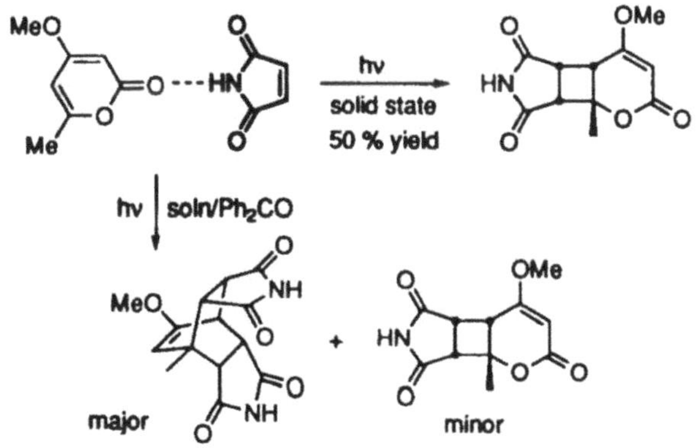

hv
solid state
50 % yield

hv | soln/Ph$_2$CO

major      +      minor

**Scheme 20**

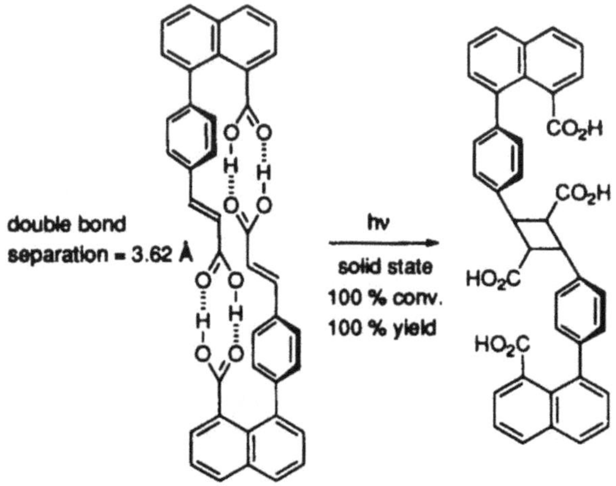

double bond
separation = 3.62 Å

hv
→
solid state
100 % conv.
100 % yield

**Scheme 21**

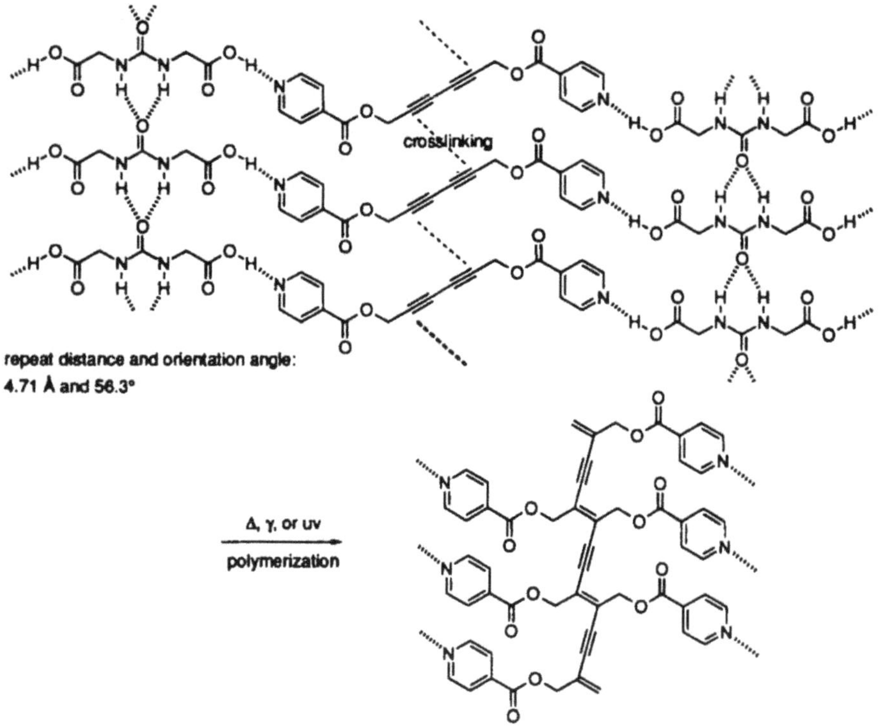

crosslinking

repeat distance and orientation angle:
4.71 Å and 56.3°

Δ, γ, or uv
→
polymerization

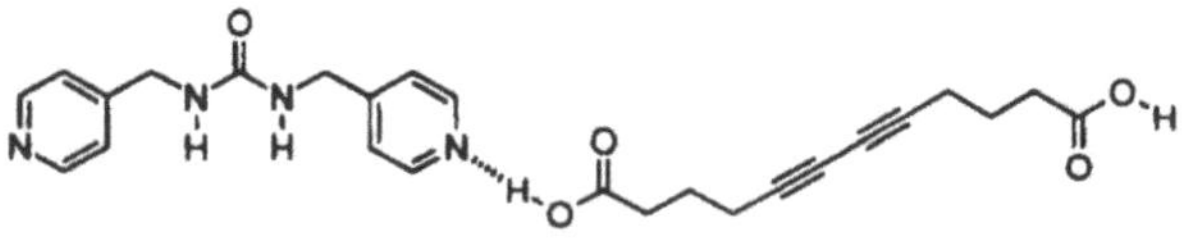

## Scheme 22

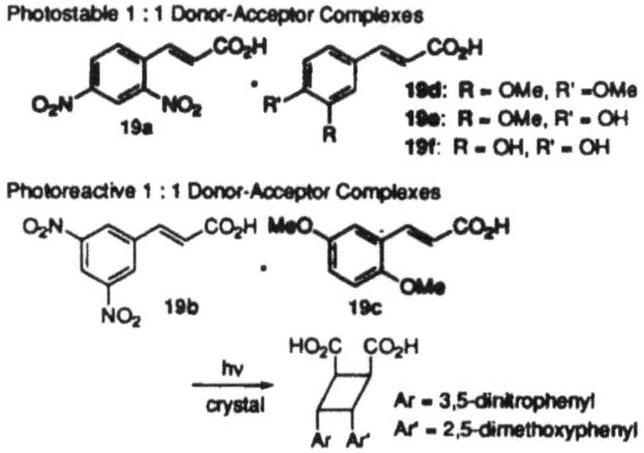

Photostable 1 : 1 Donor-Acceptor Complexes

19d: R - OMe, R' =OMe
19e: R - OMe, R' - OH
19f: R - OH, R' - OH

Photoreactive 1 : 1 Donor-Acceptor Complexes

hv
—→
crystal

Ar - 3,5-dinitrophenyl
Ar' - 2,5-dimethoxyphenyl

## Scheme 23

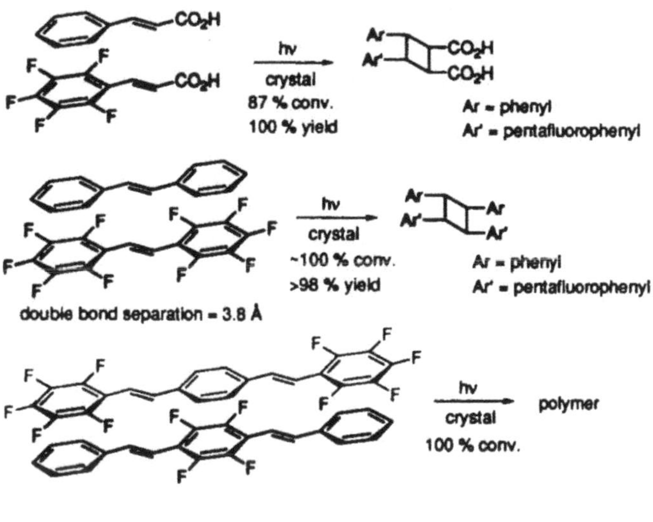

hv
—→
crystal
87 % conv.
100 % yield

Ar - phenyl
Ar' - pentafluorophenyl

hv
—→
crystal
~100 % conv.
>98 % yield

Ar - phenyl
Ar' - pentafluorophenyl

double bond separation - 3.8 Å

hv
—→
crystal
100 % conv.

polymer

(Scheme 24) [52]. The reaction was much more efficient in the solid state than in solution. When o-divinylbenzene was used, an absolute asymmetric induction leading to the chiral cycloadduct (71% and 95% ee at 15°C and −70 °C, respectively) was observed to occur via a single-crystal-to-single-crystal transformation ($P2_1 \rightarrow P2_1$).

Irradiation of crystalline charge transfer complexes of 1,2,4,5-tetracyano-

**Scheme 24**

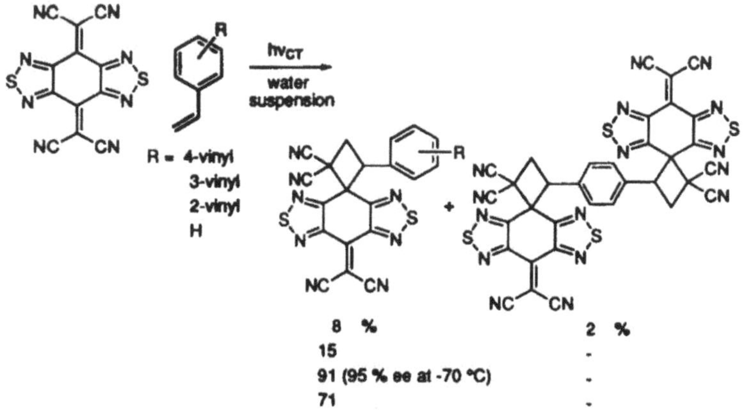

benzene (TCNB) with naphthylacetic acids gave coupling products by decarboxylative/dehydrocyanative condensation, together with methylnaphthalenes ArCH₃ by decarboxylation (Scheme 25) [53a]. Like hydrogen-bonded co-crystals (Schemes 14, 16, 17) [39–41], decarboxylation products ArCH₃ or RH were more prone to be produced in the solid state than in solution. Similar photo-condensation reactions occurred for Ar = p-tolyl, o-tolyl, p-MeO-Ph, and for 9-fluorenecarboxylic acid [53b]. For (S)-(+)-2-(6-methoxy-2-naphthyl)propanoic acid, the configuration of the asymmetric carbon was retained to some extent, although 78% of ee was lost at only 2% conversion for the donor [53c].

The pathway described in Scheme 26 is essentially the same as that pro-posed previously for the solution photolysis. The X-ray analysis of the complexes (Scheme 25, Ar = 1-naphthyl or 3-indolyl) demonstrated parallel, alternate stack-ing of D and A with the interplanar distance 3.4 Å. The C···C distance relevant to the photocondensation reaction was found to be 3.75 Å for Ar = 1-naphthyl and 4.50 Å for Ar = 3-indolyl. The longer distance for the latter explains its negligible photoreactivity [53a]. However, because the authors in this paper again [39b,40a,41] assumed a "duplicate excitation by one photon" in their mechanis-tic discussion, the details appear problematic.

Excitation of the 1:1 CT crystal of acenaphthylene and tetracyanoethylene at >500 nm afforded a [2+2] cycloadduct (Scheme 27) [54]. In contrast, the excitation in solution gave no product.

The CT excitation of the 1:2 donor–acceptor crystals of various diarylacet-ylenes (DAs) with 2,6-dichlorobenzoquinone (DB) afforded quinone methide products through [2+2] cycloaddition [55]. Selected examples are displayed in Scheme 28. Although the cocrystal c was unreactive, time-resolved (ps) diffuse

Scheme 25

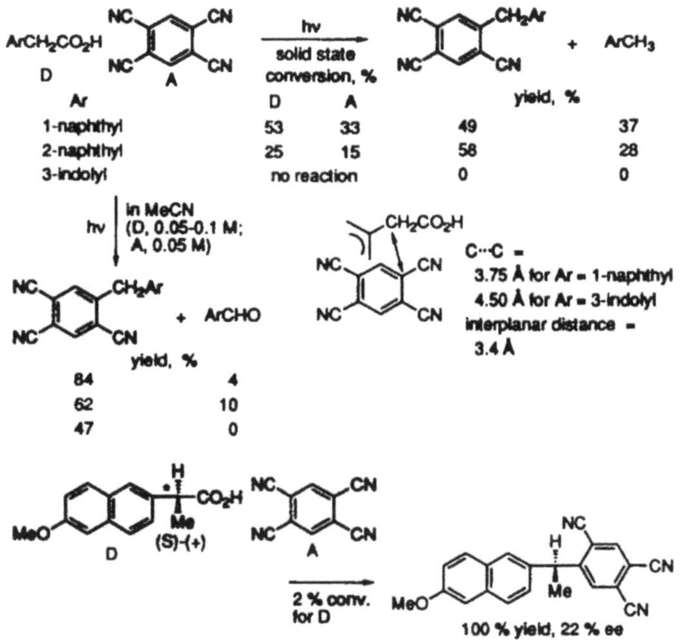
ArCH2CO2H
D
NC
CN
A
hv
solid state
conversion, %
CH2Ar
+ ArCH3
Ar
D A
yield, %
1-naphthyl 53 33 49 37
2-naphthyl 25 15 58 28
3-indolyl no reaction 0 0
hv
in MeCN
(D, 0.05-0.1 M;
A, 0.05 M)
+ ArCHO
yield, %
84 4
62 10
47 0
CH2CO2H
C···C =
3.75 Å for Ar = 1-naphthyl
4.50 Å for Ar = 3-indolyl
interplanar distance =
3.4 Å
H
CO2H
Me
MeO
(S)-(+)
2 % conv.
for D
100 % yield, 22 % ee

Scheme 26

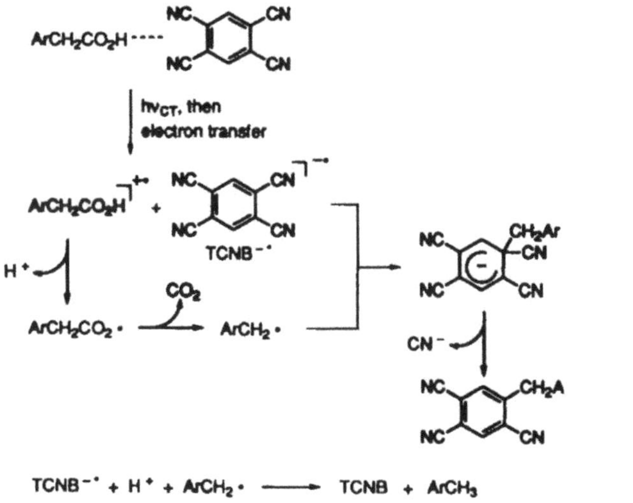
ArCH2CO2H
hvCT, then
electron transfer
TCNB−•
H+
CO2
ArCH2CO2•
ArCH2•
CH2Ar
CN−
CH2A
TCNB−• + H+ + ArCH2• ⟶ TCNB + ArCH3

Scheme 27

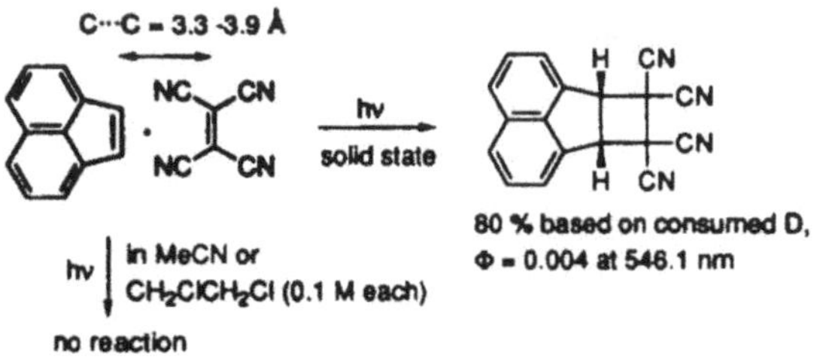
C···C = 3.3-3.9 Å
NC CN
NC CN
hv
solid state
H CN
CN
CN
H CN
80 % based on consumed D,
Φ = 0.004 at 546.1 nm
hv
in MeCN or
CH₂ClCH₂Cl (0.1 M each)
no reaction

Scheme 28

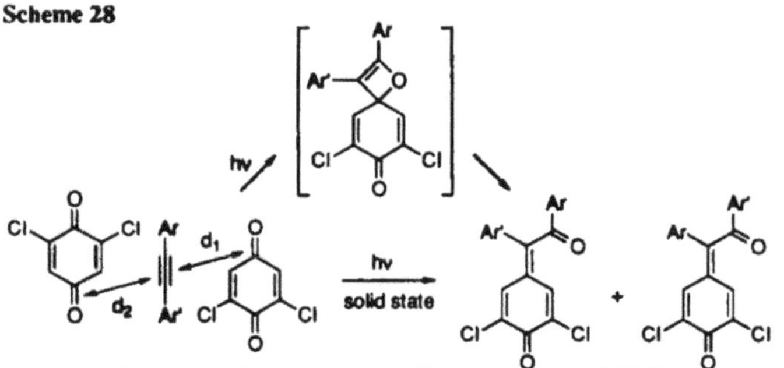
Ar
Ar'
O
Cl
Cl
O
hv
hv
solid state
d1
d2
+

| | Ar | Ar' | conv., % | yield, % | |
|---|---|---|---|---|---|
| a | 4-MePh | 3,5-diMePh | 17 | 72 (4.15 Å) | 12 (3.63 Å) |
| b | 3,5-diMePh | 3,5-diMePh | 15 | 90 (3.63 Å) | - |
| c | Ph | pentaMePh | 0 | 0 (3.84 Å) | 0 (4.34 Å) |
| | | | | ↖ $d_1$ | $d_2$ ↗ |

reflectance spectroscopy could identify the ion–radical pair intermediate [$DA^{\bullet+}$, $DB^{\bullet-}$], like in the case of the reactive cocrystal **b**. The pertinent center-to-center distances between the carbonyl group and the acetylene function ($d_1$ and $d_2$) are 4.15 Å for the major product and 3.63 Å for the minor product for **a**, 3.63 Å for **b**, and 3.84 and 4.34 Å for unreactive **c**. Consequently, the Schmidt's least-motion postulate cannot rationalize the present results. In $CH_2Cl_2$ solution, the quinone methides were readily produced from all of the donor–acceptor pairs including the pair of cocrystal **c**.

A bimolecular photoreaction between different reagents in a solid cyclodextrin phase occurred upon photolysis of crystalline ternary β-CD complexes (1 : 1 : 2 nitro/amine/β-CD) (Scheme 29) [56a]. Study by diffuse reflectance laser flash photolysis was performed and an observed transient was assigned to the radical anion of 4-nitroveratrole (X = Y = OMe) generated via photoinduced

**Scheme 29**

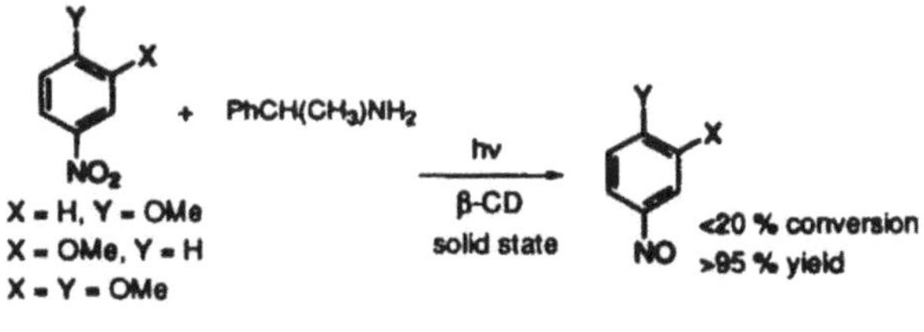

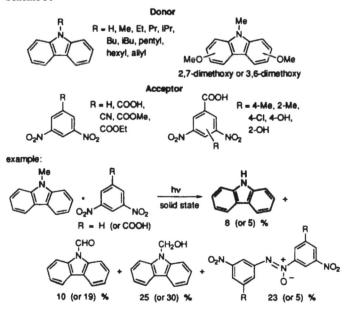

X = H, Y = OMe
X = OMe, Y = H
X = Y = OMe

+ PhCH(CH₃)NH₂

hv
β-CD
solid state

<20 % conversion
>95 % yield

electron transfer from the amine to the nitroaromatic triplet state [56b]. Many examples of stereo-, regio-, and enantioselective photoreactions in inclusion crystals have been overviewed in Ref. 1.

N-Alkylcarbazoles formed yellow to orange to deep brown 1:1 charge transfer crystals with m-dinitrobenzene, 3,5-dinitrobenzoic acid (3,5-DNBA) and its derivatives (Scheme 30) [57]. On the other hand, attempts to obtain adduct crystals of N-alkylcarbazoles with o- or p-dinitrobenzene and with 2,4- or 3,4-dinitrobenzoic acid failed. Upon solid-state photolysis, many of these adduct crystals underwent photoredox reactions and the α position of the N-alkylcarbazoles was oxidized to produce carbazole, N-acylcarbazoles, and N-(hydroxy-

**Scheme 30**

**Scheme 31**

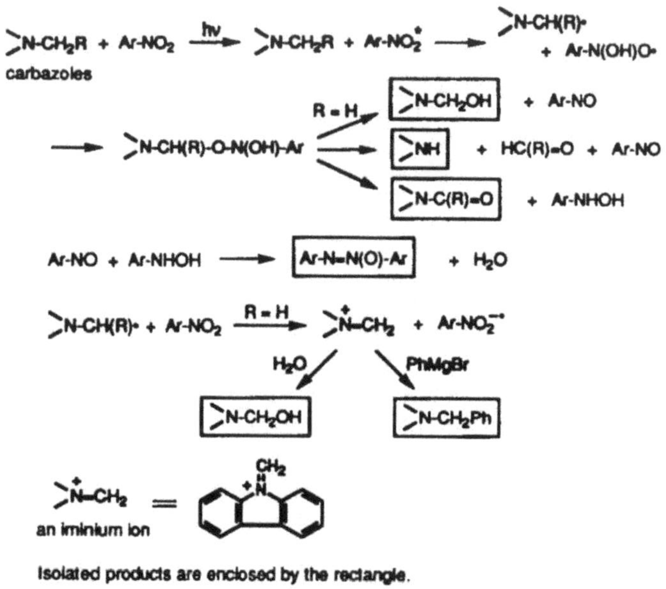

Isolated products are enclosed by the rectangle.

methyl)carbazole in addition to azoxybenzenes, as exemplified by Scheme 30. Yields are based on the employed donor (or acceptor for azoxybenzenes). In solution, the photoreactions were negligible. Possible reaction sequences are shown in Scheme 31. Intervention of an iminium ion is assumed when R = H. From the X-ray analysis of cocrystal N-methylcarbazole·3,5-DNBA, the distance between the $N$-$\alpha$-hydrogen and the nitro oxygen nearby was 2.61 Å, which is close enough for the photochemical hydrogen abstraction by the nitro group to occur. The deep brown CT crystals of 3,5-DNBA with dimethoxy-N-methylcarbazoles (strong donors) were photoinert.

The yellow to red donor–acceptor crystals of TCNB with carbazoles (9-H, 9-Me, 9-Et), naphthalenes (1-Me, 2-Me, 1-CH$_2$CN, 2-CH$_2$CN), anthracenes (9-H, 9-Me), durene, biphenyl, etc., were photostable. However, a colorless crystalline 1:2 adduct with benzylcyanide (BzCN), TCNB·BzCN, was photoreactive (BzCN is a weak donor, association constant with TCNB $K = 0.07$ M$^{-1}$ and $\lambda_{max}^{CT}$ ~ 322 nm in acetonitrile) [58]. The crystal structure (BzCN···TCNB···BzCN sandwich structure, interplane distance 3.6 Å) and the diffuse reflectance spectrum ($\lambda_{max}^{CT}$ ~ 367 nm) indicate that the adduct is a charge transfer crystal. Upon photolysis in the crystalline state, it underwent an unusual addition/rearrangement reaction to give 20, which in solution cyclized readily to an isoindole derivative 21, as shown in Scheme 32. The consecutive reaction TCNB·BzCN

**Scheme 32**

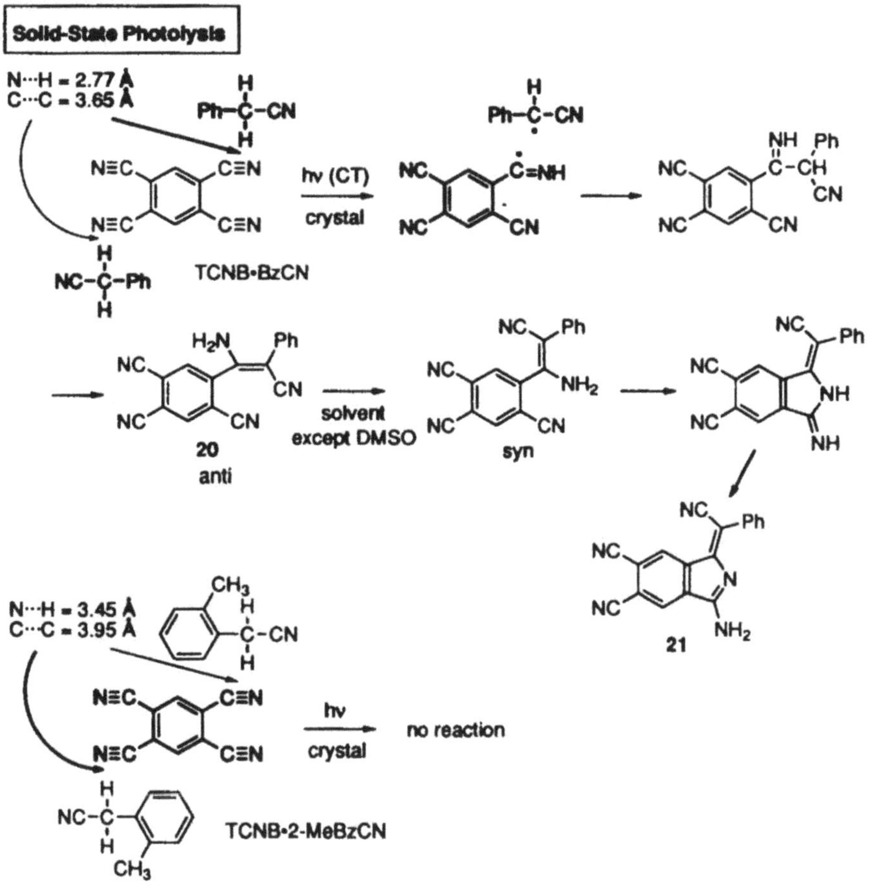

→ **20** → **21** proceeded almost quantitatively and the conversion of TCNB was 65%.

Solution irradiation of the TCNB/BzCN pair gave complex results (Scheme 33). No photoreaction occurred in acetonitrile. In contrast, irradiation of the pair in BzCN solvent, which had been distilled after pretreatment with 10% sulfuric acid, led to products completely different from those in the solid state, i.e., the elimination of hydrogen cyanide took place, resulting in the formation of **22, 23,** and **24** in 44%, 17%, and 3% yields, respectively. Irradiation in BzCN without the acid pretreatment also gave a unique result, producing **24** and a biphenyl derivative **25** [59]. Commercial BzCN is probably contaminated with amine. A

**Scheme 33**

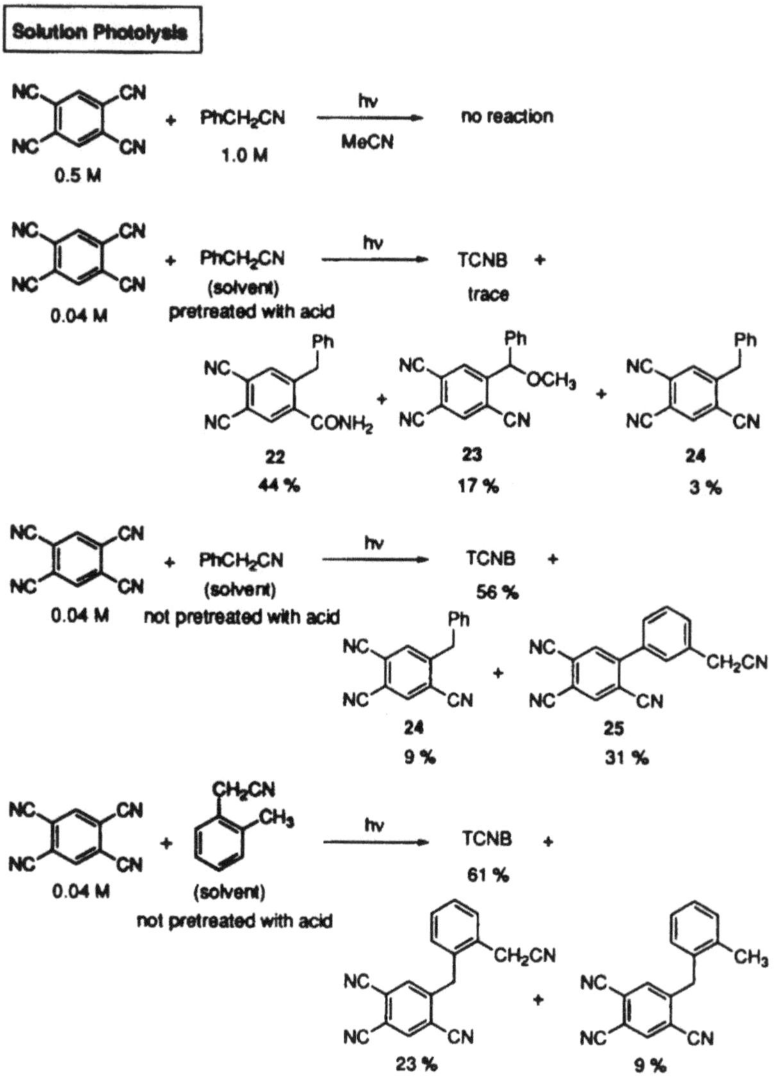

| Solution Photolysis |

NC—[benzene ring]—CN / NC—CN (0.5 M)  +  PhCH$_2$CN (1.0 M)  $\xrightarrow[\text{MeCN}]{h\nu}$  no reaction

NC—[benzene ring]—CN / NC—CN (0.04 M)  +  PhCH$_2$CN (solvent) pretreated with acid  $\xrightarrow{h\nu}$  TCNB + trace

**22** 44 %   +   **23** 17 %   +   **24** 3 %

NC—[benzene ring]—CN / NC—CN (0.04 M)  +  PhCH$_2$CN (solvent) not pretreated with acid  $\xrightarrow{h\nu}$  TCNB + 56 %

**24** 9 %   +   **25** 31 %

NC—[benzene ring]—CN / NC—CN (0.04 M)  +  [o-methylbenzyl cyanide] (solvent) not pretreated with acid  $\xrightarrow{h\nu}$  TCNB + 61 %

23 %   +   9 %

possible reaction mechanism in acid-treated BzCN is displayed in Scheme 34. It appears that the reaction was initiated by photochemical hydrogen abstraction by the cyano group both in the solid state (Scheme 32) and in the solution state (Scheme 34), but the geometry of hydrogen abstraction and the subsequent course of reaction are different due to a dramatically large effect of crystal packing.

Although 3-methylbenzylcyanide and 4-methylbenzylcyanide failed to form crystalline adducts with TCNB, 2-methylbenzylcyanide gave a 1:2 crystalline adduct TCNB·2-MeBzCN with the crystal structure similar to that of TCNB·BzCN [60]. This cocrystal, however, was photostable and this photostability was attributed to the absence of short contact between the TCNB nitrogen and the 2-MeBzCN benzylic hydrogen. The shortest N···H distance for photoinert TCNB·2-MeBzCN is 3.45 Å (C···C = 3.95 Å), considerably longer than the corresponding distance for photoreactive TCNB·BzCN, where N···H is 2.77 Å (C···C = 3.65 Å) (Scheme 32). The solution photoreaction between TCNB and

**Scheme 34**

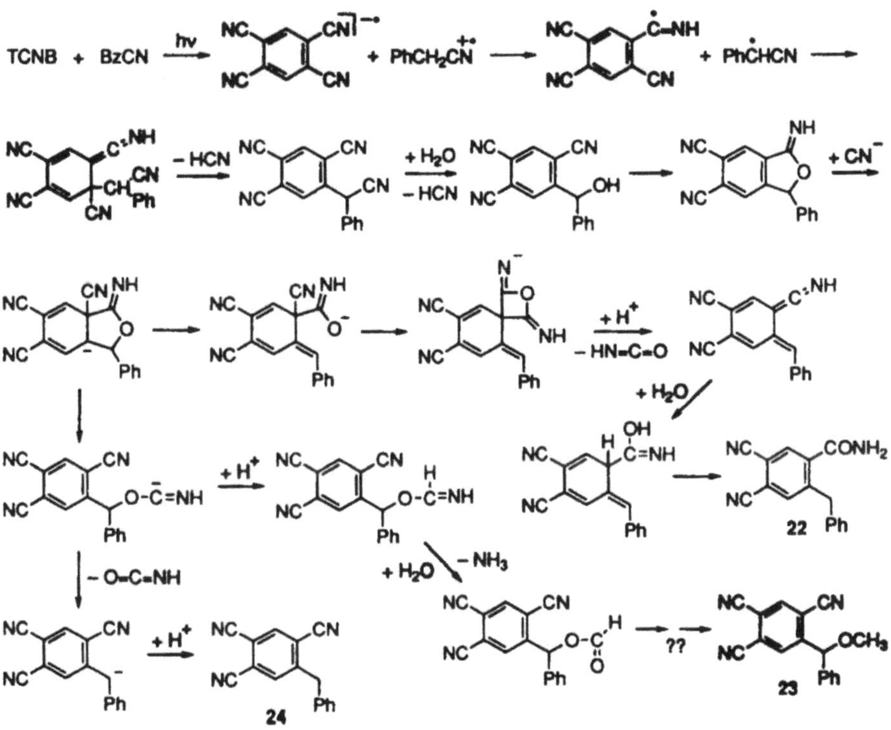

**Scheme 35**

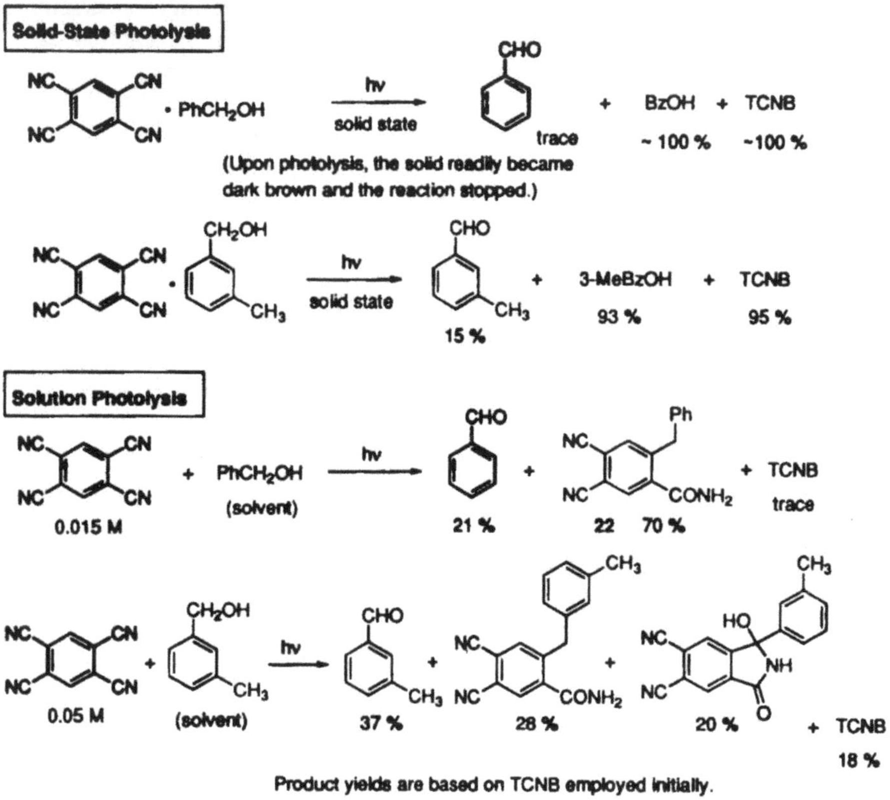

Product yields are based on TCNB employed initially.

2-Me-, 3-Me-, or 4-MeBzCN was analogous to that of the TCNB/BzCN pair, e.g., for the TCNB/2-MeBzCN pair, see Scheme 33 [59b]. However, the condensation occurred at the methyl group rather than at the cyanomethyl group and formation of biphenyl derivatives corresponding to 25 could not be found.

TCNB formed cocrystals with benzyl alcohol (BzOH) and 3-methylbenzyl alcohol (3-MeBzOH) with the 1:1 stoichiometry. Their crystal structures consist of parallel alternate stacking of the donor and acceptor molecules [61]. Although these cocrystals did not give detectable amounts of photocoupling products in the solid state, several condensation products were obtained by solution photolysis in liquid donor as solvent (Scheme 35) [59].

As the above examples demonstrate, the number of photoreactive donor–acceptor crystal is now increasing.

## VIII.  CRYSTALLINE ORGANIC SALTS

When a carboxylic acid and a nitrogen base are cocrystallized, proton transfer from the acid to base often occurs, resulting in formation of high-melting ionic crystals. The feature of crystalline organic salts is simultaneous use of hydrogen bonding interactions and ionic interactions for packing the molecules together in the crystal. Investigations of these acid–base cocrystals have indicated that a $pK_a$ difference of about 3 is required to form an ionic complex [62], although this criterion does not always hold true [63].

Scheffer and co-workers have carried out extensive research on the use of organic salts for, e.g., asymmetric synthesis (Schemes 36–38), heavy atom effects (Scheme 39), and triplet sensitization (Scheme 40) [43a,64–73]. Chiral crystals of prochiral organic acids or bases were prepared by salt formation with optically active amines or acids, respectively, and the solid-state photochemistry of the resultant salts led to asymmetric induction in prochiral acids or bases. These are illustrated for the solid-state di-$\pi$-methane photoreaction of 9,10-ethenoanthracene (dibenzobarrelene) derivatives (Schemes 36 and 37) [65–67] and for the Norrish type II photocylization of phenyl ketones (Scheme 38) [68–71]. Nonabsorbing, optically active amines or acids must be used as a chiral handle. Solution photolysis gave photoproducts with no optical activity.

For the di-$\pi$-methane reactions of the (S)-(−)-proline tert-butyl ester salt and the (S,S)-(+)-pseudoephedrine salt in Scheme 36, absolute configuration correlations between reactant and photoproduct could be elucidated by X-ray crystallography. The correct absolute configurations of the reactants and the products are as shown in the scheme. The observed selective benzo-vinyl bridging to one direction is ascribed both to reduced steric repulsion between the vinyl substituents and to better orbital overlap in the transition state [65]. It is concluded for the 11,12-dicarboxydibenzobarrelene derivatives that initial benzo–vinyl bridging is favored at the carboxylate salt-bearing vinyl carbon atom both in the solid state and in solution [66].

For the examples in Scheme 37, chiral handles were introduced into a prochichiral amine by salt formation with optically active camphorsulfonic acids [67].

L-Prolinolium $\alpha$-adamantylacetophenone-$p$-carboxylate in Scheme 38 was dimorphic. The major photoproduct from needle-shaped crystals had 97% optical purity. The optical purity of the major photoproduct from plate-like crystals was only 12%. This difference in enantiomeric excess was explained by the X-ray structures of the dimorphs. Thus, the adamantylacetophenone anion in the needle crystals adopts a single homochiral conformation, whereas in the plate form, the anion exists in both conformational chiralities [68].

The second reaction in Scheme 38 is an enantioselective single crystal-to-single crystal (topotactic) photorearrangement. The sterically more hindered endo

**Scheme 36**

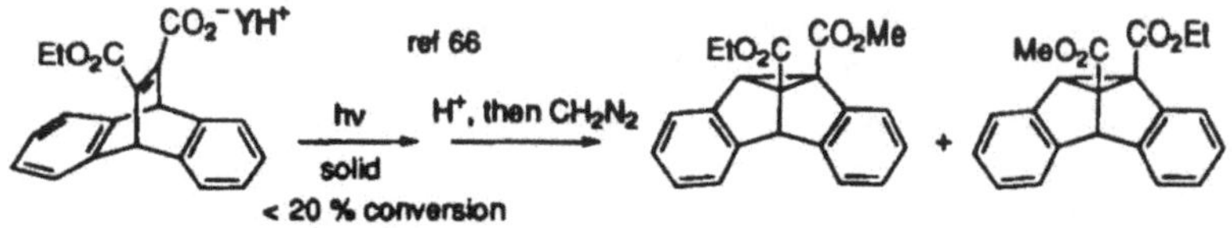

| Optically Active Amine Y | % yield | % ee | % yield | % ee |
|---|---|---|---|---|
| (R)-proline (a) | 96 | 80 (-) | 4 | 0 |
| (S)-proline (a) | 94 | 76 (+) | 6 | 0 |
| (R,S)-proline (a) | 84 | 0 | 16 | 0 |
| (S)-proline methyl ester (b) | 93 | 58 (+) | 7 | 0 |
| (S)-2-pyrrolidine methanol (c) | 100 | 37 (-) | - | - |
| (-)-strychnine (d) | 65 | 14 (+) | 35 | 0 |

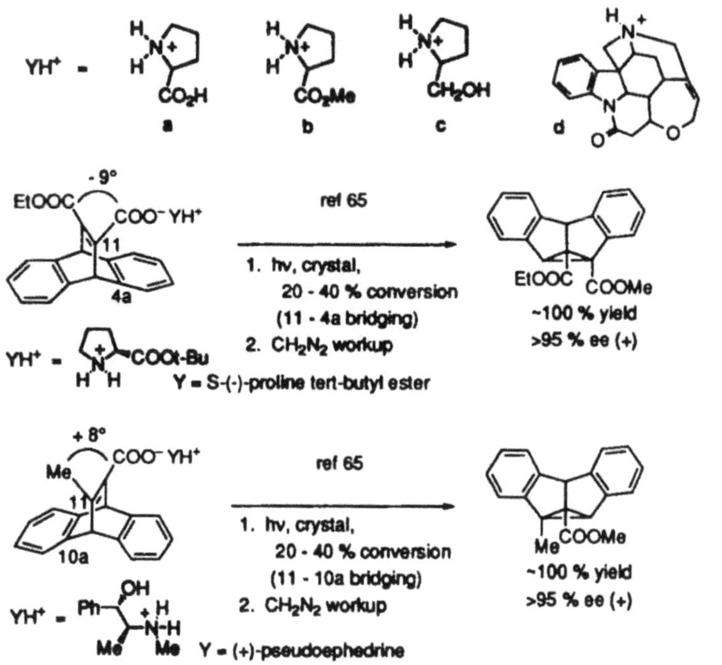

arylcyclobutanol was formed in preference to its exo isomer. Photolysis of the (S)-(−)-α-methylbenzylamine salt afforded the opposite optical antipode of cyclobutanol [69].

Direct irradiation of crystals of a β, γ-unsaturated keto acid (Scheme 39) led primarily to the 1,3-acyl shift product. Irradiation of its Li⁺, Na⁺, K⁺, Rb⁺, and Cs⁺ salts in the crystalline state, however, produced increasing amounts of

Scheme 37

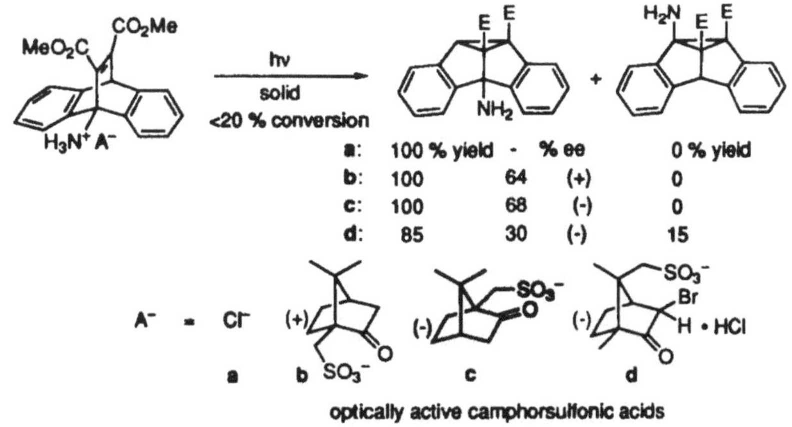
CO2Me
MeO2C
H3N+ A-
hv
solid
<20 % conversion
E E
NH2
+
H2N E E
a: 100 % yield - % ee 0 % yield
b: 100 64 (+) 0
c: 100 68 (-) 0
d: 85 30 (-) 15
A- = Cl-
(+)
SO3-
(-)
SO3-
O
(-)
SO3-
Br
H • HCl
O
a b c d
optically active camphorsulfonic acids

the oxa-di-π-methane photoproduct [72]. This result was interpreted in terms of a heavy atom–induced $S_1(n,\pi^*) \rightarrow T_1(\pi,\pi^*)$ intersystem crossing. Such an ionic heavy atom effect was not observed in aqueous solution, since the salts are dissociated.

A triplet-sensitized photoreaction in the crystalline state has been achieved by using crystalline organic salts which were prepared from cationic sensitizers ($BH^+$) and an anionic reactant (Scheme 40) [73]. Thus, upon selective excitation of the sensitizer these salts or hydrogen-bonded complex gave two dibenzosemibullvalenes through triplet–triplet energy transfer to the reactant. A dibenzocyclooctatetraene, which is a product of direct (singlet state) irradiation, was not produced.

Photochemical [2+2] cycloaddition of alkenes in the crystalline state is synthetically very useful because it usually produces only one stereoisomer predicted from the crystal structure. On the other hand, this stereospecificity of the reaction can be a disadvantage because of inaccessibility to other stereoisomers. In order to circumvent such a problem, we explored compelled orientational control of the photodimerization of particular compounds like *trans*-cinnamic acids and anthracenecarboxylic acids [74–78]. During our study, photochemistry of fluoro- and chloro-substituted *trans*-stilbene-4-carboxylic acids and their methyl esters and alkaline and alkaline earth salts in the crystalline phase was likewise studied in order to synthesize specific stereoisomers selectively (Scheme 41) [79]. Most of these stilbene compounds dimerized to give exclusively or mainly syn head-to-head cyclobutane dimers. Some were photochemically inert.

Our strategy is based on crystalline double-salt formation with diamines, which are used as a supramolecular linker to connect two acid molecules in a

**Scheme 38**

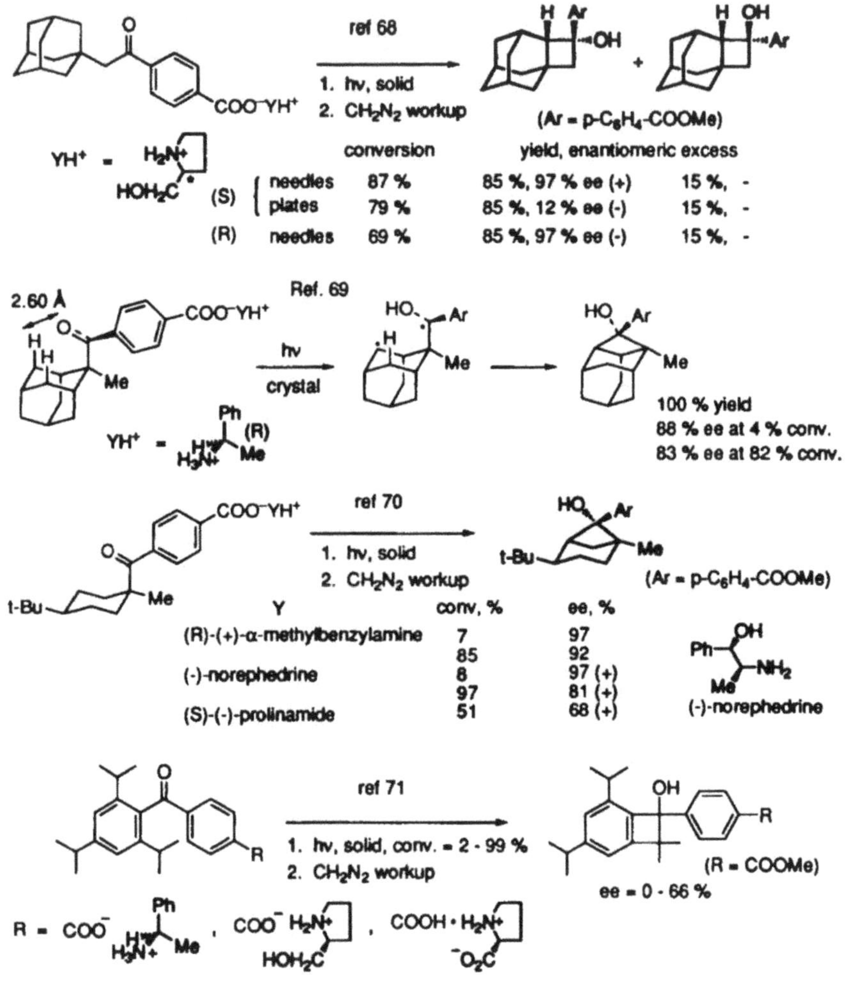

ref 68

1. hv, solid
2. CH₂N₂ workup

(Ar = p-C₆H₄-COOMe)

YH⁺ =

| | | conversion | yield, enantiomeric excess | |
|---|---|---|---|---|
| (S) | needles | 87 % | 85 %, 97 % ee (+) | 15 %, - |
| (S) | plates | 79 % | 85 %, 12 % ee (-) | 15 %, - |
| (R) | needles | 69 % | 85 %, 97 % ee (-) | 15 %, - |

2.60 Å

Ref. 69

hv
crystal

100 % yield
88 % ee at 4 % conv.
83 % ee at 82 % conv.

YH⁺ = (R)

ref 70

1. hv, solid
2. CH₂N₂ workup

(Ar = p-C₆H₄-COOMe)

| Y | conv, % | ee, % |
|---|---|---|
| (R)-(+)-α-methylbenzylamine | 7 | 97 |
| | 85 | 92 |
| (-)-norephedrine | 8 | 97 (+) |
| | 97 | 81 (+) |
| (S)-(-)-prolinamide | 51 | 68 (+) |

(-)-norephedrine

ref 71

1. hv, solid, conv. = 2 - 99 %
2. CH₂N₂ workup

(R = COOMe)

ee = 0 - 66 %

R = COO⁻ , COO⁻ H₂N⁺ , COOH·H₂N⁺

predesigned orientation. A series of diamines were investigated to steer the photo-dimerization of *trans*-cinnamic acid (t-**26**) (Scheme 42) [74]. A few polyamines were also studied for comparison. It was found that although many of the double salts showed no or low photoreactivities, those derived from gauche 1,2-diamines (t-**chxn** and c-**chxn**) were highly photoreactive, producing the β, δ, and ε dimers (see examples in Scheme 42). From simple geometrical considerations, 1,2-di-amines in a gauche conformation are expected to give rise to these three dimers. Optically active (R,R)-t-**chxn** and racemic t-**chxn** were equally effective. Al-

**Scheme 39**

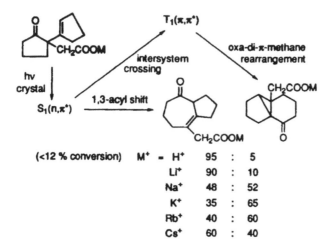

| (<12 % conversion) | M⁺ | = H⁺ | 95 | : | 5 |
|---|---|---|---|---|---|
| | | Li⁺ | 90 | : | 10 |
| | | Na⁺ | 48 | : | 52 |
| | | K⁺ | 35 | : | 65 |
| | | Rb⁺ | 40 | : | 60 |
| | | Cs⁺ | 60 | : | 40 |

though the most stable crystal modification of t-**26** is known to photodimerize into α-truxillic acid, this dimer was not produced from any of the double salts which we studied.

The double salts of ethylenediamine (**en**) with several derivatives and analogs of *trans*-cinnamic acid were investigated (Scheme 43) [75]. X-Ray studies have demonstrated that the conformation of en is gauche for the highly photoreactive double salts (t-**28**·en and t-**29**·en), whereas it is anti for less photoreactive

**Scheme 40**

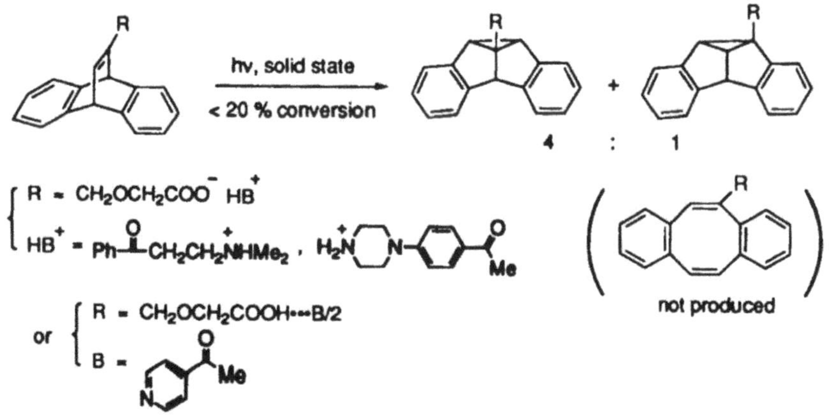

**Scheme 41**

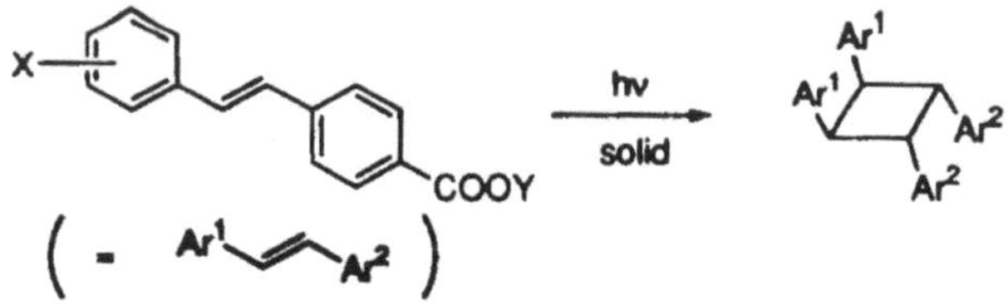

X = 2-, 3-, 4-F or Cl
Y = H, Me, Li, Na, K, Rb, Cs, Mg/2, Ca/2, Sr/2, Ba/2

t-**27**·en or photoinert t-**26**·en. The products were predominantly β-truxinic dimers. This is of value because the β-truxinic dimer of 2-methoxycinnamic acid (t-**27**) cannot be obtained by irradiation of its homocrystals. Perhaps, it should also be mentioned that although the t-**27** molecules in the double salt t-**27**·en are separated (Cα···Cα′ = 5.76 Å, Cβ···Cβ′ = 5.35 Å) much more than Schmidt's critical distance (4.2 Å), they can still undergo photodimerization.

As displayed in Scheme 44 [76], irradiation of double salts prepared from anthracene-9-propionic acid (9-AP) with (R,R)- or (S,S)-t-**chxn** readily afforded the thermally unstable head-to-head dimer as a sole product rather than the stable head-to-tail dimer which is the predominant photoproduct in solution. In contrast, the homocrystal of 9-AP and the double salt 9-AP·**en** were photostable in the solid state.

From the above results (Schemes 42–44), it appears that *trans*- and *cis*-cyclohexane-1,2-diamines (t-**chxn** and c-**chxn**) or gauche 1,2-diamines are useful linkers for preparing double salts of unusual photoreactivity. Evidently, photoreactive overlap configurations are more likely to happen to the gauche conformation of 1,2-diamine. There are, however, limitations. For example, the double salt 9-AP·(±)-t-**chxn**, which crystallized with 1 mol of included MeOH, yielded virtually no head-to-head dimer but only a small yield of the usual head-to-tail dimer of 9-AP [76].

In the case of double salts of anthracene-9-carboxylic acid (9-AC), which usually cyrstallized with included water and/or solvent, the 9-AC⁻ component in the crystal underwent decarboxylation and reduction along with dimerization (Scheme 45) [77]. Although the head-to-tail dimer and a mixed dimer were obtained, the head-to-head one was not produced. The reactivity was influenced by the amount of the included solvent, e.g., $(9\text{-}AC^-)_2(c\text{-}\mathbf{chxn}H_2^{2+})(EtOH)_{2.1}$ was photostable, while $(9\text{-}AC^-)_2(c\text{-}\mathbf{chxn}H_2^{2+})(EtOH)_{0.5}$ gave six photoproducts. Incidentally, the elusive head-to-head dimer was selectively obtained by solid-state irradiation of the 9-AC homocrystal as a thermally unstable product [78].

Divalent metals were used instead of diamines as a linker of two acid

## Scheme 42

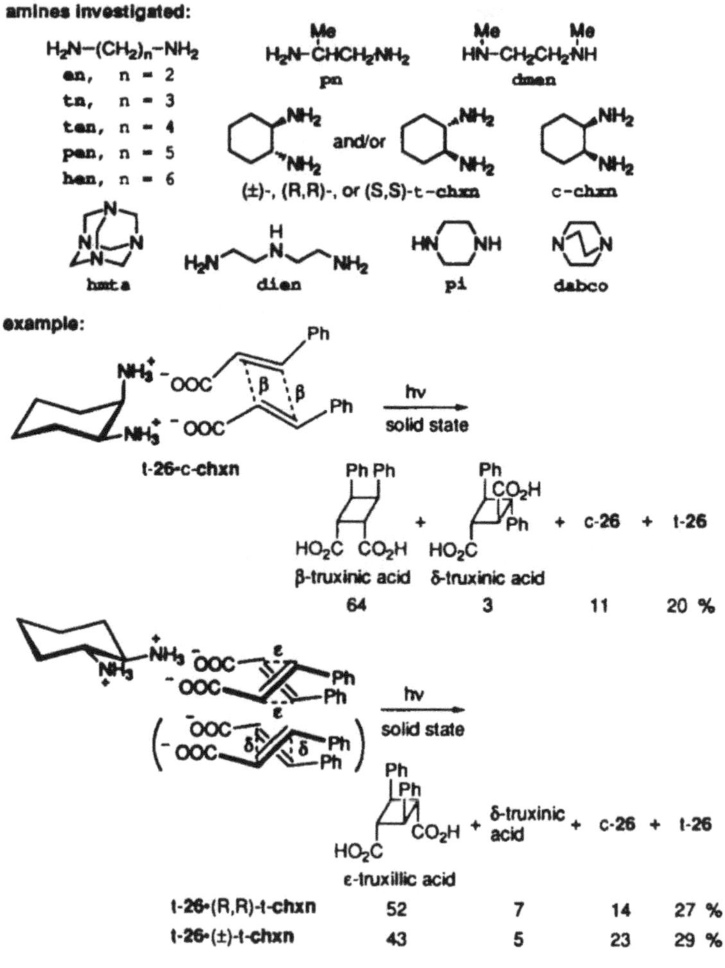

amines investigated:

$H_2N-(CH_2)_n-NH_2$
en, n = 2
tn, n = 3
ten, n = 4
pen, n = 5
hen, n = 6

Me
|
$H_2N-CHCH_2NH_2$
pn

Me      Me
|        |
$HN-CH_2CH_2NH$
dmen

(±)-, (R,R)-, or (S,S)-t-chxn    and/or

c-chxn

hmta

H
|
$H_2N$ N $NH_2$
dien

HN  NH
pi

N  N
dabco

example:

$NH_3^+$ ⁻OOC   Ph
             β  β
$NH_3^+$ ⁻OOC   Ph

$\xrightarrow{h\nu}$ solid state

t-26-c-chxn

Ph Ph

HO_2C CO_2H    HO_2C
β-truxinic acid    δ-truxinic acid
64              3          11       20 %

Ph
|CO_2H
Ph    + c-26 + t-26

$-NH_3$ ⁻OOC ε
$NH_3^+$ ⁻OOC   Ph
                Ph
(⁻OOC δ δ Ph Ph)

$\xrightarrow{h\nu}$ solid state

Ph
|Ph
CO_2H + δ-truxinic + c-26 + t-26
HO_2C         acid

ε-truxillic acid

| | | | | |
|---|---|---|---|---|
| t-26-(R,R)-t-chxn | 52 | 7 | 14 | 27 % |
| t-26-(±)-t-chxn | 43 | 5 | 23 | 29 % |

molecules. These metal-containing salts also crystallized together with water (and solvent) in such a manner as $M^{2+}(t\text{-PhCH}=\text{CHCOO}^-)_2(H_2O)_n$ and $M^{2+}(9\text{-AC}^-)_2(H_2O)_n(\text{solvent})_m$ (M = Mg, Ca, Sr, Ba, Zn). Their solid-state photoreactivities again depended on the amount of included water or solvent, i.e., the included solvent or water hampered the photoreactivity (e.g., Schemes 45 and 46) [80].

By using dicarboxylic acids as a non-covalent intermolecular linker, a compelled orientational control of the solid-state [2+2] photodimerization of *trans-*

## Scheme 43

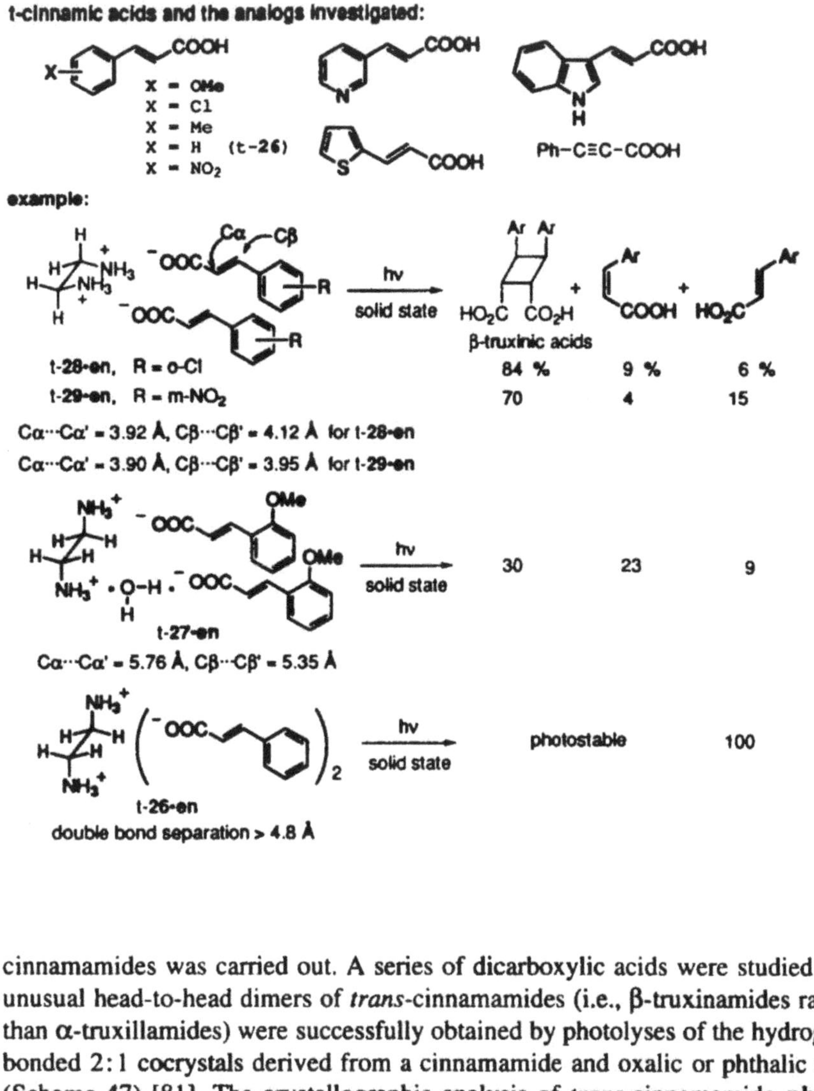

t-cinnamic acids and the analogs investigated:

X = OMe
X = Cl
X = Me
X = H   (t-26)
X = NO₂

Ph–C≡C–COOH

example:

                                                    β-truxinic acids

t-28-en, R = o-Cl                                   84 %        9 %        6 %
t-29-en, R = m-NO₂                                  70          4          15

Cα···Cα' = 3.92 Å, Cβ···Cβ' = 4.12 Å for t-28-en
Cα···Cα' = 3.90 Å, Cβ···Cβ' = 3.95 Å for t-29-en

t-27-en                                             30          23         9

Cα···Cα' = 5.76 Å, Cβ···Cβ' = 5.35 Å

t-26-en                                             photostable            100

double bond separation > 4.8 Å

cinnamamides was carried out. A series of dicarboxylic acids were studied and unusual head-to-head dimers of *trans*-cinnamamides (i.e., β-truxinamides rather than α-truxillamides) were successfully obtained by photolyses of the hydrogen-bonded 2:1 cocrystals derived from a cinnamamide and oxalic or phthalic acid (Scheme 47) [81]. The crystallographic analysis of *trans*-cinnamamide·pht (t-30·pht) has shown that two amide molecules are brought to an overlap configuration via a supramolecular linker **pht**. The two olefinic double bonds, however, are anti and not parallel (C···C distances, 3.82 and 4.85 Å). Therefore, a crankshaft–like dynamic conformational change [82] is thought to occur in the crystal prior to cycloaddition to β-truxinamide (Scheme 48). It is to be noted that in

**Scheme 44**

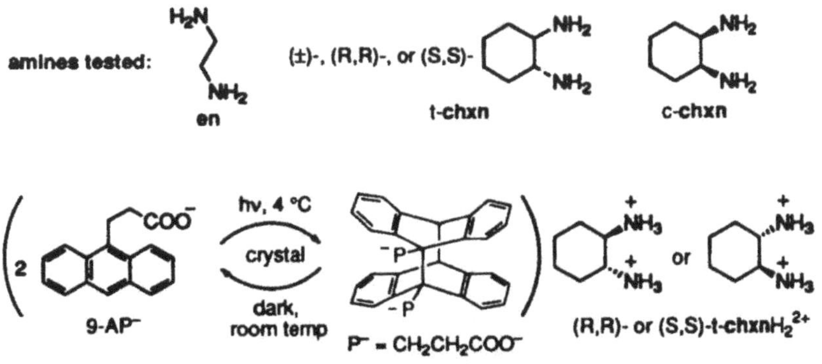

amines tested: (en, t-chxn, c-chxn structures)

(±)-, (R,R)-, or (S,S)- t-chxn

9-AP⁻        hv, 4 °C / crystal ⇌ dark, room temp        (R,R)- or (S,S)-t-chxnH₂²⁺

P⁻ = CH₂CH₂COO⁻

Forward and backward transformations were nearly quantitative;
conversion to the dimer at quasi-photostationary state at 4 °C > 55 %.

such a case the prediction of the product stereochemistry on the basis of the crystal structure is not direct.

The structural features of t-30·ox are 1.) the layered crystal structure and 2.) the presence of disorder at the ethylene carbons [81]. The former feature is considered to be the clue responsible for the compulsory reaction control. As pictured in Scheme 49, molecules of t-30 are arranged in a syn head-to-head fashion in the layer and are thus suitable for cycloaddition leading to β-truxinamide. The latter feature indicates that the ethylene bond moiety of t-30 is particularly susceptible to torsional vibration in the crystal lattice, as was likewise suggested for t-30·pht (Scheme 48).

Fumaric acid is photostable in the solid state. Its ammonium salt, however, photoreacted to give the syn [2+2] cyclobutane dimer (Scheme 50) [83]. N-(1-Octadecyl)-4-stilbazolium p-chloro- or p-bromobenzene sulfonate underwent quantitative photodimerization to give the syn head-to-head dimer (Scheme 51) [84]. The structure of the dimers from these photodimerization reactions indicates that a compulsory control of the stereochemistry by steric and coulombic repulsion was unsuccessful in these cases. We are now looking for other crystallographic strategies to synthesize the all-*trans*-cyclobutane dimer of fumaric acid.

In general, *trans,cis*-photoisomerization of olefin double bond is inefficient in the solid state. On the contrary, the isomerization from trans to cis isomers was achieved relatively easily for several kinds of salts (Schemes 50 [83] and 52 [85]). It is concluded from the crystal structure determinations that the isomerizability does not correlate with the cavity in the crystal, but with the structure

Scheme 45

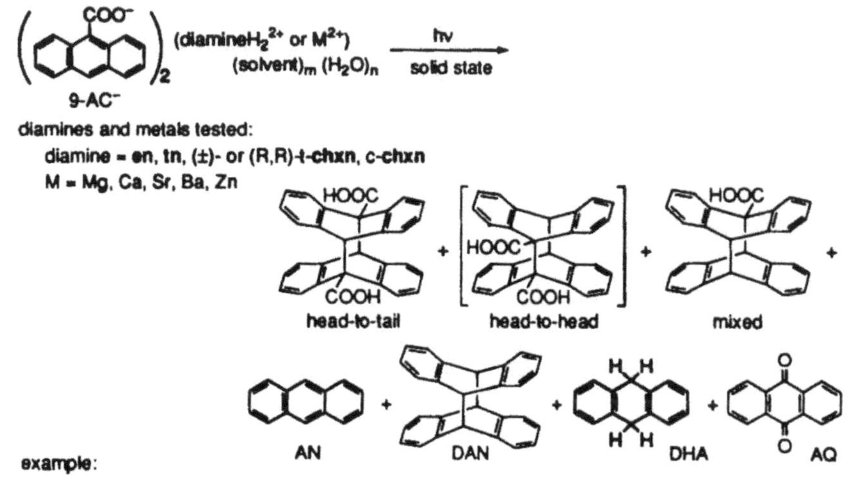
COO⁻
(diamineH2 2+ or M2+)
(solvent)m (H2O)n
hν
solid state
9-AC⁻
diamines and metals tested:
diamine = en, tn, (±)- or (R,R)-t-chxn, c-chxn
M = Mg, Ca, Sr, Ba, Zn
HOOC
COOH
head-to-tail
head-to-head
mixed
AN
DAN
DHA
AQ
example:

| | h-to-t | mixed | AN | DAN | DHA | AQ | 9-AC |
|---|---|---|---|---|---|---|---|
| $(9\text{-}AC^-)_2(c\text{-}chxnH_2^{2+})(EtOH)_{2.1}$ | | | = photostable | | | 2 | 98 % |
| $(9\text{-}AC^-)_2(c\text{-}chxnH_2^{2+})(EtOH)_{0.5}$[a] | 16 | 5 | 5 | 2 | 2 | 2 | 67 |
| $(9\text{-}AC^-)_2Zn^{2+}(MeOH)(H_2O)_2$ | | | = photostable | | | 3 | 97 |
| $(9\text{-}AC^-)_2Zn^{2+}(H_2O)_{0.2}$[a] | 7 | 42 | 3 | 3 | 0 | 3 | 34 |

[a]Dried in vacuo at 70 °C.

of the hydrogen bond network, which tends to be preserved in the course of the isomerization [85].

An interesting dimerization reaction was reported for the crotonic acid salt (Scheme 53) [86].

## IX. SOLID MIXTURES

Despite a large number of reports on the solid-state bimolecular photoreactions (photodimerization) of cinnamic acids [2], bimolecular photoreactions between different organic compounds in the solid state had been little studied when the present author undertook investigations about the bimolecular photoreaction of solid mixtures [87]. I began this study because with no special equipment needed

**Scheme 46**

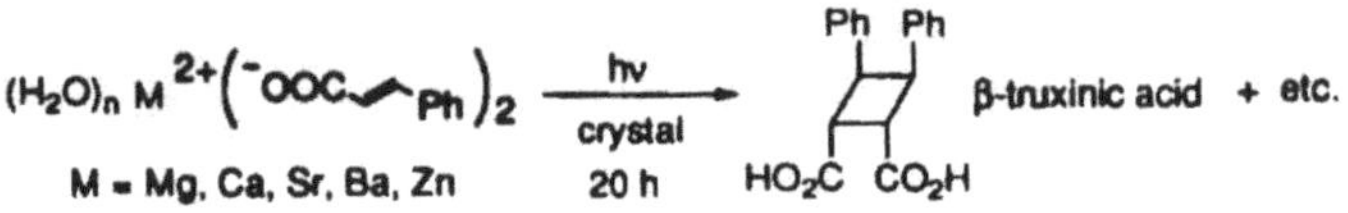

$(H_2O)_n\ M^{2+}\left(^-OOC\diagdown\!\diagup Ph\right)_2 \xrightarrow[\substack{\text{crystal}\\ \text{20 h}}]{h\nu}$    [structure: Ph Ph / HO₂C CO₂H cyclobutane]   β-truxinic acid + etc.

M = Mg, Ca, Sr, Ba, Zn

example:

| M | n | dimer | | | | cinnamic acid | |
|---|---|---|---|---|---|---|---|
| | | β | δ | α | ε | cis | trans |
| Ca | 0.2[a] | 61 | 9 | 0 | 4 | 9 | 7 % |
| | 1.0 | 3 | — photostable | | | | 97 |
| Zn | 0[a] | 22 | 14 | 0 | 5 | 15 | 39 |
| | 2.0 | photostable | | | | | 100 |

[a]Dried at 70 °C in vacuo.

**Scheme 47**

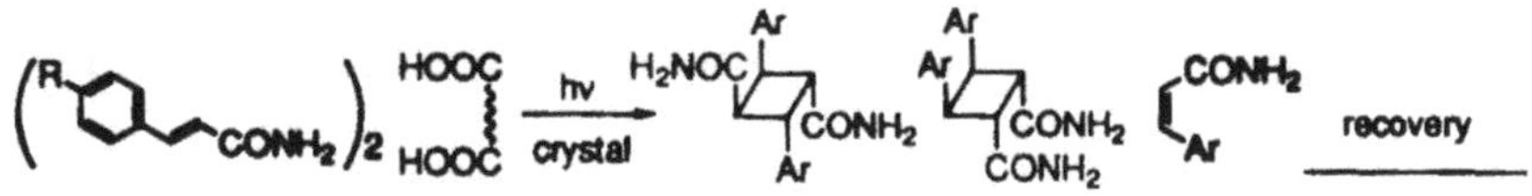

$\left(R\diagdown\!\bigcirc\!\diagup CONH_2\right)_2 \ \ \substack{HOOC\\HOOC} \xrightarrow[\text{crystal}]{h\nu}$   [α-truxillamide structure: H₂NOC, Ar, CONH₂, Ar]   [β-truxinamide structure: Ar, Ar, CONH₂, CONH₂]   [structure: CONH₂, Ar]   recovery

| acid component | α-truxillamide | β-truxinamide | | recovery | |
|---|---|---|---|---|---|
| | | | | acid | amide |
| R = H (t-30)   HOOC–COOH ox | 6 | 42 | 8 | ne | 26 |
|  HOOC⌒COOH su | photostable | | | 100 | 100 |
| 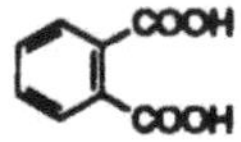 (o-phthalic) COOH COOH pht | 3 | 37 | 5 | 100 | 51 |
|  HOOC⌒COOH fu | 1 | 5 | 5[a] | 45 | 39 |
| R = p-Cl   HOOC–COOH ox | 2 | 86 | 0 | ne | 5 |
| R = p-Me   HOOC–COOH ox | 0 | 43 | 0[b] | ne | 1 |
| R = H, p-Me, or p-Cl   none | Only the α-truxillamides were quantitatively formed. | | | | |

[a]Other products: [structure: H₂NOC, HOOC, Ph, COOH] 50 %, [structure: COOH, COOH] 5 %   [b]Other products: polymer

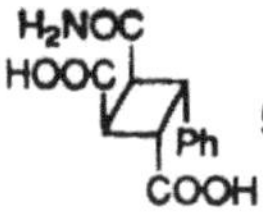

ne = not estimated

**Scheme 48**

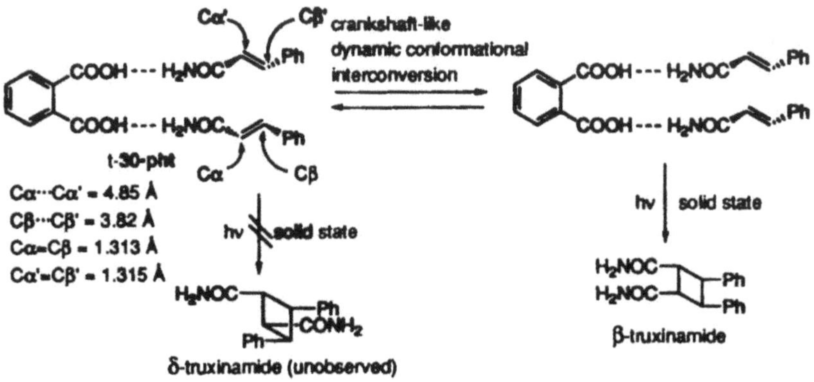

Cα'–Cβ' crankshaft-like dynamic conformational interconversion

Cα'···Cα' = 4.85 Å
Cβ···Cβ' = 3.82 Å
Cα–Cβ = 1.313 Å
Cα'–Cβ' = 1.315 Å

t-30-pht

hν  solid state

δ-truxinamide (unobserved)

hν | solid state

β-truxinamide

it could be easily carried out at any place and at low cost. To begin with, we reported [2+2] photocycloaddition and photoaddition reactions [87–92]. Solid mixtures were usually prepared by melting together the components followed by cooling the melt or by grinding the components together in an agate mortar and pestle. A solid sample prepared by either method afforded a similar result. The reaction selectivity observed in the solid state was different from that in solution, as described below.

An efficient [2+2] photoaddition of benzophenone to substituted imidaz-

**Scheme 49**

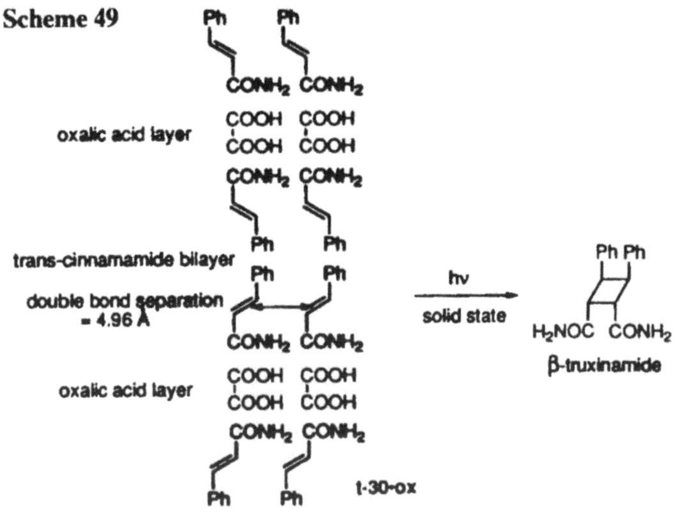

oxalic acid layer

trans-cinnamamide bilayer

double bond separation = 4.96 Å

oxalic acid layer

hν
solid state

β-truxinamide

t-30-ox

Scheme 50

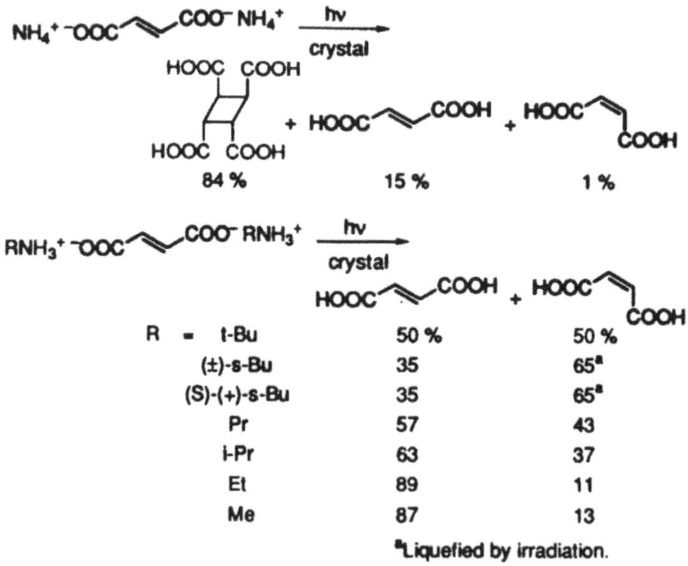
NH4+ -OOC COO- NH4+
hv
crystal
HOOC COOH
HOOC COOH
+ HOOC COOH
+ HOOC COOH
84 %
15 %
1 %
RNH3+ -OOC COO- RNH3+
hv
crystal
HOOC COOH
+ HOOC COOH
R = t-Bu 50 % 50 %
(±)-s-Bu 35 65a
(S)-(+)-s-Bu 35 65a
Pr 57 43
i-Pr 63 37
Et 89 11
Me 87 13
aLiquefied by irradiation.

oles giving oxetanes was found to occur in acetonitrile or benzene solution [87]. The photoaddition likewise went on for substituted benzophenones with their triplet energies less than 70 kcal/mol. In the state of a solid mixture prepared by grinding or cooling the melt, on the other hand, the photoaddition occurred only for the combination of benzophenone and 1-methyl-2,4,5-triphenylimidazole (Scheme 54) [87]. It appears that only for this case, the molecules are arranged in proper geometries favorable for the oxetane formation to occur.

Because of the topochemical restriction of the solid state, bimolecular pho-

Scheme 51

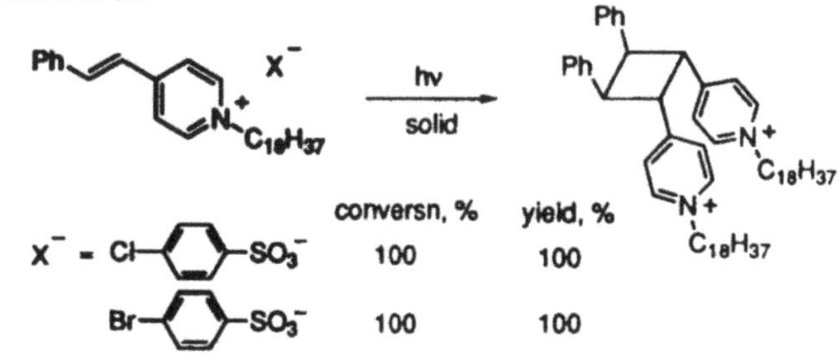
Ph
X-
N+
C18H37
hv
solid
Ph
Ph
N+
C18H37
N+
C18H37
conversn, %
yield, %
X- = Cl SO3- 100 100
Br SO3- 100 100

**Scheme 52**

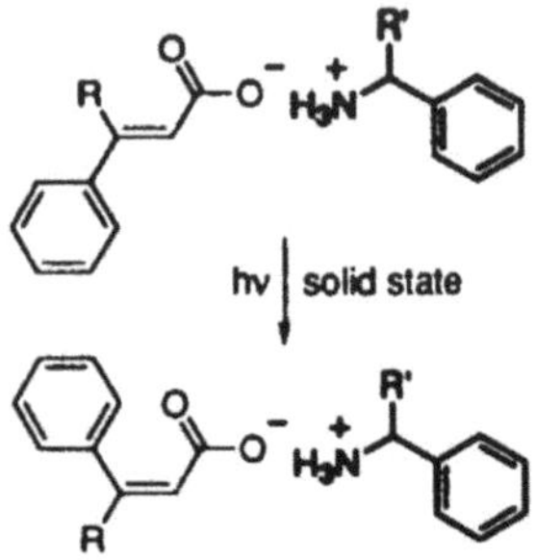
R
O
O⁻
H₃N⁺
R'
hν
solid state
R
O
O⁻
H₃N⁺
R'

| R | R' | E : Z |
|---|---|---|
| Me | Me | 6 : 94 |
| Me | Ph | 33 : 67 |
| Et | Me | 20 : 80 |
| Et | Ph | 92 : 8 |

**Scheme 53**

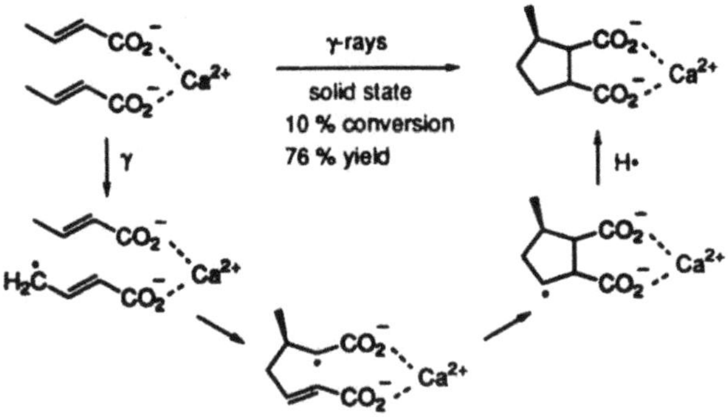
γ-rays
solid state
10 % conversion
76 % yield
γ
H•
Ca²⁺

**Scheme 54**

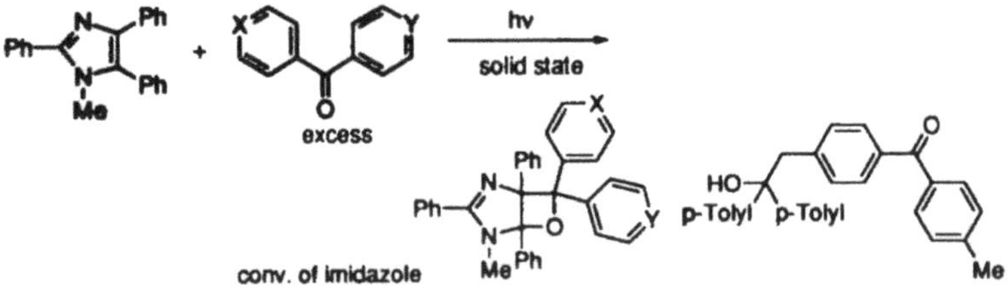
hν
solid state
excess
Ph
N
Me
X
Y
O
HO
p-Tolyl
p-Tolyl
Me

| | conv. of imidazole | | |
|---|---|---|---|
| X = Y = HC | 47 % | 100 % yield | - |
| X = Y = MeC | 1 | no oxetane formed | 15 % yield |
| X = MeC, Y = HC | 8 | no oxetane formed | - |
| X = HC, Y = N | 6 | no oxetane formed | - |

(In solution, 85-100 % yield of oxetane in every case)

**Scheme 55**

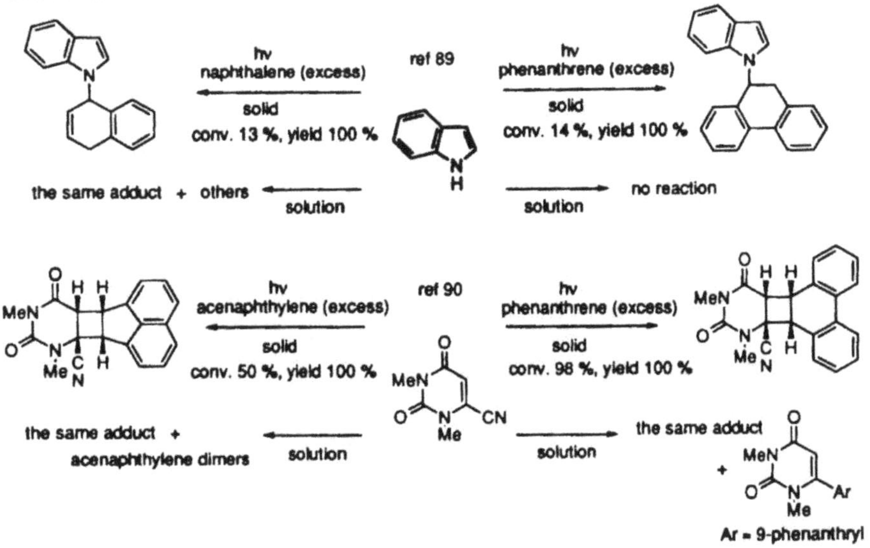

toreactions may occur more selectively in the solid state than in solution. Reactions between indole and naphthalene or phenanthrene and reactions between 6-cyano-1,3-dimethyluracil and acenaphthylene or phenanthrene are the examples (Scheme 55) [89,90]. These selectivities may be due to the presence of certain regular patterns of intermolecular interactions even in a crystalline mixture of different compounds. PXD and DSC analyses revealed that the crystalline mixture of indole and naphthalene forms a molecular compound, while the crystalline mixture of indole and phenanthrene is a simple mixture of each microcrystal [89b].

Workers at Nankai University and Ryukoku University continued this work and studied surprisingly many solid mixtures. They reported photocondensation [93] and hydrogen abstraction reactions [95,96] in addition to photoaddition reactions [94]. Irradiation of solid mixtures, which were prepared from 5-formyl-1,3-dimethyluracil and indole by resolidifying the melt, gave a condensation product, whereas 1,3-dimethyluracil was the sole product in solution (Scheme 56) [93]. Similar photocondensation products were obtained from reactions of 5-formyl-1,3-dimethyluracil and 4-hydroxybenzaldehyde with antipyrine, from reactions of 4-hydroxybenzaldehyde, 4-methoxycarbonylbenzaldehyde, and terephthaldicarboxaldehyde with indole, and from other combinations, regardless of the medium (solid or solution) (Scheme 56).

The crystalline mixture of carbazole and *trans*-stilbene (1:1 to 1:4, by

**Scheme 56**

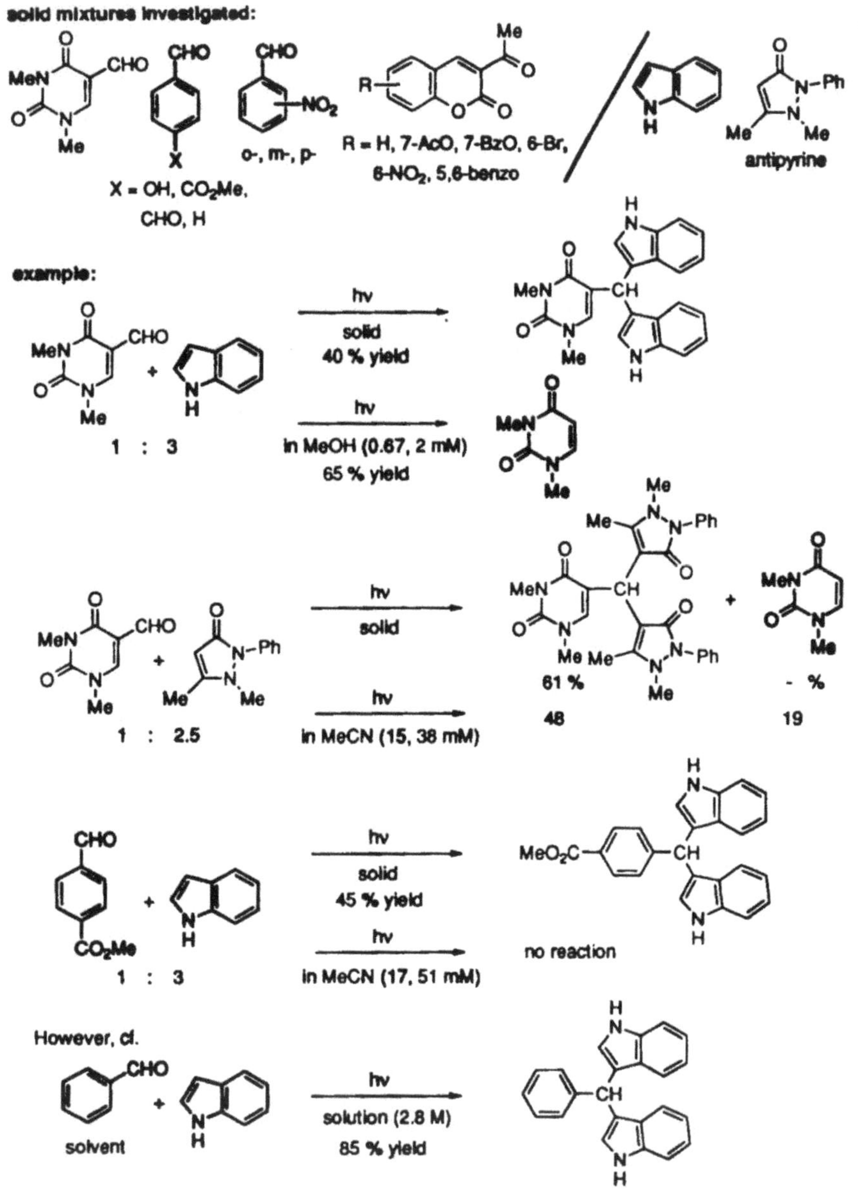

solid mixtures investigated:

R = H, 7-AcO, 7-BzO, 6-Br,
6-NO₂, 5,6-benzo

o-, m-, p-

X = OH, CO₂Me,
CHO, H

antipyrine

example:

$\xrightarrow[\text{40 \% yield}]{\text{hv solid}}$

$\xrightarrow[\substack{\text{in MeOH (0.67, 2 mM)} \\ \text{65 \% yield}}]{\text{hv}}$

1 : 3

$\xrightarrow[\text{solid}]{\text{hv}}$

$\xrightarrow[\text{in MeCN (15, 38 mM)}]{\text{hv}}$

1 : 2.5

61 %          48          - %          19

$\xrightarrow[\text{45 \% yield}]{\text{hv solid}}$

$\xrightarrow[\text{in MeCN (17, 51 mM)}]{\text{hv}}$        no reaction

1 : 3

However, cf.

solvent

$\xrightarrow[\substack{\text{solution (2.8 M)} \\ \text{85 \% yield}}]{\text{hv}}$

cooling the melt) afforded a N-H photoadduct in 25–57% yield (Scheme 57) [94]. It is confirmed that the smaller the crystallite sizes of each compound are, the faster the reaction is. This suggests that the photoreaction proceeded at the interface of the two crystallites. By using indole, phenothiazine, or diphenylamine instead of carbazole, similar photoaddition reactions occurred. In an acetonitrile solution, the same reactions occurred.

Solid mixtures (by cooling the melt) of benzoquinones with polymethyl-benzenes were studied (Scheme 58) [95]. It was shown by single-crystal X-ray diffraction, PXD, DSC, IR, and phase diagram construction that the combination of duroquinone and durene formed a 2:1 molecular compound. This molecular compound underwent photochemical intermolecular hydrogen abstraction to give durohydroquinone and two adducts. Other combinations also underwent similar

**Scheme 57**

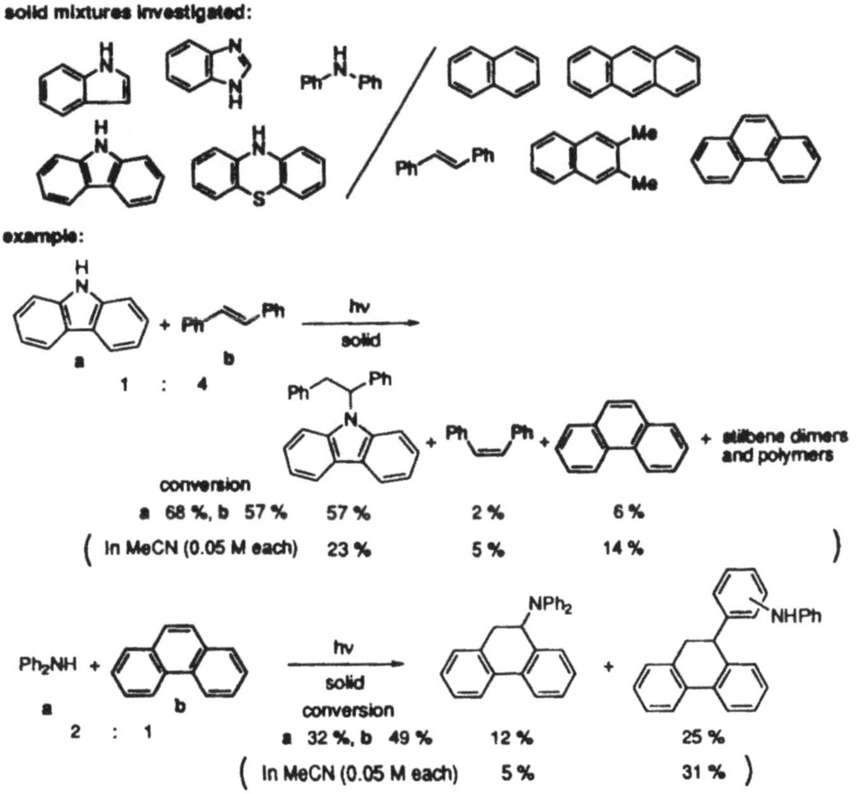

solid mixtures investigated:

example:

Scheme 58

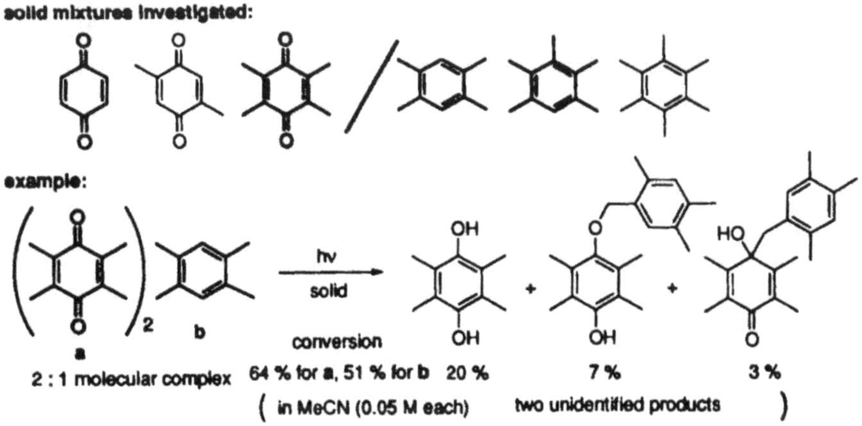
solid mixtures investigated:
example:
a
b
2
hv
solid
conversion
OH
OH
OH
OH
HO
O
O
2 : 1 molecular complex
64 % for a, 51 % for b
20 %
7 %
3 %
( in MeCN (0.05 M each) two unidentified products )

reactions, but the molecular compound formations were not observed in these cases.

The solid mixtures (by cooling the melt) between benzophenones and hydrogen donors were irradiated to afford hydrogen transfer products (Scheme 59) [96]. For example, benzophenone and durene (2:1 molar ratio) gave benzopina-

Scheme 59

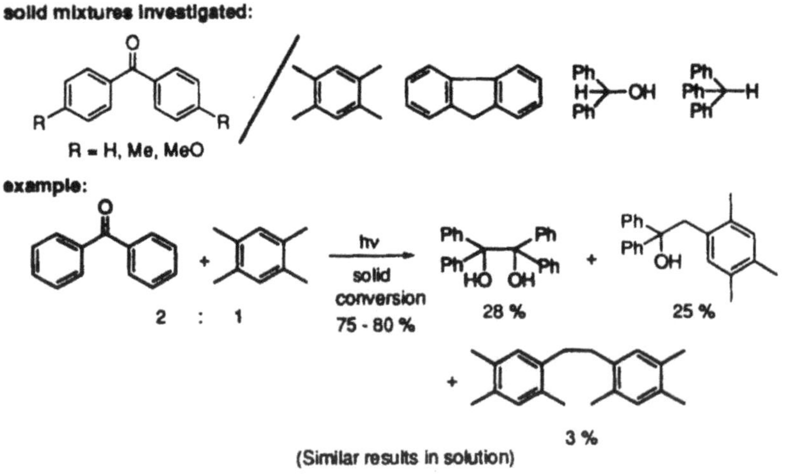
solid mixtures investigated:
R = H, Me, MeO
example:
hv
solid
conversion
75 - 80 %
2 : 1
Ph
HO
OH
OH
28 %
25 %
3 %
(Similar results in solution)

**Scheme 60**

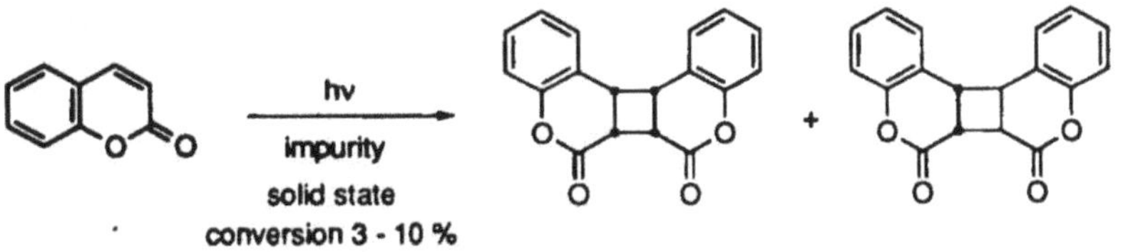

added impurity = antipyrine, phenanthrene, 6-cyano-1,3-dimethyluracil, 5-bromo-1,3-
dimethyluracil, 5-iodo-1,3-dimethyluracil,

col, a coupling product and a durene dimer with 20–25% recoveries of starting materials. Photolyses in solution gave the same products.

They reported even the effect of added impurities: coumarin, which is photoinert in the solid state, photodimerized when admixed (by cooling the melt) with various foreign substances such as antipyrine (Scheme 60) [97]. Compounds which are photoreactive in solution may frequently be photoinert in the solid state owing to the rigid crystalline environment. Addition of external substances disrupts the crystal lattice and the compounds recover their photoreactivity. This is very commonly observed.

The products from thermolysis or photolysis of p-benzoquinone or 1,4-naphthoquinone with p-chlorothiophenol were investigated (Scheme 61) [98].

Photocoupling of 5-bromouracil to L-tryptophan in frozen aqueous solution or in the solid state was reported (Scheme 62) [99]. The solid mixture was prepared by evaporation of an aqueous solution or by freeze-drying.

A well-ground mixture of fullerene ($C_{60}$) and 9-methylanthracene (1:1) was

**Scheme 61**

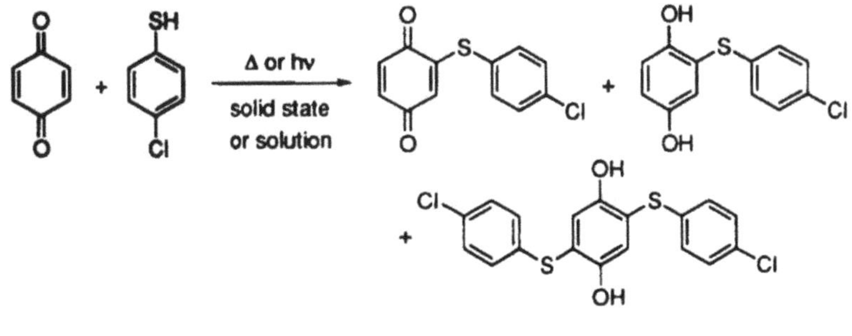

**Scheme 62**

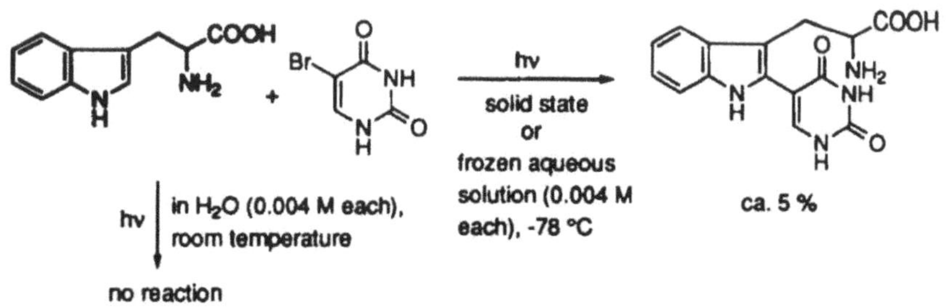

irradiated in the solid state to give the Diels-Alder adducts without any formation of 9-methylanthracene dimers (Scheme 63) [100]. In contrast, unsubstituted an-thracene did not react with $C_{60}$ under the same conditions. The Diels-Alder reac-tion appears to proceed via photoinduced electron transfer to excited $C_{60}$ and the higher ionization potential of anthracene than 9-methylanthracene was thought to be responsible for their difference in reactivity.

Photosensitized decomposition of many pesticides (dieldin, aldrin, isodrin, endrin, and others) was investigated (Scheme 64) [101].

Irradiation of a 1:1 (weight) solid mixture of m-phenylenediacrylic acid and picramide with >430 nm light gave a cyclobutane oligomer of m-PDA (~80% conversion) (Scheme 65). This photooligomerization reaction of m-PDA must be sensitized by picramide via a T-T energy transfer, since the light was absorbed only by picramide and m-PDA phosphorescence was observed. Simi-larly, 1,2-benzoanthraquinone and 2-nitrofluorene also sensitized the reaction, although they were less effective than picramide [102].

Mechanistic details of bimolecular photoreactions occurring in solid mixtures are vague. Workers at Ryukoku University attacked this problem

**Scheme 63**

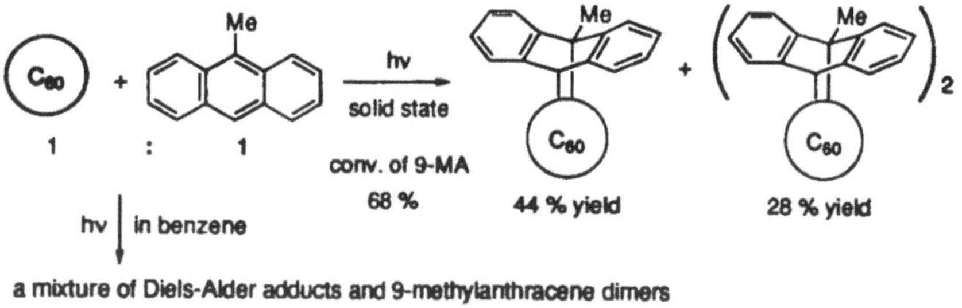

a mixture of Diels-Alder adducts and 9-methylanthracene dimers

**Scheme 64**

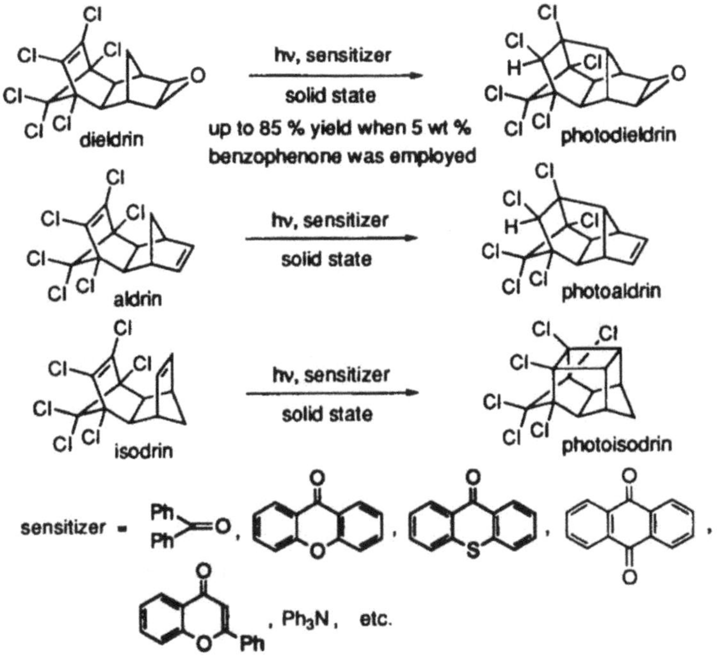

[93b,94a,94c]. On the basis of PXD, IR, and DSC analyses, they claim that each combination of indole/naphthalene, carbazole/anthracene, or 5-formyl-1,3-dimethyluracil/antipyrine forms a molecular compound, while a solid mixture between diphenylamine and phenanthrene, diphenylamine and *trans*-stilbene, carbazole and *trans*-stilbene, indole and phenanthrene, indole and *trans*-stilbene, phenothiazine and *trans*-stilbene, or 4-hydroxybenzaldehyde and indole is a simple mechanical mixture. In the latter case, the photoreaction was presumed to occur at the interface of each crystallite. It was shown for carbazole/*trans*-stilbene that the smaller the crystallite sizes were, the faster the photoaddition was [94a].

**Scheme 65**

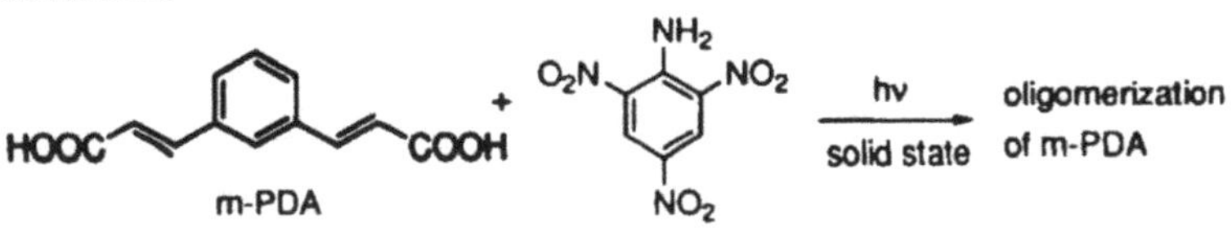

In the former case, it was assumed to proceed within the crystal lattice of the molecular compound. However, in a series of combinations, a simple mixture and a molecular compound gave the same types of photoproduct. Furthermore, essentially the same photoreaction occurred in solution. Thus, there is still much to be done to elucidate the progress of mixed bimolecular photoreactions in the solid state.

The above type of study cannot be designed or planned, and is tedious. The workers at Ryukoku challenged this task, to which few organic photochemists had paid attention. Unfortunately, they left this exploration suddenly and changed their focus to hydrogen-bonded cocrystals and charge transfer crystals, studying the traditional crystal structure–reactivity relationship (see Secs. VI and VII). As a result, my original purpose of discovering both a guide to rational design of the system and a spectacular reaction characteristic of the solid mixture which contrasts with solution reactions is yet to be realized.

## X.  PERSPECTIVES

Organic photoreactions in the solid state are strictly controlled by the crystal structure. As a result, at least in principle, the solid-state photoreactivities and selectivities must be freely controllable by the high-level design of the crystal structure by manipulating a range of supramolecular interactions. Studies on compulsory control of unimolecular and bimolecular photochemical reactions in mixed molecular crystals and solid mixtures are a primary step toward this ultimate goal. In contrast to one-component crystals, the organic photochemistry of such multicomponent crystals is still very young and hence has an abundant serendipity and possibility. More investigations in this field are required.

## ACKNOWLEDGMENT

This review was made possible by the support of the Grant-in-Aid for Scientific Research from the Japanese Government (Nos. 07231213, 08221213, 10132229, 10440215). I also acknowledge Iketani Science and Technology Foundation, Murata Science Foundation, Shorai Foundation for Science and Technology, and Nagase Science and Technology Foundation.

**Appendix 1.**  Typical Supramolecular Interactions.

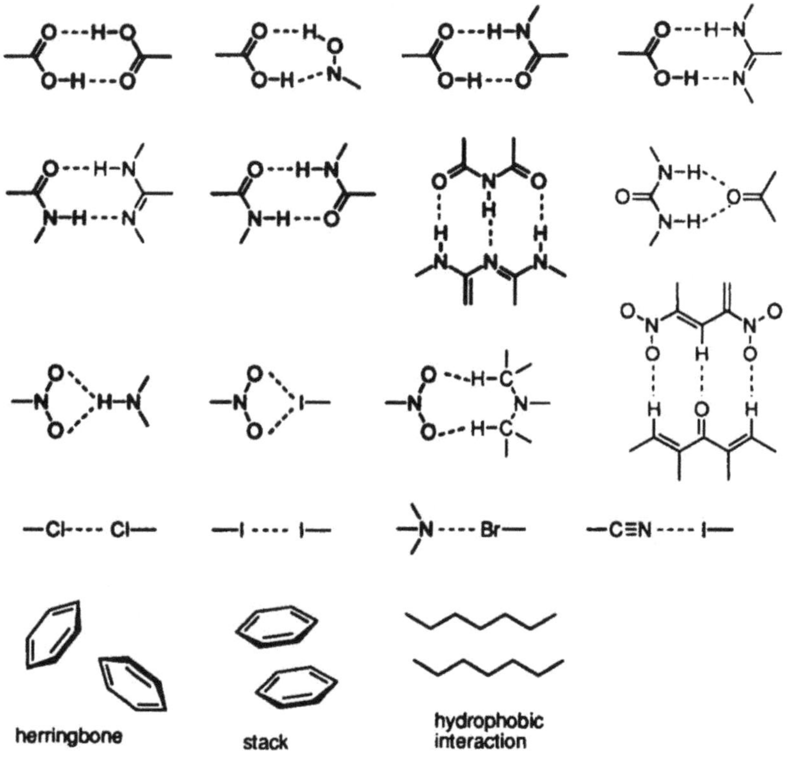

herringbone          stack          hydrophobic
                                     interaction

**Appendix 2.** Numerous Examples of Crystalline Molecular Complexes.

[p-aminobenzoic acid·3,5-dinitrobenzoic acid, 1:1]
Etter, M. C.; Frankenbach, G. M. *Chem. Mater.*
1989, *1*, 10-12. Harris, K. D. M.; Hollingsworth,
M. D. *Nature* 1989, *341*, 19.

[diacetamide·hydroquinone, 2:1]
Etter, M. C.; Reutzel, S. M. *J. Am. Chem. Soc.* 1991, *113*,
2586-2598.

[diacetamide-4-hydroxybenzoic acid, 1:1]

[1,4-dicyanobutane·urea, 1:1]
Hollingsworth, M. D.; Brown, M. E.; Santarsiero,
B. D.; Huffman, J. C.; Goss, C. R. *Chem. Mater.*
1994, *6*, 1227-1244. Hollingsworth, M. D.;
Santarsiero, B. D.; Oumar-Maharnat, H.; Nichols,
C. J. *Chem. Mater.* 1991, *3*, 23-25.

[1,3-bis(m-nitrophenyl)urea·cyclohexanone, 1:1]
[1,3-bis(m-nitrophenyl)urea·N,N-dimethyl-p-nitroaniline, 1:1]
Etter, M. C.; Urbanczyk-Lipkowska, Z.; Zia-Erbahimi, M.;
Panunto, T. W. *J. Am. Chem. Soc.* 1990, *112*, 8415-8426.

[benzamide·succinic acid, 2:1]
Leiserowitz, L. *Acta Cryst.* 1976, *B32*, 775-802.

[benzoic acid·acetophenone oxime, 1:1]   [benzamide·benzaldehyde oxime, 1:1]
Maurin, J. K.; Winnika-Maurin, M.; Paul, I. C.; Curtin, D. Y. *Acta Cryst.* 1993, *B49*, 90-96.

## Appendix 2. Continued

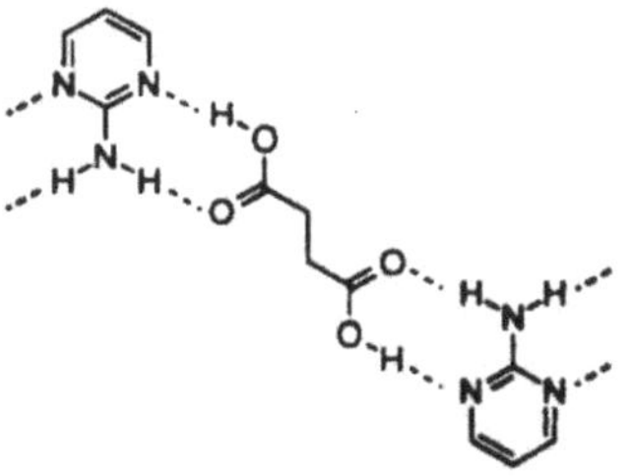

[2-aminopyrimidine-succinic acid, 1:1]
Etter, M. C.; Adsmond, D. A. *J. Chem. Soc.,
Chem. Commun.* 1990, 589-591.

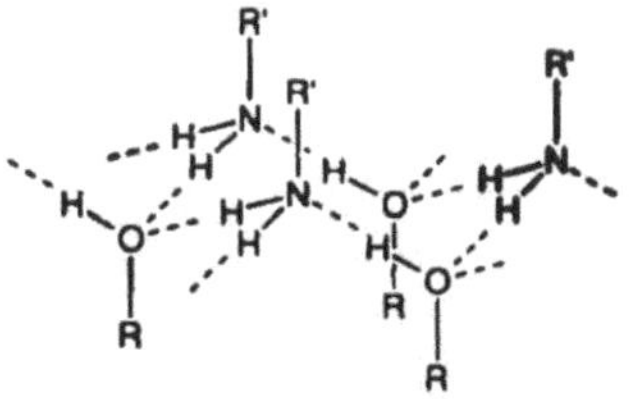

three-dimensional diamondoid networks
[hydroquinone-p-phenylenediamine, 1:1]
Ermer, O.; Eling, A. *J. Chem. Soc., Perkin Trans. 2*
1994, 925-944.

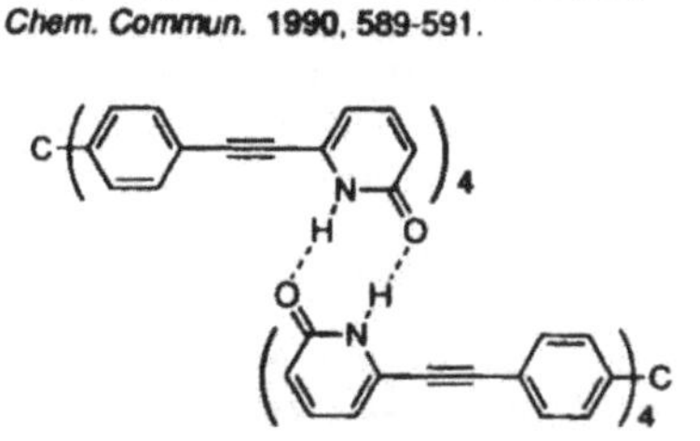

three-dimensional diamondoid lattice
[tetrapyridone tecton-butyric acid, 1:2]

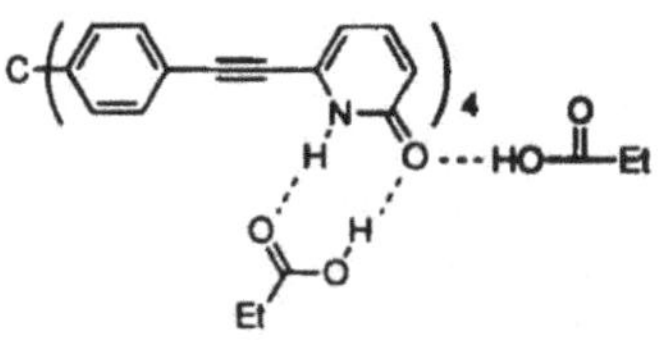

[tetrapyridone tecton-propionic acid, 1:8]

Simard, M.; Su, D.; Wuest, J. D. *J. Am. Chem. Soc.* 1991, *113*, 4696-4698. cf. Brunet, P.; Simard, M.;
Wuest, J. D. *J. Am. Chem. Soc.* 1997, *119*, 2737-2738; Zaworotko, M. J. *Nature* 1997, *386*, 220-221.

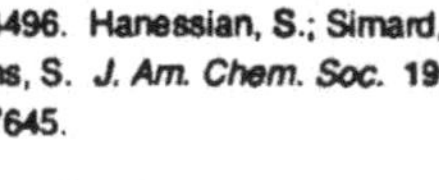

left-handed trihelicate structure

[(S,S)-trans-1,2-diaminocyclohexane-(S,S)-
trans-1,2-cyclohexanediol, 1:1]
Hanessian, S.; Gomtsyan, A.; Simard, M.;
Roelens, S. *J. Am. Chem. Soc.* 1994, *116*,
4495-4496. Hanessian, S.; Simard, M.;
Roelens, S. *J. Am. Chem. Soc.* 1995, *117*,
7630-7645.

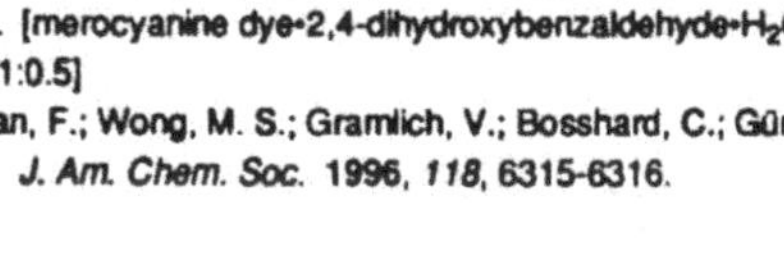

[merocyanine dye-2-amino-5-nitrophenol, 1:1]
Pan, F.; Wong, M. S.; Gramlich, V.; Bosshard, C.; Günter,
P. *J. Chem. Soc., Chem. Commun.* 1996, 1557-1558.
cf. [merocyanine dye-2,4-dihydroxybenzaldehyde-H₂O,
1:1:0.5]
Pan, F.; Wong, M. S.; Gramlich, V.; Bosshard, C.; Günter,
P. *J. Am. Chem. Soc.* 1996, *118*, 6315-6316.

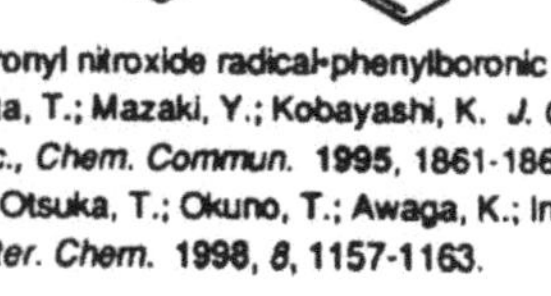

[nitronyl nitroxide radical-phenylboronic acid, 1:1]
Akita, T.; Mazaki, Y.; Kobayashi, K. *J. Chem.
Soc., Chem. Commun.* 1995, 1861-1862.
cf. Otsuka, T.; Okuno, T.; Awaga, K.; Inabe, T. *J.
Mater. Chem.* 1998, *8*, 1157-1163.

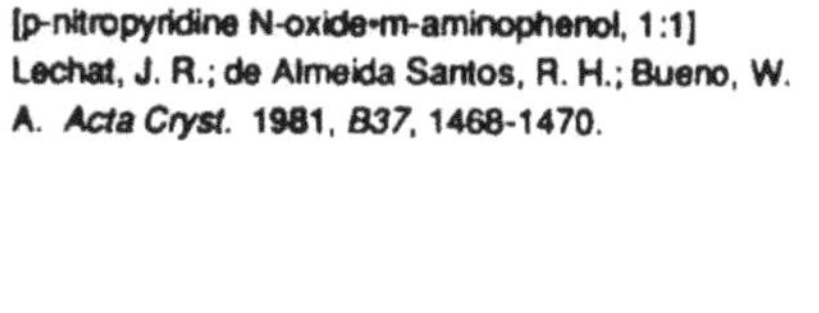

[p-nitropyridine N-oxide-m-aminophenol, 1:1]
Lechat, J. R.; de Almeida Santos, R. H.; Bueno, W.
A. *Acta Cryst.* 1981, *B37*, 1468-1470.

## Appendix 2. Continued

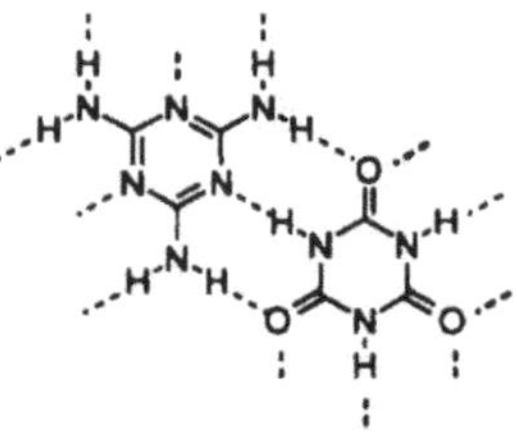

[melamine–cyanuric acid, 1:1]

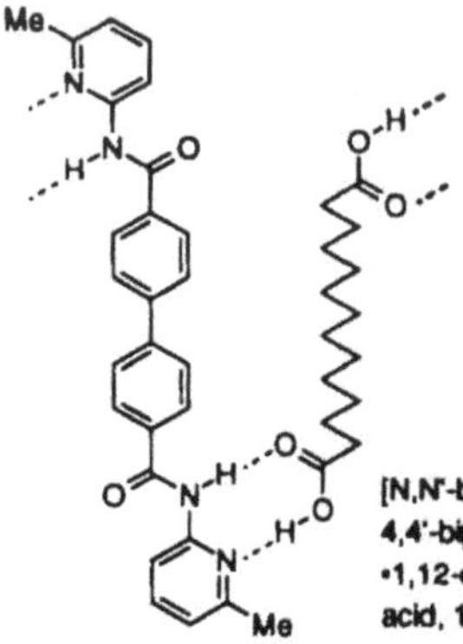

linear tape

X = H, F, Cl, Br, I, CH₃, CF₃

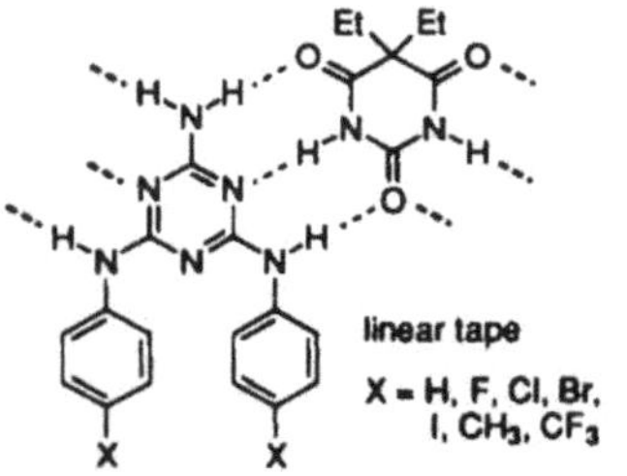

[N,N'-bis(p-X-phenyl)melamine–5,5-diethylbarbituric acid, 1:1]

[N,N'-bis[2-(6-methyl)pyridyl]-4,4'-biphenyldicarboxamide •1,12-dodecanedicarboxylic acid, 1:1]

Garcia-Tellado, F.; Goswami, S.; Chang, S. K.; Geib, S. J.; Hamilton, A. D. *J. Am. Chem. Soc.* **1990**, *112*, 7393-7394.
Garcia-Tellado, F.; Geib, S. J.; Goswami, S.; Hamilton, A. D. *J. Am. Chem. Soc.* **1991**, *113*, 9265-9269.
cf. Bielawski, C.; Chen, Y.-S.; Zhang, P.-J.; Prest, P.-J.; Moore, J. S. *J. Chem. Soc., Chem. Commun.* **1998**, 1313-1314.

A crinkled tape when X = CO₂Me and a rosette when X = t-Bu. Zerkowsky, J. A.; Seto, C. T.; Wierda, D. A.; Whitesides, G. M. *J. Am. Chem. Soc.* **1990**, *112*, 9025-9026. Zerkowsky, J. A.; Seto, C. T.; Whitesides, G. M. *J. Am. Chem. Soc.* **1992**, *114*, 5473-5475. Zerkowsky, J. A.; MacDonald, J. C.; Seto, C. T.; Wierda, D. A.; Whitesides, G. M. *J. Am. Chem. Soc.* **1994**, *116*, 2382-2391. Zerkowsky, J. A.; Whitesides, G. M. *J. Am. Chem. Soc.* **1994**, *116*, 4298-4304. Zerkowsky, J. A.; Mathias, J. P.; Whitesides, G. M. *J. Am. Chem. Soc.* **1994**, *116*, 4305-4315.

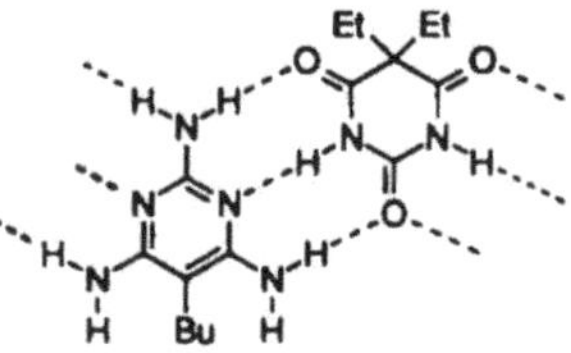

[2,4,6-triamino-5-butylpyrimidine–5,5-diethylbarbituric acid, 1:1]

Lehn,J.-M.; Mascal, M.; DeCian,A.; Fischer, *J. Chem. Soc., Chem. Commun.* **1990**, 479-481; *J. Chem. Soc., Perkin Trans. 2* **1992**, 461-467.

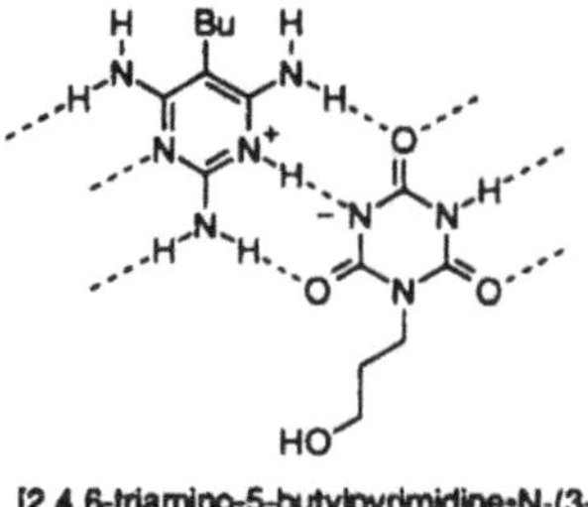

[2,4,6-triamino-5-butylpyrimidine–N-(3-hydroxypropyl)cyanuric acid, 1:1]

Mascal, M.; Fallon, P. S.; Batsanov, A. S.; Heywood, B. R.; Champ, S.; Colclough, M. *J. Chem. Soc., Chem. Commun.* **1995**, 805-806. Mascal, M.; Hext, N. M.; Warmuth, R.; Moore, M. H.; Turkenburg, J. P. *Angew. Chem., Int. Ed. Engl.* **1996**, *35*, 2204-2206.

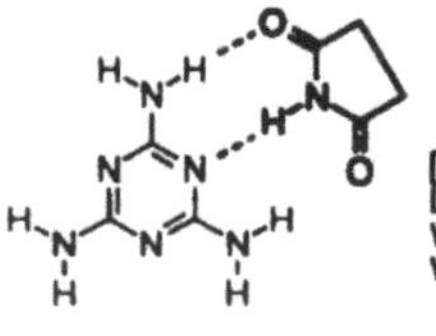

[melamine–succinimide, 1:1]
Lange, R. F. M.; Beijer, F. H.; Sijbesma, R. P.; Hooft, R. W. W.; Kooijman, H.; Spek, A. L.; Kroon, J.; Meijer, E. W. *Angew. Chem., Int. Ed. Engl.* **1997**, *36*, 969-971.

## Appendix 2. Continued

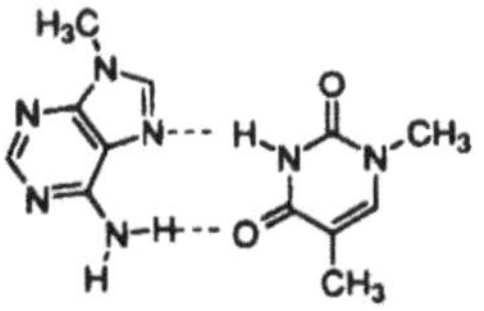

[9-methyladenine-1-methylthymine, 1:1]
Hoogsteen mode
Ref 8

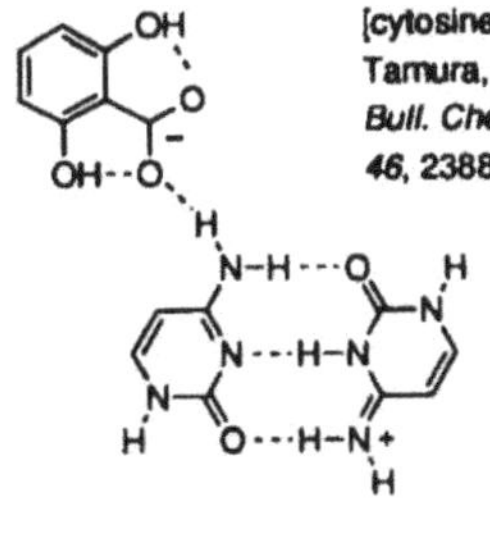

[cytosine-resorcylic acid, 2:1]
Tamura, C.; Sato, S.; Hata, T.
*Bull. Chem. Soc. Jpn.* 1973,
*46*, 2388-2394.

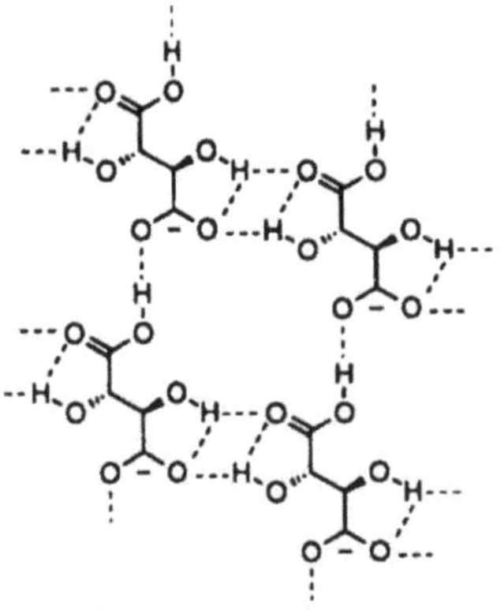

((S)-1-phenylethylammonium cations are not shown)

[(S)-1-phenylethylammonium hydrogen-meso-tartrate]
Aakeröy, C. B.; Hitchcock, P. B.; Seddon, K. R. *J.
Chem. Soc., Chem. Commun.* 1992, 553-555.
Aakeröy, C. B.; Hitchcock, P. B. *J. Mater. Chem.*
1993, *3*, 1129-1135.

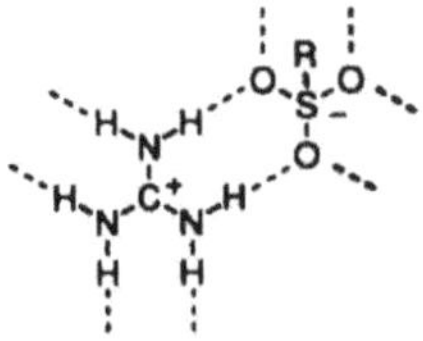

R = Me, Et, Pr, Bu, CF₃, (1S)-(+)-10-camphor,
Ph, 1-naph, 2-naph

R = Me, Et, Pr, Bu, $CF_3$, (1S)-(+)-10-camphor,
Ph, 1-naph, 2-naph

[guanidinium alkane- and arenesulfonates]
Russell, V. A.; Etter, M. C.; Ward, M. D. *J. Am.
Chem. Soc.* 1994, *116*, 1941-1952. Swift, J.
A.; Pivovar, A. M.; Reynolds, A. M.; Ward, M. D.
*J. Am. Chem. Soc.* 1998, *120*, 5887-5894.

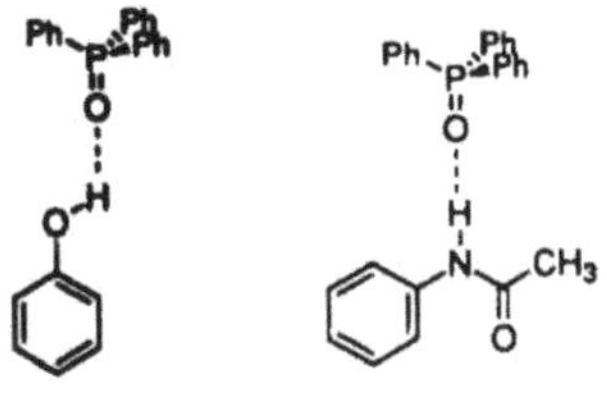

[triphenylphosphine oxide-phenol, 1:1]
[TPPO-acetanilide, 1:1]
Ref 11

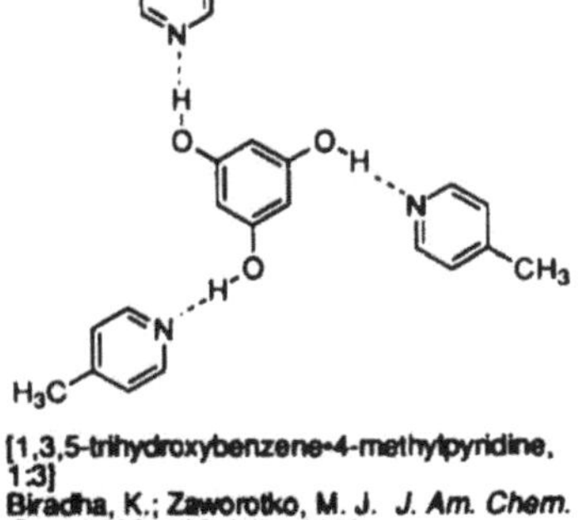

[1,3,5-trihydroxybenzene-4-methylpyridine,
1:3]
Biradha, K.; Zaworotko, M. J. *J. Am. Chem.
Soc.* 1998, *120*, 6431-6432.

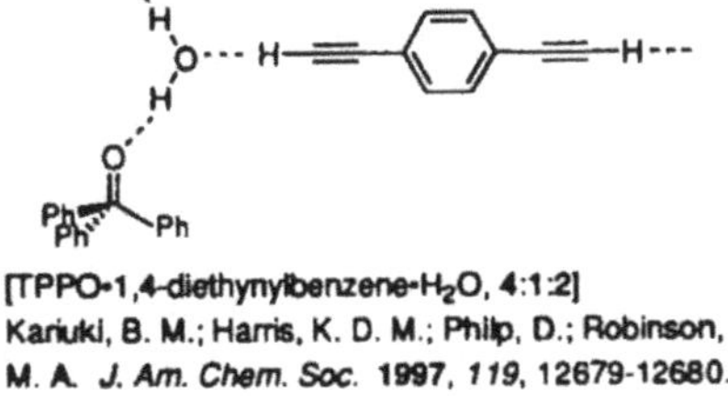

[TPPO-1,4-diethynylbenzene-H₂O, 4:1:2]
Kariuki, B. M.; Harris, K. D. M.; Philp, D.; Robinson, J.
M. A. *J. Am. Chem. Soc.* 1997, *119*, 12679-12680.

# Appendix 2.  Continued

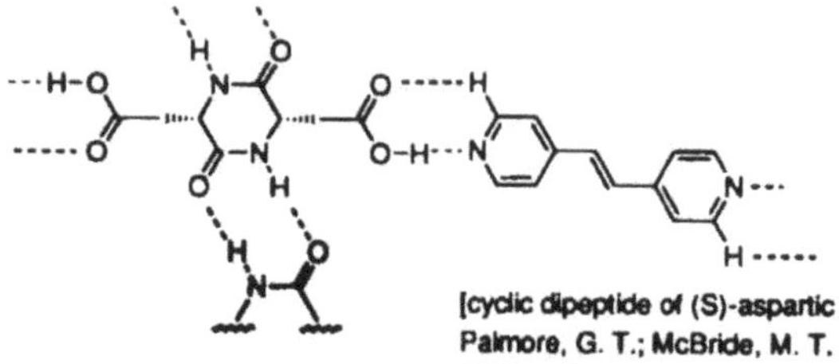

[cyclic dipeptide of (S)-aspartic acid-(E)-1,2-bis(4-pyridyl)ethene, 1:1]
Palmore, G. T.; McBride, M. T. *J. Chem. Soc., Chem. Commun.*
1998, 145-146.

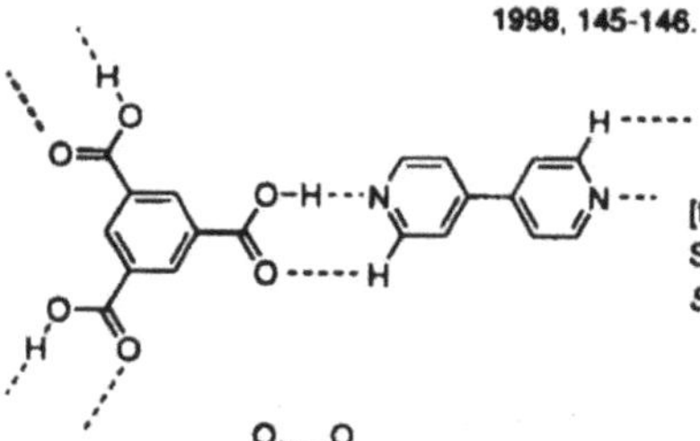

[trimesic acid-4,4'-bipyridine, 1:1.5]
Sharma, C. V. K.; Zaworotko, M. J. *J. Chem.
Soc., Chem. Commun.* 1996, 2655-2656.

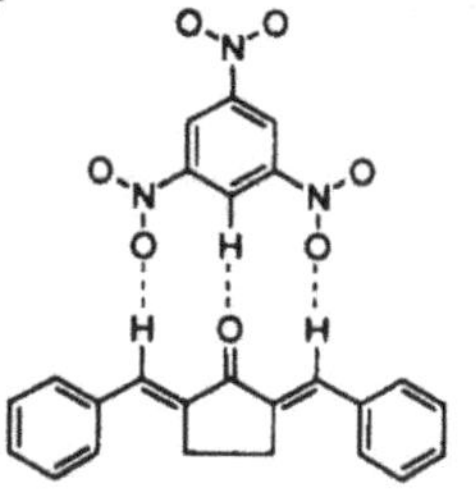

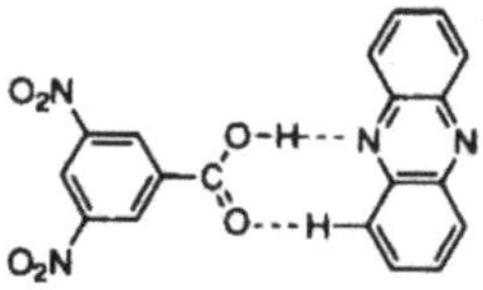

[3,5-dinitrobenzoic acid-phenazine, 1:1]
Pedireddi, V. R.; Jones, W.; Chorlton, A. P.; Docherty,
R. *J. Chem. Soc., Chem. Commun.* 1996, 997-998.

[2,5-dibenzylidenecyclopentanone-1,3,5-trinitrobenzene, 1:2]
Biradha, K.; Sharma, C. V. K.; Panneerselvam, K.; Shimoni, L.; Carrell, H. L.; Zacharias, D.
E.; Desiraju, G. R. *J. Chem. Soc., Chem. Commun.* 1993, 1473-1475. Biradha, K.; Nangia,
A.; Desiraju, G. R.; Carrell, C. J.; Carrell, H. L. *J. Mater. Chem.* 1997, 7, 1111-1122.

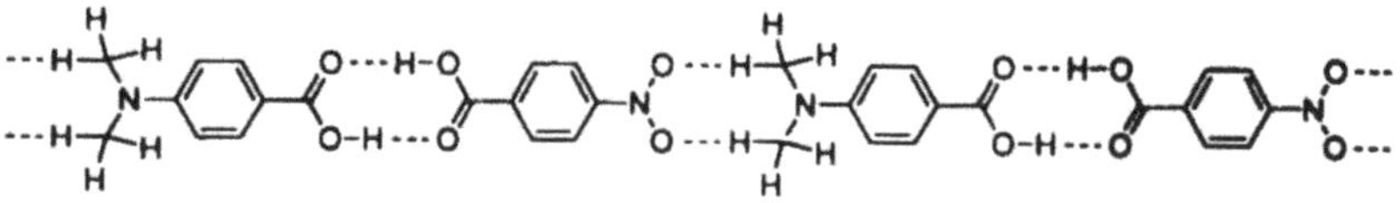

[4-nitrobenzoic acid-4-(N,N-dimethylamino)benzoic acid, 1:1]
Sharma, C. V. K.; Panneerselvam, K.; Pilati, T.; Desiraju, G. R. *J. Chem. Soc., Chem. Commun.* 1992,
832-833. Sharma, C. V. K.; Desiraju, G. R. *J. Chem. Soc., Perkin Trans. 2* 1994, 2345-2352. Sharma, C.
V. K.; Panneerselvam, K.; Pilati, T.; Desiraju, G. R. *J. Chem. Soc., Perkin Trans. 2* 1993, 2209-2216.

## Appendix 2. Continued

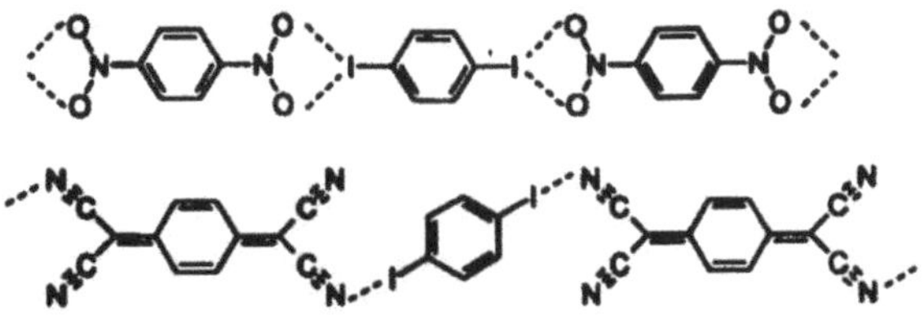

[1,4-diiodobenzene•1,4-dinitrobenzene, 1:1]
[1,4-diiodobenzene•7,7,8,8-tetracyanoquinodimethane, 1:1]
Allen, F. H.; Goud, B. S.; Hoy, V. J.; Howard, J. A. K.; Desiraju, G. R. *J. Chem. Soc., Chem. Commun.* 1994, 2729-2730. Thalladi, V. R.; Goud, B. S.; Hoy, V. J.; Allen, F. H.; Howard, J. A. K.; Desiraju, G. R. *J. Chem. Soc., Chem. Commun.* 1996, 401-402. Ranganathan, A.; Pedireddi, V. R. *Tetrahedron Lett.* 1998, *39*, 1803-1806.

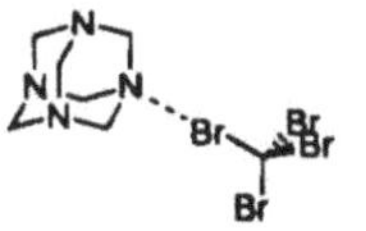

three-dimensional diamondoid lattice

[hexamethylenetetramine•carbon tetrabromide, 1:1]
Reddy, D. S.; Craig, D. C.; Rae, A. D.; Desiraju, G. R. *J. Chem. Soc., Chem. Commun.* 1993, 1737-1739. Reddy, D. S.; Craig, D. C.; Desiraju, G. R. *J. Am. Chem. Soc.* 1996, *118*, 4090-4093. Reddy, D. S.; Craig, D. C.; Desiraju, G. R. *J. Chem. Soc., Chem. Commun.* 1994, 1457-1458.

a

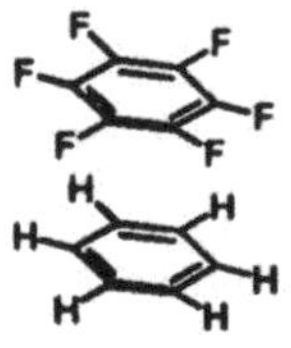

[benzene•hexafluorobenzene, 1:1]
Williams, J. H.; Cockcroft, J. K.; Fitch, A. N. *Angew. Chem., Int. Ed. Engl.* 1992, *31*, 1655-1657. Ref 37a.

b

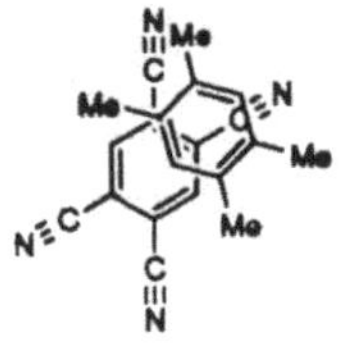

[durene•1,2,4,5-tetracyanobenzene, 1:1]
Lefebvre, J.; Miniewicz, A.; Kowal, R. *Acta Cryst.* 1989, *C45*, 1372-1376.

c

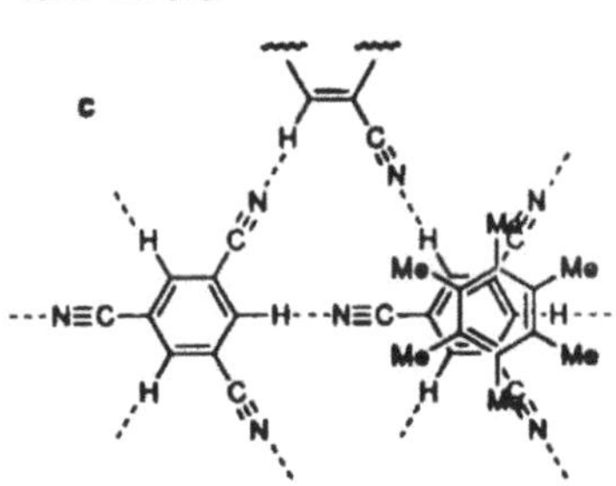

layer structure (HMB layer is not shown.)

[1,3,5-tricyanobenzene•hexamethylbenzene, 1:1]
Reddy, D. S.; Goud, B. S.; Panneerselvam, K.; Desiraju, G. R. *J. Chem. Soc., Chem. Commun.* 1993, 663-664. Reddy, D. S.; Craig, D. C.; Desiraju, G. R. *J. Am. Chem. Soc.* 1996, *118*, 4090-4093.
cf. [Trimethylisocyanurate•1,3,5-trinitrobenzene, 1:1] similar layer structure
Thalladi, V. R. ; Panneerselvam, K.; Carrell, C. J.; Carrell, H. L.; Desiraju, G. R. *J. Chem. Soc., Chem. Commun.* 1995, 341-342.

## Appendix 2.   Continued

d

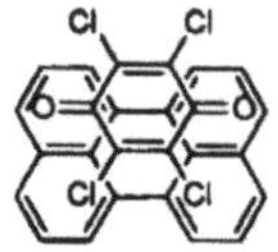

[perylene-p-chloranil, 1:1]
Kozawa, K.; Uchida, T. *Acta Cryst.* 1983, *C39*,
1233-1235.

e

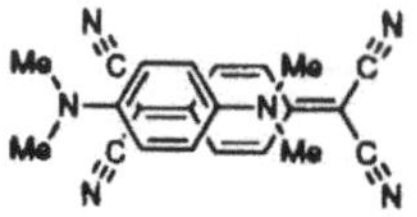

[7,7,8,8-tetracyanoquinodimethane-N,N,N',N'-
tetramethyl-p-phenylenediamine, 1:1]
Hanson, A. W. *Acta Cryst.* 1965, *19*, 610-613.

f                                                          two polymorphs

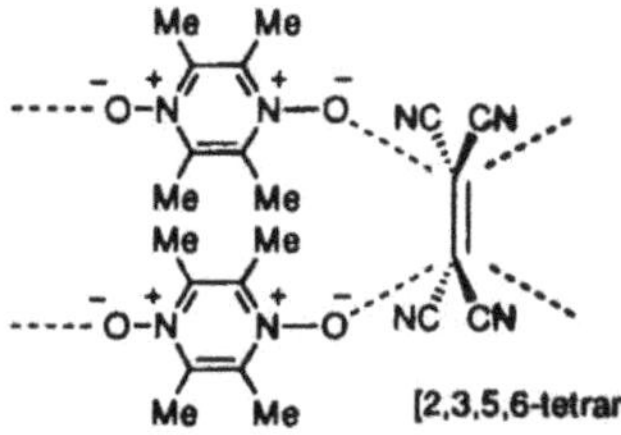

[2,3,5,6-tetramethylpyrazine-N,N'-dioxide·tetracyanoethylene, 2:1]

Greer, M. L.; McGee, B. J.; Rogers, R. D.; Blackstock, S. C. *Angew. Chem., Int. Ed. Engl.* 1997, *36*,
1864-1867. Greer, M. L.; Duncan, J. R.; Duff, J. L.; Blackstock, S. C. *Tetrahedron Lett.* 1997, *38*,
7665-7668. Blackstock, S. C.; Poehling, K.; Greer, M. L. *J. Am. Chem. Soc.* 1995, *117*, 6617-
6618; *ibid.* 1997, *119*, 1498. Greer, M. L.; Blackstock, S. C. *J. Org. Chem.* 1996, *61*, 7895-7903.
Greer, M. L.; Blackstock, S. C. *J. Am. Chem. Soc.* 1997, *119*, 11343-11344.

g

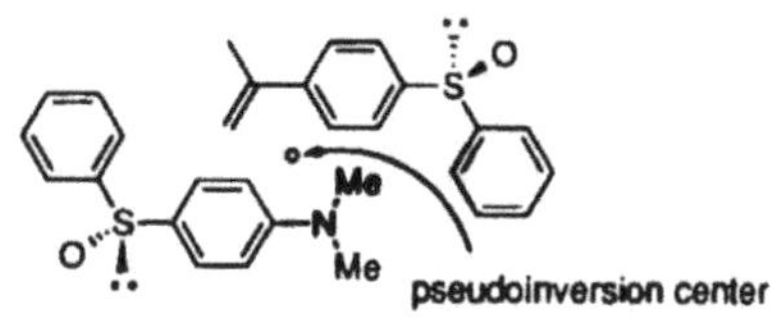

pseudoinversion center

[(S)-p-(isopropenyl)diphenyl sulfoxide-(R)-p-
(dimethylamino)diphenyl sulfoxide, 1:1]
Davis, R. E.; Whitesell, J. K.; Wong, M.-S.;
Chang, N.-L. In ref 17d, pp 63-106. Whitesell,
J. K.; Davis, R. E.; Wong, M.-S.; Chang, N.-L.
*J. Am. Chem. Soc.* 1994, *116*, 523-527.

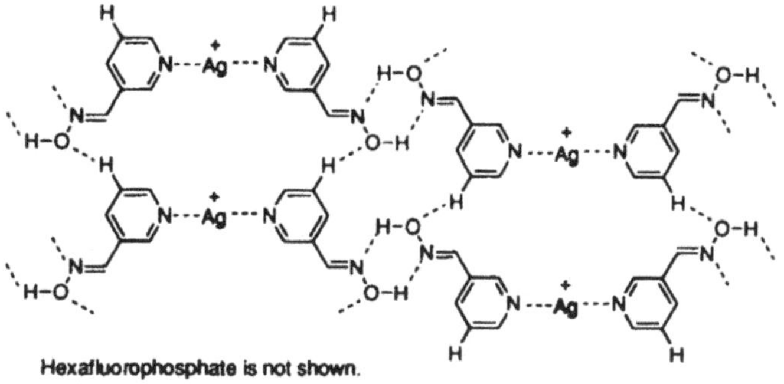

Hexafluorophosphate is not shown.
[bis(3-aldoximepyridine)silver(I) hexafluorophosphate]
Aakeröy, C. B.; Beatty, A. M.; Leinen, D. S. *J. Am. Chem. Soc.* 1998, *120*, 7383-7384.
Aakeröy, C. B.; Beatty, A. M. *J. Chem. Soc., Chem. Commun.* 1998, 1067-1068.

## REFERENCES

1. Ito, Y. *Synthesis* **1998**, 1–32.
2. General reviews on organic solid-state photochemistry: (a) Green, B.S.; Arad-Yellin, R.; Cohen, M.D. *Top. Stereochem.* **1986**, *16*, 131–218. (b) Ramamurthy, V.; Venkatesan, K. *Chem. Rev.* **1987**, 87, 433–481. (c) Scheffer, J.R.; Garcia-Garibay, M.; Nalamasu, O. *Org. Photochem.* **1987**, *8*, 249–347. (d) Scheffer, J.R.; Pokkuluri, P.R. In *Photochemistry in Organized and Constrained Media*; Ramamurthy, V., Ed.; VCH: New York, 1991, pp 185–246. (e) Venkatesan, K.; Ramamurthy, V. In *Photochemistry in Organized and Constrained Media*; Ramamurthy, V., Ed.; VCH: New York, 1991, pp 133–184. (f) Singh, N.B.; Singh, R.J.; Singh, N.P. *Tetrahedron* **1994**, *50*, 6441–6493.
3. Kitaigorodsky, A.I. Mixed Crystals; Springer-Verlag: Berlin, 1984; pp 17–48.
4. (a) Patil, A.O.; Curtin, D.Y.; Paul, I.C. *J. Am. Chem. Soc.* **1984**, *106*, 348–353. (b) Patil, A.O.; Curtin, D.Y.; Paul, I.C. *J. Chem. Soc., Perkin Trans. 2* **1986**, 1687–1692.
5. Caira, M.R.; Nassimbeni, L.R.; Wildervanck, A.F. *J. Chem. Soc., Perkin Trans. 2* **1995**, 2213–2216.
6. Cohen, M.D.; Cohen, R.; Lahav, M.; Nie, P.L. *J. Chem. Soc., Perkin Trans. 2* **1973**, 1095–1100.
7. Etter, M.C.; Frankenbach, G.M.; Bernstein, J. *Tetrahedron Lett.* **1989**, *30*, 3617–3620.
8. Etter, M.C.; Reutzel, S.M.; Choo, C.G. *J. Am. Chem. Soc.* **1993**, *115*, 4411–4412 and references cited therein.
9. Pedireddi, V.R.; Jones, W.; Chorlton, A.P.; Docherty, R. *J. Chem. Soc., Chem. Commun.* **1996**, 987–988.
10. Huang, K.-S.; Britton, D.; Etter, M.C.; Byrn, S.R. *J. Mater. Chem.* **1997**, *7*, 713–720.
11. Etter, M.C.; Baures, P.W. *J. Am. Chem. Soc.* **1988**, *110*, 639–640.
12. Toda, F.; Tanaka, K.; Miyamoto, H.; Miyahara, I.; Hirotsu, K. *J. Chem. Soc., Perkin Trans. 2* **1997**, 1877–1885.
13. Hasegawa, M.; Kinbara, K.; Adegawa, Y.; Saigo, K. *J. Am. Chem. Soc.* **1993**, *115*, 3820–3821.
14. Maekawa, Y.; Lim, P.-J.; Saigo, K.; Hasegawa, M. *Macromolecules* **1991**, *24*, 5752–5755.
15. Kinbara, K.; Adegawa, Y.; Saigo, K.; Hasegawa, M. *Bull. Chem. Soc. Jpn.* **1993**, *66*, 1204–1210.
16. Maekawa, Y.; Kato, S.; Hasegawa, M. *J. Am. Chem. Soc.* **1991**, *113*, 3867–3872.
17. Many reviews are available as listed below.
    (a) Etter, M.C. *Acc. Chem. Res.* **1990**, *23*, 120–126. (b) Etter, M.C. *J. Phys. Chem.* **1991**, *95*, 4601–4610. (c) Russell, V.A.; Ward, M.D. *Chem. Mater.* **1996**, 8, 1654–1666. (d) *The Crystal as a Supramolecular Entity*; Perspectives in Supramolecular Chemistry, Vol. 2; Desiraju, G.R., Ed.; Wiley: Chichester, 1996. (e) Desiraju, G.R. *Angew. Chem., Int. Ed. Engl.* **1995**, *34*, 2311–2327. (f) Desiraju, G.R. *J. Chem. Soc., Chem. Commun.* **1997**, 1475–1481. (g) Desiraju, G.R. *Crystal Engineering: The Design of Organic Solids*; Materials Science Monographs, Vol. 54; Elsevier:

New York, 1989. (h) *Solid-State Supramolecular Chemistry: Crystal Engineering*; Comprehensive Supramolecular Chemistry, Vol. 6; MacNicol, D.D.; Toda, F.; Bishop, R., Eds.; Pergamon: Oxford, 1996. (i) Desiraju, G.R. *Acc. Chem. Res.* **1996**, *29*, 441–449. (j) Bernstein, J.; Davis, R.E.; Shimoni, L.; Chang, N.-L. *Angew. Chem., Int. Ed. Engl.* **1995**, *34*, 1555–1573. (k) Aakeröy, C.B.; Seddon, K.R. *Chem. Soc. Rev.* **1993**, 397–407. (l) Aakeröy, C.B. *Acta Cryst.* **1997**, *B53*, 569–586. (m) MacDonald, J.C.; Whitesides, G.M. *Chem. Rev.* **1994**, *94*, 2383–2420. (n) Fan, E.; Vicent, C.; Geib, S.J.; Hamilton, A.D. *Chem. Mater.* **1994**, *6*, 1113–1117. (o) Subramanian, S.; Zaworotko, M.J. *Coord. Chem. Rev.* **1994**, *137*, 357–401. (p) Zaworotko, M.J. *Chem. Soc. Rev.* **1994**, 283–288. (q) Moore, J.S.; Lee, S. *Chem. Ind. (London)* **1994**, 556–560. (r) Wong, M.S.; Bosshard, C.; Günter, P. *Adv. Mater.* **1997**, *9*, 837–842. (s) Thalladi, V.R.; Brasselet, S.; Weiss, H.-S.; Bläser, D.; Katz, A.K.; Carrell, H.L.; Boese, R.; Zyss, J.; Nangia, A.; Desiraju, G.R. *J. Am. Chem. Soc.* **1998**, *120*, 2563–2577.

18.  In addition, a number of carboxylic acid cocrystals, of which some are of course chiral, are being reported in Aust. J. Chem. and by Japanese workers. For example, Lynch, D.E.; Latif, T.; Smith, G.; Byriel, K.A.; Kennard, C.H.L.; Parsons, S. *Aust. J. Chem.* **1998**, *51*, 403–408. Smith, G.; Lynch, D.E.; Kennard, C.H.L.; Byriel, K.A. *Aust. J. Chem.* **1998**, *51*, 437–439. Hamilton, D.G.; Lynch, D.E.; Byriel, K.A.; Kennard, C.H.L.; Sanders, J.K.M. *Aust. J. Chem.* **1998**, *51*, 441–444. Koshima, H.; Matsuura, T. *Yuki Gosei Kagaku Kyokai Shi* **1998**, *56*, 268–279 and 466–477.

19.  (a) Hunter, C.A.; Sanders, J.K.M. *J. Am. Chem. Soc.* **1990**, *112*, 5525–5534. (b) Hunter, C.A. *Angew. Chem., Int. Ed. Engl.* **1993**, *32*, 1584–1586. (c) Dahl, T. *Acta Chem. Scand.* **1994**, *48*, 95–106.

20.  Kitaigorodsky, A.I. *Molecular Crystals and Molecules*; Academic Press: New York, 1973.

21.  (a) Munakata, M.; Wu, L.P.; Kuroda-Sowa, T. *Bull. Chem. Soc. Jpn.* **1997**, *70*, 1727–1743. (b) Dance, I. In ref 17d, pp 137–233. (c) Burrows, A.D.; Chan, C.-W.; Chowdhry, M.M.; McGrady, J.E.; Mingos, D.M.P. *Chem. Soc. Rev.* **1995**, 329–339. (d) Zaworotko, M.J. *Angew. Chem., Int. Ed. Engl.* **1998**, *37*, 1211–1213. (e) Dagani, R. *Chem. Eng. News* **1998**, June 8, 35–46. (f) Braga, D.; Grepioni, F.; Desiraju, G.R. *Chem. Rev.* **1998**, *98*, 1375–1405.

22.  Hung, J.D.; Lahav, M.; Luwisch, M.; Schmidt, G.M.J. *Isr. J. Chem.* **1972**, *10*, 585–599.

23.  Cohen, M.D.; Cohen, R. *J. Chem. Soc., Perkin Trans. 2* **1976**, 1731–1735.

24.  Theocharis, C.R.; Desiraju, G.R.; Jones, W. *J. Am. Chem. Soc.* **1984**, *106*, 3606–3609.

25.  (a) Elgavi, A.; Green, B.S.; Schmidt, G.M.J. *J. Am. Chem. Soc.* **1973**, *95*, 2058–2059. (b) Warshel, A.; Shakked, Z. *J. Am. Chem. Soc.* **1975**, *97*, 5679–5684.

26.  Green, B.S.; Lahav, M.; Rabinovich, D. *Acc. Chem. Res.* **1979**, *12*, 191–197.

27.  (a) Addadi, L.; Lahav, M. *J. Am. Chem. Soc.* **1978**, *100*, 2838–2844; *J. Am. Chem. Soc.* **1979**, *101*, 2152–2156; *Pure Appl. Chem.* **1979**, *51*, 1269–1284. (b) Addadi, L.; van Mil, J.; Lahav, M. *J. Am. Chem. Soc.* **1982**, *104*, 3422–3429. (c) van Mil, J.; Addadi, L.; Gati, E.; Lahav, M. *J. Am. Chem. Soc.* **1982**, *104*, 3429–3434. (d) van Mil, J.; Addadi, L.; Lahav, M.; Leiserowitz, L. *J. Chem. Soc., Chem. Commun.* **1982**, 584–587.

28. Vaida, M.; Shimon, L.J.W.; van Mil, J.; Ernst-Cabrera, K.; Addadi, L.; Leiserowitz, L.; Lahav, M. *J. Am. Chem. Soc.* **1989**, *111*, 1029–1034.

29. (a) Vaida, M.; Shimon, L.J.W.; Weisinger-Lewin, Y.; Frolow, F.; Lahav, M.; Leiserowitz, L.; McMullan, R.K. *Science* **1988**, *241*, 1475–1479. (b) Shimon, L.J.W.; Vaida, M.; Frolow, F.; Lahav, M.; Leiserowitz, L.; Weisinger-Lewin, Y.; McMullan, R.K. *Faraday Discuss.* **1993**, *95*, 307–327.

30. (a) Sarma, J.A.R.P.; Desiraju, G.R. *J. Chem. Soc., Perkin Trans. 2* **1987**, 1187–1193. (b) Sarma, J.A.R.P.; Desiraju, G.R. *J. Am. Chem. Soc.* **1986**, *108*, 2791–2793. (c) Sarma, J.A.R.P.; Desiraju, G.R. *J. Chem. Soc., Chem. Commun.* **1984**, 145–147.

31. Dhurjati, M.S.K.; Sarma, J.A.R.P.; Desiraju, G.R. *J. Chem. Soc., Chem. Commun.* **1991**, 1702–1703.

32. Berkovic, G.E.; Ludmer, Z. *J. Am. Chem. Soc.* **1982**, *104*, 4280–4282.

33. Garcia-Garibay, M.; Scheffer, J.R.; Trotter, J.; Wireko, F. *Tetrahedron Lett.* **1987**, *28*, 1741–1744.

34. (a) Chmielewski, J.; Lewis, J.J.; Lovell, S.; Zutshi, R.; Savickas, P.; Mitchell, C.A.; Subramony, J.A.; Kahr, B. *J. Am. Chem. Soc.* **1997**, *119*, 10565–10566. (b) Kahr, B.; Jang, S.-H.; Subramony, J.A.; Kelley, M.P.; Bastin, L. *Adv. Mater.* **1996**, *8*, 941–944. (c) Mitchell, C.A.; Lovell, S.; Thomas, K.; Savickas, P.; Kahr, B. *Angew. Chem., Int. Ed. Engl.* **1996**, *35*, 1021–1023, and references cited therein.

35. Garcia-Garibay, M.A.; Shin, S.H.; Chao, I.; Houk, K.N.; Khan, S.I. *Chem. Mater.* **1994**, *6*, 1297–1306.

36. (a) Ito, Y.; Yasui, S.; Yamauchi, J.; Ohba, S.; Kano, G. *J. Phys. Chem. A* **1998**, *102*, 5415–5420. (b) Fukushima, S.; Ito, Y.; Hosomi, H.; Ohba, S. *Acta Cryst. Sect.* **1998**, *B54*, 895–906. (c) Yamauchi, J.; Yasui, S.; Ito, Y. *Chem. Lett.* **1998**, 137–138. For other papers about solid-state photocyclization of 2,4,6-triisopropylbenzophenones, see Ito, Y.; Matsuura, T.; Fukuyama, K. *Tetrahedron Lett.* **1988**, *29*, 3087–3090: Ito, Y.; Kano, G.; Nakamura, N. *J. Org. Chem.* **1998**, *63*, 5643–5647. Hosomi, H.; Ito, Y.; Ohba, S. *Acta Cryst. Sect.* **1998**, *B54*, 907–911.

37. (a) Vieth, H.H.; Mache, V.; Stehlik, D. *J. Phys. Chem.* **1979**, *83*, 3435. (b) Colpa, J.P.; Prass, B.; Stehlik, D. *Chem. phys. Lett.* **1984**, *107*, 469. (c) Prass, B.; Colpa, J.P.; Stehlik, D. *J. Chem. Phys.* **1988**, *88*, 191; *Chem. Phys. Lett.* **1989**, *136*, 187. (d) Yamauchi, S.; Terazima, M.; Hirota, N. *J. Phys. Chem.* **1985**, *89*, 4804–4808. (e) Hoshi, N.; Yamauchi, S.; Hirota, N. *J. Phys. Chem.* **1990**, *94*, 7523–7529. (f) Hoshi, N.; Hara, K.; Yamauchi, S.; Hirota, N. *J. Phys. Chem.* **1991**, *95*, 2146–2150. (g) Winkler, I.C.; Hanson, D.M. *J. Am. Chem. Soc.* **1984**, *106*, 923–925. (h) Saitow, K.; Endo, K.; Katoh, R.; Kotani, M. *Chem. Phys. Lett.* **1994**, *229*, 323–327. (i) Berkovic, G.E.; Ludmer, Z. *J. Chem. Soc., Chem. Commun.* **1984**, 232–233. (j) Lazarev, G.G.; Kuskov, V.L.; Lara, F.; García, F.; Rieker, A.; Tordo, P. *Chem. Phys. Lett.* **1993**, *212*, 319–325.

38. Ito, Y.; Scheffer, J.R. Unpublished result.

39. (a) Koshima, H.; Ding, K.; Matsuura, T. *J. Chem. Soc., Chem. Commun.* **1994**, 2053–2054. (b) Koshima, H.; Ding K.; Chisaka, Y.; Matsuura, T.; Miyahara, I.; Hirotsu, K. *J. Am. Chem. Soc.* **1997**, *119*, 10317–10324.

40. (a) Koshima, H.; Ding, K.; Miura, T.; Matsuura, T. *J. Photochem. Photobiol. A: Chem.* **1997**, *104*, 105–112. (b) Koshima, H.; Nakagawa, T.; Matsuura, T. *Tetrahedron Lett.* **1997**, *38*, 6063–6066.

41. Koshima, H.; Ding, K.; Chisaka, Y.; Matsuura, T. *J. Am. Chem. Soc.* **1996**, *118*, 12059–12065.
42. Ref 26 and references cited therein.
43. (a) Leibovitch, M.; Olovsson, G.; Scheffer, J.R.; Trotter, J. *Pure Appl. Chem.* **1997**, *69*, 815–823. (b) Sakamoto, M. *Chem. Eur. J.* **1997**, *3*, 684–689.
44. Koshima, H.; Wang, Y.; Matsuura, T.; Miyahara, I.; Mizutani, H.; Hirotsu, K.; Asahi, T.; Masuhara, H. *J. Chem. Soc., Perkin Trans. 2* **1997**, 2033–2038.
45. (a) Asahi, T.; Matsuo, Y.; Masuhara, H. *Chem. Phys. Lett.* **1996**, *256*, 525–530. (b) Asahi, T.; Matsuo, Y.; Masuhara, H.; Koshima, H. *J. Phys. Chem.* **1997**, *101*, 612–616.
46. Obata, T.; Shimo, T.; Somekawa, K. 74th Annu. Meet. Chem. Soc. Jpn. **1998**, Abstracts II, p. 1433.
47. (a) Feldman, K.S.; Campbell, R.F. *J. Org. Chem.* **1995**, *60*, 1924–1925. (b) Feldman, K.S.; Campbell, R.F.; Saunders, J.C.; Ahn, C.; Masters, K.M. *J. Org. Chem.* **1997**, *62*, 8814–8820.
48. (a) Kane, J.J.; Liao, R.-F.; Lauher, J.W.; Fowler, F.W. *J. Am. Chem. Soc.* **1995**, *117*, 12003–12004. (b) Ball, P. *Nature* **1996**, *381*, 648–650. (c) Schauer, C.L.; Matway, E.; Fowler, F.W.; Lauher, J.W. *J. Am. Chem. Soc.* **1997**, *119*, 10245–10246.
49. Sarma, J.A.R.P.; Desiraju, G.R. *J. Chem. Soc., Perkin Trans. 2* **1985**, 1905–1912.
50. (a) Desiraju, G.R.; Sharma, C.V.K.M. *J. Chem. Soc., Chem. Commun.* **1991**, 1239–1241. (b) Sharma, C.V.K.; Panneerselvam, K.; Shimoni, L.; Katz, H.; Carrell, H.L.; Desiraju, G.R. *Chem. Mater.* **1994**, *6*, 1282–1292.
51. Coates, G.W.; Dunn, A.R.; Henling, L.M.; Ziller, J.W.; Lobkovsky, E.B.; Grubbs, R.H. *J. Am. Chem. Soc.* **1998**, *120*, 3641–3649.
52. (a) Suzuki, T.; Fukushima, T.; Yamashita, Y.; Miyashi, T. *J. Am. Chem. Soc.* **1994**, *116*, 2793–2803. (b) Suzuki, T. *Pure Appl. Chem.* **1996**, *68*, 281–284.
53. (a) Koshima, H.; Ding, K.; Chisaka,Y.; Matsuura, T.; Ohashi, Y.; Musaka, M. *J. Org. Chem.* **1996**, *61*, 2352–2357. (b) Koshima, H.; Ding, K.; Miyahara, I.; Hirotsu, K.; Kanazaki, M.; Matsuura, T. *J. Photochem. Photobiol. A: Chem.* **1995**, *87*, 219–223. (c) Koshima, H.; Ding, K.; Chisaka, Y.; Matsuura, T. *Tetrahedron: Asymmetry* **1995**, *6*, 101–104.
54. Haga, N.; Nakajima, H.; Takayanagi, H.; Tokumaru, K. *J. Chem. Soc., Chem. Commun.* **1997**, 1171–1172.
55. Bosch, E.; Hubig, S.M.; Lindeman, S.V.; Kochi, J.K. *J. Org. Chem.* **1998**, *63*, 592–601.
56. (a) Mir, M.; Marquet, J.; Cayón, E. *Tetrahedron Lett.* **1992**, *33*, 7053–7056. (b) Mir, M.; Wilkinson, F.; Worrall, D.R.; Bourdelande, J.L.; Marquet, J. *J. Photochem. Photobiol. A: Chem.* **1997**, *111*, 241–247.
57. (a) Ito, Y.; Asaoka, S.; Saito, I.; Ohba, S. *Tetrahedron Lett.*, **1994**, *35*, 8193–8196. (b) Ito, Y.; Asaoka, S.; Kokubo, K.; Ohba, S.; Fukushima, S. *Mol. Cryst. Liq. Cryst.* **1998**, *313*, 125–134.
58. Ito, Y.; Endo, S.; Ohba, S. *J. Am. Chem. Soc.* **1997**, *119*, 5974–5975.
59. (a) Ito, Y.; Nakabayashi, H. To be published. (b) Ito, Y.; Nakabayashi, H.; Saito, I. 74th Annu. Meet. Chem. Soc. Jpn. **1998**, Abstracts II, p. 697.
60. Hosomi, H.; Ohba, S.; Ito, Y.; Nakabayashi, H. *Acta Cryst.* **1997**, *C53*, IUC9700032.

61. Ito, Y.; Ohba, S. To be published.
62. Etter, M.C.; Adsmond, D.A. *J. Chem. Soc., Chem. Commun.* **1990**, 589–591.
63. Liao, R.-F.; Lauher, J.W.; Fowler, F.W. *Tetrahedron* **1996**, *52*, 3153–3162.
64. Gamlin, J.N.; Jones, R.; Leibovitch, M.; Patrick, B.; Scheffer, J.R.; Trotter, *J. Acc. Chem. Res.* **1996**, *29*, 203–209.
65. Gudmundsdottir, A.D.; Scheffer, J.R.; Trotter, J. *Tetrahedron Lett.* **1994**, *35*, 1397–1400.
66. Gudmundsdottir, A.D.; Scheffer, J.R. *Tetrahendron Lett.* **1990**, *31*, 6807–6810.
67. Gudmundsdottir, A.D.; Scheffer, J.R. *Photochem. Photobiol.* **1991**, *54*, 535–538.
68. Jones, R.; Scheffer, J.R.; Trotter, J.; Yang, J. *Tetrahedron Lett.* **1992**, *33*, 5481; *Acta Cryst.* **1994**, *B50*, 601–607.
69. Leibovitch, M.; Olovsson, G.; Scheffer, J.R.; Trotter, J. *J. Am. Chem. Soc.* **1997**, *119*, 1462–1463.
70. Leibovitch, M.; Olovsson, G.; Sundarababu, G.; Ramamurthy, V.; Scheffer, J.R.; Trotter, J. *J. Am. Chem. Soc.* **1996**, *118*, 1219–1220.
71. (a) Koshima, H.; Maeda, A.; Masuda, N.; Matsuura, T.; Hirotsu, K.; Okada, K.; Mizutani, H.; Ito, Y.; Fu, T.Y.; Scheffer, J.R.; Trotter, J. *Tetrahedron: Asymmetry* **1994**, *5*, 1415–1418. (b) Fu, T.Y.; Scheffer, J.R.; Trotter, J. *Acta Cryst.* **1997**, *C53*, 1259–1262.
72. Borecka, B.; Gudmundsdottir, A.D.; Olovsson, G.; Ramamurthy, V.; Scheffer, J.R.; Trotter, J. *J. Am. Chem. Soc.* **1994**, *116*, 10322–10323.
73. Gamlin, J.N.; Olovsson, G.; Pitchumani, K.; Ramamurthy, V.; Scheffer, J.R.; Trotter, J. *Tetrahedron Lett.* **1996**, *37*, 6037–6040.
74. Ito, Y.; Borecka, B.; Scheffer, J.R.; Trotter, J. *Tetrahedron Lett.*, **1995**, *36*, 6083–6086.
75. Ito, Y.; Borecka, B.; Olovsson, G.; Trotter, J.; Scheffer, J.R. *Tetrahedron Lett.*, **1995**, *36*, 6087–6090.
76. Ito, Y.; Olovsson, G. *J. Chem. Soc., Perkin Trans. 1*, **1997**, 127–133.
77. Ito, Y. *Mol. Cryst. Liq. Cryst.*, **1996**, *277*, 247–253.
78. Ito, Y.; Fujita, H. *J. Org. Chem.*, **1996**, *61*, 5677–5680.
79. Brune, H.-A.; Debaerdemaeker, T.; Günther, U.; Schmidtberg, G.; Ziegler, U. *J. Photochem. Photobiol. A: Chem.* **1994**, *83*, 113–128.
80. Ito, Y. To be published.
81. (a) Ito, Y.; Ohba, S. To be published. (b) Ito, Y. *Ann. Symp. Photochem. Jpn.* **1996**, Abstracts, p. 250.
82. (a) Ogawa, K.; Sano, T.; Yoshimura, S.; Takeuchi, Y.; Toriumi, K. *J. Am. Chem. Soc.* **1992**, *114*, 1041–1051. (b) Ogawa, K.; Harada, J.; Tomoda, S. *Acta Cryst.* **1995**, *B51*, 240–248. (c) Harada, J.; Ogawa, K.; Tomoda, S. *Acta Cryst.* **1997**, *B53*, 662–672.
83. (a) Ito, Y. To be published; 74th Annu. Meet. Chem. Soc. Jpn. **1998**, Abstracts II, p. 1434. (b) Hosomi, H.; Ito, Y.; Ohba, S. *Acta Cryst.* **1998**, *C54*, 142–145.
84. Quina, F.H.; Whitten, D.G. *J. Am. Chem. Soc.* **1977**, *99*, 877–883.
85. Kinbara, K.; Kai, A.; Maekawa, Y.; Hashimoto, Y.; Naruse, S.; Hasegawa, M.; Saigo, K. *J. Chem. Soc., Perkin Trans. 2* **1996**, 247–253.
86. Cho, T.H.; Chaudhuri, B.; Snider, B.B.; Foxman, B.M. *J. Chem. Soc., Chem. Commun.* **1996**, 1337–1338.

87. (a) Ito, Y.; Jiben, M.; Suzuki, S.; Kusunaga, Y.; Matsuura, T. *Tetrahedron Lett.* **1985**, 26, 2093–2096. (b) Ito, Y.; Kusunaga, Y.; Tabata, K.; Arai, H.; Meng, J.; Matsuura, T. *J. Appl. Polym. Sci.*, **1993**, 50, 1989–1998.

88. Ito, Y. In *Photochemistry on Solid Surfaces*; Anpo, M.; Matsuura, T., Eds.; Elsevier: Amsterdam, 1989; pp 469–480.

89. (a) Meng, J.; Ito, Y.; Matsuura, T. *Tetrahedron. Lett.* **1987**, 28, 6665–6666. (b) Meng, J.; Wang, W.; Wang, H.; Matsuura, T.; Koshima, H.; Sugimoto, I.; Ito, Y. *Photochem. Photobiol.* **1993**, 57, 597–602.

90. Meng, J.; Zhu, Z.; Wang, R.; Yao, X.; Ito, Y.; Ihara, H.; Matsuura, T. *Chem. Lett.* **1990**, 1247–1248.

91. Ito, Y. *Mol. Cryst. Liq. Cryst.* **1992**, 219, 29–36.

92. For other photoreactions of the solid mixtures, see Ito, Y.; Aoki,Y.; Matsuura, T.; Kawatsuki, N.; Uetsuki, M. *J. Appl. Polym. Sci.*, **1991**, 42, 409–415 and Ito, Y. *Oxidation Commun.* **1992**, 15, 149–155.

93. (a) Meng, J.; Fu, D.; Gao, Z.; Wang, R.; Wang, H.; Saito, I.; Kasatani, R.; Matsuura, T. *Tetrahedron* **1990**, 46, 2367–2370. (b) Meng, J.; Wang, W.; Xiong, G.; Wang, Y.; Fu, D.; Du, D.; Wang, R.; Wang, H.; Koshima, H.; Matsuura, T. *J. Photochem. Photobiol. A: Chem.* **1993**, 74, 43–49. (c) Meng, J.; Du, D.-M.; Xiong, G.-X.; Wang, W.-G.; Wang, Y-M.; Koshima, H.; Matsuura, T. *J. Heterocycl. Chem.* **1994**, 31, 121–124. (d) Du, D.M.; Wang, Y.M.; Meng, J.B.; Zhou, X.Z. *Chem. J. Chin. Univ.* **1996**, 17, 252–254. (e) Du, D.M.; Wang, Y.M.; Xiong, G.X.; Wang, W.G.; Meng, J.B. *Chem. Res. Chin. Univ.* **1995**, 11, 111–116.

94. (a) Koshima, H.; Ichimura, H.; Matsuura, T. *Chem. Lett.* **1994**, 847–848. (b) Koshima, H.; Ichimura, H.; Hirotsu, K.; Miyahara, I.; Wang, Y.; Matsuura, T. *J. Photochem. Photobiol. A: Chem.* **1995**, 85, 225–229. (c) Koshima, H.; Matsuura, T. *Mol. Cryst. Liq. Cryst.* **1996**, 277, 55–62. (d) Meng, J.; Wang, W.; Wang, Y.; Wang, H.; Koshima, H.; Matsuura, M. *Mol. Cryst. Liq. Cryst.* **1994**, 242, 403–411.

95. (a) Koshima, H.; Yao, X.; Wang, H.; Wang, R.; Matsuura, T. *Tetrahedron Lett.* **1994**, 35, 4801–4804. (b) Koshima, H.; Chisaka, Y.; Wang, Y.; Yao, X.; Wang, H.; Wang, R.; Maeda, A.; Matsuura, T. *Tetrahedron* **1994**, 48, 13617–13630.

96. (a) Koshima, H.; Bittl, D.P.H.; Miyoshi, F.; Wang, Y.; Matsuura, T. *J. Photochem. Photobiol. A: Chem.* **1995**, 86, 171–176. (b) Koshima, H.; Matsuura, T. *J. Photochem. Photobiol. A: Chem.* **1996**, 100, 85–91.

97. Meng, J.; Fu, D.; Yao, X.; Wang, R.; Matsuura, T. *Tetrahedron* **1989**, 45, 6979–6986.

98. Wang, Y.M.; Du, D.M.; Chen, X.M.; Meng, J.B.; Zhang, H.P. *Chem. Res. Chin. Univ.* **1996**, 17, 235–239.

99. (a) Saito, I.; Ito, S.; Matsuura, T.; Hélène, C. *Photochem. Photobiol.* **1981**, 33, 15–19. (b) Koshima, H.; Yamashita, K.; Matsuura, T.; *Photomed. Photobiol.* **1995**, 17, 109–110.

100. Mikami, K.; Matsumoto, S.; Tonoi, T.; Okubo, Y.; Suenobu, T.; Fukuzumi, S. *Tetrahedron Lett.* **1998**, 39, 3733–3736.

101. (a) Burton, W.B.J. *Chem. Educ.* **1979**, 56, 483. (b) Ivie, G.W.; Casida, J.E. *J. Agric. Food Chem.* **1971**, 19, 405–409 and 410–416.

102. Nakanishi, F.; Nakanishi, H.; Kato, M.; Tawata, M.; Hattori, S. *J. Appl. Polym. Sci.* **1981**, 26, 3505–3510.

# 2

# Asymmetric Photochemical Reactions in Solution

**Simon R. L. Everitt and Yoshihisa Inoue**

Inoue Photochirogenesis Project, ERATO, Japan Science and Technology Corporation and Osaka University, Toyonaka, Japan

## I.  INTRODUCTION AND BACKGROUND

The asymmetric synthesis of molecules in the ground state (thermal processes) is increasingly becoming the normal mode of synthesis rather than the exception, and it is now usually possible to prepare compounds with excellent degrees of enantio- or diastereoselectivity using modern synthetic methods [1–10]. It is apparent that the level of interest in asymmetric photochemistry has not driven the available procedures to their ultimate limits, and a great deal of work remains to be carried out in this area if stereocontrol is to become as high as more "conventional" thermal processes. However, it would not be unreasonable to describe asymmetric photochemistry as a relatively young field. Indeed, diastereodifferentiating photochemical reactions induced by chiral auxiliaries were not extensively explored until the mid-1970s. We reviewed the literature up to mid-1991 in "Asymmetric photochemical reactions in solution" [11] and we have chosen to use the same title for this chapter in an attempt to show areas in which developments have taken place. It is from that point in time onward, through mid-1998, that the current work is focused. However, a description of the associated historical background and a large number of pertinent references may be found in our earlier work. The reader may also be interested in several somewhat older reviews on this area [12], including the much cited review by Rau, which still has a good deal of importance to the modern photochemist. Although brief commentaries on theory will be made in the introductory comments for each section, the reader should note that the scope of this chapter is not sufficient for an in-depth analysis, and should instead refer to the original articles for more detailed descriptions.

Advantages of photochemical processes can be summarized by the following points. 1.) Light drives the reaction, which leads to the involvement of electronically excited state(s). 2.) Selective excitation of specific components in the system is made possible through the correct choice of exciting wavelength. 3.) The substrate in the excited state usually exhibits a reactivity which is completely different for that observed for thermal processes, and thus novel/strained compounds unique to photochemical reactions can be produced. 4.) Intermolecular interaction in the excited state is also unique. The electronic excitation permits charge transfer interactions between donor–acceptor pairs with appropriate redox potentials. 5.) Because of the high energy introduced by the absorption of photons in the UV/Vis region, only low activation energies are needed, making low-temperature experiments possible. 6.) Photoreactions often provide a direct and efficient route to one-step syntheses of thermally inaccessible or difficult-to-attain compounds. 7.) In the context of asymmetric induction, an inherent advantage of the photochemistry is the use of circularly polarized light (CPL) as a physical source of chirality. However, photosystems have inherent disadvantages too; lifetimes of the electronically excited state are very short ($\tau \cong$ ps–$\mu$s), making intermolecular interaction as well as detection/observation of any transient species difficult, and only weak, noncovalent interactions are involved in the excited state, which can sometimes lead to poor reaction control.

Several reviews on various subjects contained within the scope of our title have been published in the literature that we have considered, and these will be brought to the reader's attention in the relevant parts of this work. In a very general sense, however, it has become apparent that the field of asymmetric photochemistry is maturing very rapidly, and several processes now are considered to be standard technique if a particular structure is to be synthesized, e.g., the [2+2] photocycloaddition for the preparation of the four-membered ring. Several mechanisms have recently been elucidated, such as the Schenck reaction [13], which has been explained thoroughly by Adam and co-workers, and it is possible to use this understanding to design precursors for products which then can be produced with exceptional degrees of stereocontrol. Multiple chiral centers are often generated in a single-step asymmetric photoprocess, which is also useful to the synthetic chemist in the preparation of complex products. Photochemical processes also have the advantage that irradiation can be carried out over a range of temperatures without affecting the reaction mechanism, although control of both reaction temperature [14] and, more recently, pressure [15] have been shown to offer a good method for directing the stereoselectivity of certain photoprocesses, in which even switching of product chirality is possible using the same photosensitizer.

Solid-state asymmetric photoprocesses of chiral crystals of achiral molecules (often referred to as "absolute" asymmetric syntheses) will not be described in detail in this chapter, although examples will be used in the text where appropriate. It is worth noting that these processes proceed with excellent levels of asymmetry, and a wide range of different reactions have been carried out. It is also possible to include a small amount of photosubstrate into the crystal of a chiral yet photochemically inert molecule, and use the rigid, highly preorganized environment to induce high levels of optical activity to the photoproducts.

In the following pages we will seek to give a general overview of the recent literature that falls into the category of "asymmetric photochemistry in solution," highlighting what we consider to be of particular interest to the reader, while aiming to thoroughly reference the rest of the available material.

## II. CLASSIFICATION

The classification of asymmetric photochemical reactions is generalized in Table 1. This quick reference serves to highlight areas of this review that have proven to be the most successful in terms of generating optically active products, as well as demonstrating very rapidly the means by which the optical activity is imparted to the product. We will discuss each of these sections in turn in this chapter.

In principle, it is possible for an asymmetric photochemical process to be carried out in any phase, from solid through to gas, but the full range of conceivable reactions has not been demonstrated experimentally. No gas phase asymmetric photochemistry has been reported to date, and this is probably a result of the very low numbers of deactivating collisions that will occur, thus allowing the relaxation of a photoexcited state through to a vibrationally excited ground state, leading to decomposition products that are very similar to those from the thermal reaction. However, in the solution state, it is business as usual for the photochemist, and a wide variety of reactions have been carried out, the majority of which have the chiral information that is incorporated into the product relayed through either a chiral auxiliary or, more rarely, through interaction of the substrate with a chiral complexing agent or chiral sensitizer. It is only in photoreactions in the solid state that optical activity can be generated in a photoproduct without the need for preexisting chiral information in the system. In terms of efficiency in the propagation of chiral information, it is clear that solid-state asymmetric photochemistry offers the best results if the system produces appropriate crystals. However, we are primarily concerned with solution state processes, and we will only

**Table 1** Classification of Asymmetric Photochemical Reactions

| Media | Chiral source | Excitation mode | Differentiation mechanism | Chiral source/ substrate ratio | Optical yield |
|---|---|---|---|---|---|
| Gas | | | | | No examples |
| Solution | Circularly polar- ized light | Direct | Absolute asymmetric synthesis | 0 (CPL needed) | Low-moderate |
| | Sensitizer | Sensitization | Enantiodifferentiation | ≪1 | Low-high |
| | Complexing agent | Direct | Diastereodifferentiation | <1 | Moderate-high |
| | Chiral auxiliary | Direct | Diastereodifferentiation | 1 | Moderate-high |
| | Supramolecule | Direct/sensitization | Enantiodifferentiation | ≤1 | Low-moderate |
| | Solvent | Direct | Enantiodifferentiation | ≫1 | Low |
| Molecular assembly | Liquid crystal | Direct | Enantiodifferentiation | ≫1 | Low |
| | Micelle | | | | No examples |
| | Polymer matrix | Direct | Enantiodifferentiation | ≫1 | Low-moderate |
| Solid | Host cavity | Direct | Enantiodifferentiation | 1 | Low-high |
| | Host lattice | Direct | Enantiodifferentiation | >1 | Moderate-high |
| | Crystal lattice | Direct | Enantiodifferentiation | 1 | Moderate-high |
| | Modified clay | Direct | Enantiodifferentiation | >1 | Moderate |

briefly describe solid-state reactions in the context of chiral complexing agents (see Sec. IV). Work concerning the use of zeolites as a reaction medium has also appeared recently in the literature [16–18], but this also falls outside the scope of this chapter.

Optical activity generation using CPL relies on an optically selective excitation, which electronically activates one enantiomer in a racemic mixture, leading to either photodestruction or photoderacemization. The generation of optically active products through interaction with a chiral photosensitizer requires the input of only a small amount of chiral information, and this type of process is quite efficient in terms of the amplification of chiral centers. Indeed, the low concentrations of sensitizer that are often required, coupled with good turnover numbers, renders this process very attractive in these terms.

Chiral complexing agents represent the next level down since, although in theory the amount of chiral information that must be programmed into a system before or after photochemistry is carried out can be quite low (if turnover of the chiral host/prochiral photoreactive guest is possible), it is quite often necessary to use the "host" in much greater than stoichiometric quantities. Of course, if the chiral host can be recovered and recycled, the process becomes quite attractive from the point of view of conservation/amplification of chiral centers. Asymmetric photochemical reactions that use a chiral auxiliary as the driving force behind the stereochemistry of the forming chiral centers unavoidably require at least one equivalent of chiral center per substrate molecule, and quite often, more than one chiral center is required to give good control over the reaction. This renders the use of chiral auxiliaries the least efficient method for the generation of new chiral centers. However, it is through the introduction of a chiral auxiliary that high levels of stereocontrol can be applied to asymmetric photoprocesses, resulting in exceptionally high optical purities (op's). For this reason the use of induction through a chiral auxiliary is by far the most popular of the four classes of reaction that we have described.

We will describe each of these four main areas in order of decreasing chirality amplification efficiency. It is perhaps a reflection on the state of the art that the least efficient processes in terms of the amount of chiral information contained in starting material compared to that of final product, i.e., solution state asymmetric photoprocesses with chiral auxiliaries, are by far the most common examples in the current literature. However, this is probably a result of the increasing ease with which experimental techniques are applied, and also the increasing number of available reactions/substrates amenable to the photoprocess as the level of chiral amplification is reduced. It will remain the challenge of the modern photochemist to circumvent these limitations to enable chirality generation from substrates with almost no optical activity, which must still be considered a very demanding goal.

## III.  DIRECT ASYMMETRIC PHOTOCHEMISTRY WITH CPL

Arguably the most efficient methods for the transfer of chiral information with a photochemical system described in this review involve "absolute" asymmetric photochemistry induced by CPL. In this photoprocess, an optically inactive, racemic substrate is irradiated by a left- or right-handed light source ($l$- or $r$-CPL). Differential interactions of this chiral light source with each enantiomer lead to either enantioselective destruction, or a shift in the photoequilibrium between the two enantiomers. Both of these processes lead to optically active products. Because of these photochirogenesis features, direct asymmetric photochemistry with CPL has been implicated in the generation of enantiomerically enriched material in prebiotic stages of the universe [19], and as such has a great deal of associated interest in the scientific community. However, in this chapter we shall focus only on the chemical aspects of the use of CPL.

The technique of absolute asymmetric photochemistry was first conceived by le Bel [20] and van't Hoff [21] in the late 19th century but has until recently been difficult to put into practice. (For a description of the historical development of the CPL technique, see Inoue [11] and references contained therein.) These difficulties may be a result of problems associated with generating photons of sufficient energy/narrow wavelength distribution, or because of the relatively few compounds with a significantly large anisotropy factor ($g$), which is a prerequisite for successful asymmetric photochemistry with CPL. The $g$ factor, which was first defined by Kuhn [22,23], is the relative difference of the molar extinction coefficients of an optically pure compound toward $l$- and $r$-CPL at a given wavelength [see Eq. (1)].

$$g = (\varepsilon_l - \varepsilon_r)/\varepsilon = \Delta\varepsilon/\varepsilon \qquad\qquad (1)$$

where $\varepsilon = (\varepsilon_l - \varepsilon_r)/2$, and therefore $0 \leq g < 2$. It is most common for $g$ factors to be well below unity for compounds such as aromatics, which possess fully allowed $\pi,\pi^*$ transition(s) (typically $g \cong 10^{-3}$–$10^{-4}$; e.g., 4-hydroxyphenylglycine has a $g$ factor of $8.8 \times 10^{-5}$ at 266 nm [24,25]). However, much larger $g$ factors are encountered for carbonyl compounds; these posses a partially forbidden $n,\pi^*$ transition, which has a very small $\varepsilon$ value (usually 10–50 $M^{-1}$ $cm^{-1}$), and $g$ factors can, in extreme cases, reach values as high as 0.24 (for trans-bicyclo[4.3.0]nonan-8-one at 313 nm) and 0.30 (for tricyclo[4.4.0.0$^{3,8}$]decan-2-one at 288 nm) [26]. It is of vital importance to note that the $g$ factor is wavelength-dependent, and thus a monochromatic source of CPL is most desirable for absolute asymmetric photochemistry, as this prevents a reduction in the $g$ factor, through averaging out over a range of wavelengths. In this area, dramatic improvements in available technology, allowing access to tunable, high-intensity, monochromatic sources of CPL, have made possible some of the advances in this area that we shall describe below.

Photochemical asymmetric fixation, a technique by which the asymmetric element is "fixed" through a photoreaction induced by CPL irradiation, has not been discussed in the recent literature. However, for the sake of completeness, we will briefly describe the process here. This technique normally requires that the thermal racemization process is fast enough to keep the substrate racemic during the photolysis and that the excited state racemizes much more slowly than the relaxation to the product. It is, of course, essential that there be an appreciable $g$ factor at the irradiation wavelength. Under these conditions, the product's op is governed simply by the ratio of the extinction coefficients at the irradiation wavelength. Only a few reactions have been reported to satisfy these criteria, and none of these more recently than the mid 1970s [11, and refs. therein].

## A. Photodestruction/Photoproduction

The enantiodifferentiating interaction of $l$- or $r$-CPL with a racemic mixture can result in the destruction of each enantiomer at a different rate, resulting in an enhancement ("production") of optical purity in the more slowly reacting species. In order for this to be possible, the process must be carried out under the following conditions: 1.) $\Delta\varepsilon$ must be nonzero; 2.) an irreversible photoreaction must occur; 3.) there must be no thermal or photochemical racemization of the substrate; and ideally 4.) the product must not absorb at the irradiating wavelength. When these criteria are met, direct asymmetric photodestruction/production using CPL can be considered as two independent parallel processes which occur at different rates. Although the destruction process is inherently accompanied by a production of some product (either through small modification of the substrate, or more usually through decomposition, accompanied by the extrusion of gaseous products, such as $CO$, $CO_2$, $H_2$, $H_2O$, etc.), it is not usual to isolate the potentially optically active photoproducts, and the conversion and op of the substrate are usually taken as a measure of the extent of reaction. Extensive theoretical treatments and kinetic analyses of the enantioselective photodestruction process have been written [27–29]. A limiting feature of this process is that to obtain a high op it is necessary to closely approach 100% destruction. This is demonstrated by Eq. (2) for the optical rotation ($\alpha$) of the remaining enantiomer during the photodestruction process, where $\Gamma$ is the optical activity of the pure enantiomer, $g$ is the anisotropy factor, and $x$ is the extent of photodestruction:

$$\alpha = \Gamma(g/2)(1 - x) \ln [1/(1 - x)] \qquad (2)$$

This equation is illustrated for different $g$ factors in Fig. 1. Kuhn et al. [22] highlighted this fact, and we can see from the plot of optical purity versus % conversion that even for high $g$ factors, an exceptionally high level of destruction must be carried out. If we consider that $g$ factors usually fall far below unity, then we begin to see a failing of this method. For a more typical (theoretical)

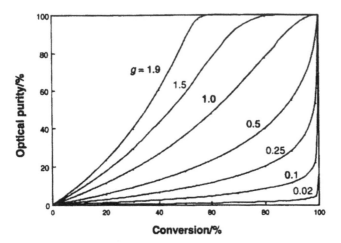

**Figure 1** Theoretical optical purity of remaining substrate as a function of % conversion for a range of $g$ factors ($0 \leq g < 2$).

example, a molecule with a $g$ factor of 0.04 affords an op of 2.0% at 63.2% destruction. This increases to 6.0% at 95% destruction and even at 99.995% destruction the op has only risen to 20% [22].

Recently, Shimizu et al. [30] reported an efficient enantioselective destruction of the L isomer of a D/L-tartaric acid mixture, using focused $r$-CPL from an XeF excimer laser. Initial attempts to destroy optically pure samples of each enantiomer resulted in a decrease of the L-tartaric acid sample by ~14% when irradiated with $1 \times 10^4$ J of focused $r$-CPL at 351 nm, and photodestruction products such as CO, $CO_2$, $H_2$, etc., were observed. Unusually, the D isomer (**1**) was found to remain almost unchanged upon irradiation with the same light source. Thus, treatment of a racemic mixture with $r$-CPL afforded a sample enriched in **1**, which had an op of ~7.5%, a result which is exceptionally high for this type of process (see Scheme 1). It was observed that when unfocused $r$-CPL was used, there was hardly any decrease in the op of the L-tartaric acid (0.11%),

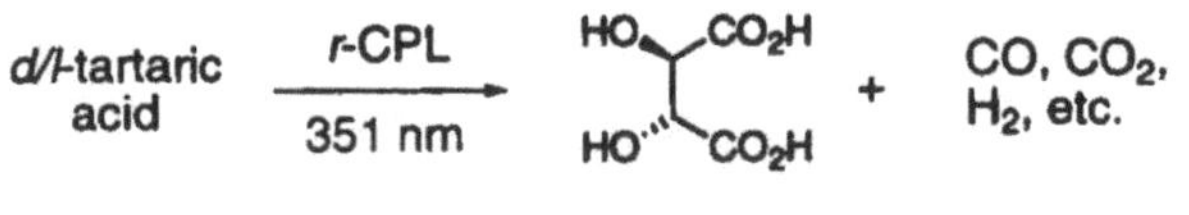

$$\textit{d/l-}\text{tartaric acid} \xrightarrow[\text{351 nm}]{r\text{-CPL}} \quad \text{HO}\underset{\text{HO}^{\prime\prime}}{\overset{\text{CO}_2\text{H}}{\bigwedge}}\text{CO}_2\text{H} \quad + \quad \text{CO, CO}_2, \text{H}_2, \text{etc.}$$

**1 (14% ee)**

**Scheme 1**  Enantioselective photodestruction with CPL.

and the authors point out that this source was approximately 150 times weaker in intensity than the focused source. This would highlight the need for a focused laser source in order to enhance the nonresonant absorption in addition to a sharp irradiating wavelength to maximize the $\Delta\varepsilon$. Unfortunately, this paper contains no mention of the $g$ factor or the % destruction carried out in this reaction, although it is implied that there was no interaction whatsoever with the L isomer in this instance. It is also interesting to consider that there is no appreciable absorption band for tartaric acid at 351 nm. The photodecomposition must therefore be made possible through nonresonant excitation, or simply through thermal (or plasma) reaction induced by the focused laser light.

Using the enantioselective binding by the transport protein bovine serum albumin (BSA) which has two binding sites for (R)- and (S)-1,1'-bi-2-naphthol (**2** and **3**) [31], Zandomeneghi reported that non-CPL irradiation of the (R) and (S) enantiomers (**2** and **3**) complexed to BSA led to highly enantiodifferentiating photolysis [31]. When a racemic mixture of the substrate was dissolved in an aqueous solution containing BSA (0.12:0.17 mM), one binding site was found to be occupied in an approximately 2:1 ratio of **2** and **3**, and a new absorption band appeared at 355–420 nm. If the level of occupancy was set to below one substrate molecule per BSA, the first site was shown to be almost completely occupied by **3**. Preferential photodestruction of **3** was possible using this system. At 57% photodestruction, the enantiomeric excess (ee) of remaining **2** was reported as a remarkable 89%. This large ee was attributed to a modified $g$ factor, which was calculated to be 1.63 ± 0.04 at 365 nm—three orders of magnitude higher than Kuhn's $g$ factor for the same molecule. Although this is not strictly a "pure" absolute asymmetric photodestruction, as the chirality enhancement is also a function of the chiral complexing agent BSA, it is a remarkable achievement in a field which is considered to hold little practical value as far as asymmetric synthesis is concerned.

## B.  Photoderacemization

From a synthetic viewpoint, it is perhaps more interesting to apply the inherent chirality of $l$- and $r$-CPL to the process of photochemical deracemization, in which the racemic substrate is photoexcited, causing the enantiomers to interconvert to given an unequal mixture which ultimately becomes a photostationary state (pss). Unlike the previous example of asymmetric destruction/production, the total amount of material in the system ($[R] + [S]$) does not change throughout the process, and the pss will always be the same for a given substrate at the same excitation wavelength, regardless of the op of the starting material. It is possible to imagine two distinct situations where such a process can be applied: 1.) the photoderacemization of a chiral substrate, whereby the excited enantiomers directly interconvert, or 2.) the intervention of a common, prochiral intermediate,

which relaxes to the ground state enantiomer. In each case, any observed enantiomeric excess is related only to $\Delta\varepsilon$ of each enantiomer toward the CPL at the
exciting wavelength, since the deactivating reverse reactions cannot be enantiodifferentiating. For an overview of the kinetics involved in both of these processes, the reader is drawn to the works of Stevenson [32], Rau [27], and Inoue
[11]. Interestingly, it is possible to evaluate the anisotropy factor from the optical
purity at the pss [see Eq. (3)], and from this equation we can see that a high $g$
factor is imperative for good ee's:

$$ee_{pss} = ([R] - [S])/([R] + [S]) = g/2 \qquad (3)$$

Historically, very few examples of photoderacemization induced by CPL
have been reported, and this is also true for the recent literature. Feringa, who
has reported extensive work on the use of sterically overcrowded olefins as potential optical data storage devices, has used CPL to carry out the "dynamic control
and amplification of molecular chirality" with the aim of generating a photo-
switchable device that can be "read" by nondestructive means [33]. The requirements for an optical data storage device that relies on photoisomerization are
defined by Feringa et al. as follows [34]: 1.) the two states (0 and 1) must be
thermally stable, and not interconvert at room temperature; 2.) the system must
be fatigue-resistant, undergoing many cycles without material degradation; 3.)
the two states must be readily interconverted in a controlled way; and 4.) the
information must be readable by nondestructive means. It was proposed that two
different pss's would be formed when a mixture of **4** and **5** was irradiated with
*l*- or *r*-CPL. Thus, the racemic mixture was irradiated with each type of CPL at
313 nm, where $\Delta\varepsilon$ value was found to be at its maximum (Scheme 2). Alternating
between the two light sources gave an alternating pair of pss's which had op's of

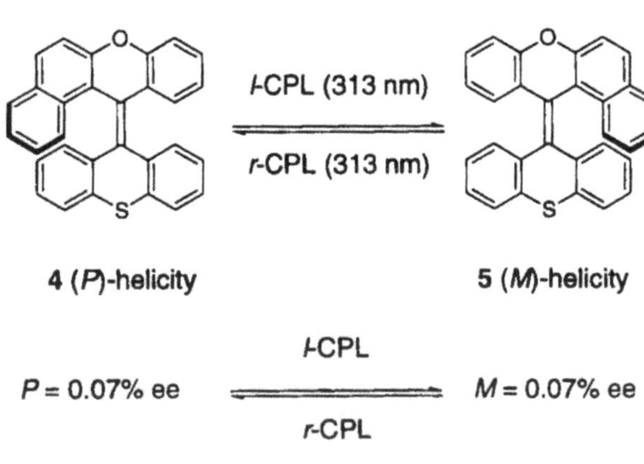

| | *l*-CPL (313 nm) | |
| 4 (*P*)-helicity | | 5 (*M*)-helicity |
| | *r*-CPL (313 nm) | |

| | *l*-CPL | |
| *P* = 0.07% ee | | *M* = 0.07% ee |
| | *r*-CPL | |

**Scheme 2**  Photoderacemization.

0.07% and −0.07%. These values were considered disappointingly low, simply because the g factor was inherently small for the allowed transition and the CPL source used had a bandwidth of 10 nm, thus diluting the effects that could be observed. This system showed good fatigue resistance over several cycles. In the same paper, the use of this switching system to control the phase of a liquid crystal was reported. 4′-(Pentyloxy)-4-biphenylcarbonitrile, when doped with 20 wt% **4**, was found to undergo a phase transition from nematic to cholesteric upon irradiation with 313 nm CPL, and the helicity of this phase could be controlled, with r-CPL affording the right-handed cholesteric phase and vice versa. In the closing comments, it is pointed out that an increase in the helical twisting power and/or anisotropy factor of the chromophore should decrease the amount of dopant needed to carry out this transformation.

When a photostationary state is set up by the irradiation of a sample, it is often possible that this pss has some degree of optical activity. Feringa has also reported the switching of substrate chirality by unresolved light, in which optically active photoresponsive sterically overcrowded alkenes (**6** and **7**) were selectively isomerized using two different wavelengths, to give two pss's which have different helicity [35–38]. The switching processes, which are shown in Scheme 3, were improved by adding donor–acceptor substituents to the molecules. These molecules were added as dopants in liquid crystals [37], enabling phase switching from nematic to cholesteric, the helicity of which was controlled by the predominant helicity of the pss. When the amino group was protonated, the system became "frozen" and no photoisomerization occurred.

Another example of the use of a photoresolution process with the aim of controlling a liquid crystal–based optical switch was reported by Zhang and Schuster [39]. In this paper, the ketone **8**, which has a g factor of 0.0502 at

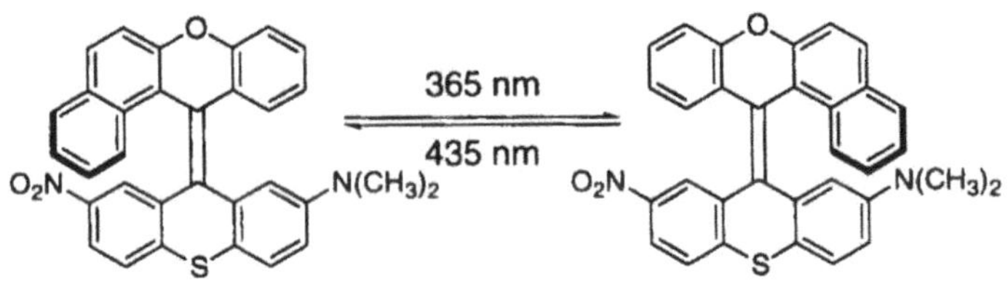

Scheme 3  Chirality switching using light.

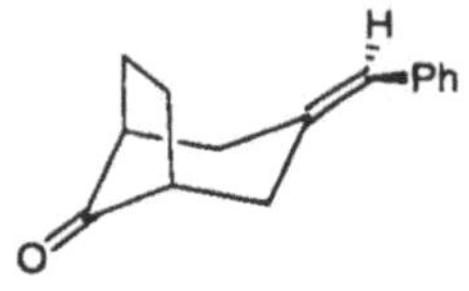

8

$g = 0.0502$ (313 nm)

**Figure 2**  A substrate subjected to photoderacemization.

313 nm, was calculated as capable of forming a pss with an op of 2.5% when irradiated with CPL. Excitation of **8** resulted in intersystem crossing to the ketone triplet, and energy transfer leads to the styryl triplet state, followed by an interconversion of the enantiomers through rotation about the double bond. When an 8 mM solution of **8** was irradiated with CPL through a 305 nm cutoff filter for 47 h, an enantiomeric excess of 1.6% was observed, thus demonstrating the success of the photoresolution (see Fig. 2).

The above examples involve the switching of two enantiomers through photoirradiation. Using an alternative procedure, Inoue et al. have used *l*- and *r*-CPL to enantioselectively photodestroy a racemic mixture of (*E*)-cyclooctene (**9**) to achiral (*Z*)-cyclooctene (**10**), leaving an optically active mixture of (*R*)-(−)- and (*S*)-(+)-(*E*)-cyclooctene [(*R*)-(−)-**9** and (*S*)-(+)-**9**] [19]. This was achieved using 190-nm light produced by a helical undulator (an insertion device for synchrotron), which gives intense, monochromatic CPL down to the soft X-ray region. This photoreaction (see Scheme 4) caused almost 70% conversion of **9** to the (*Z*) isomer and the optical activity of the remaining (*E*) isomer gradually increased to reach a plateau. Although the op's reported in Inoue's system were very low (−0.12% for *r*-CPL and +0.12% for *l*-CPL), the fact that all of the optical activity in the system is derived from the CPL source is vitally important, and it is to be hoped that further groups will follow this lead.

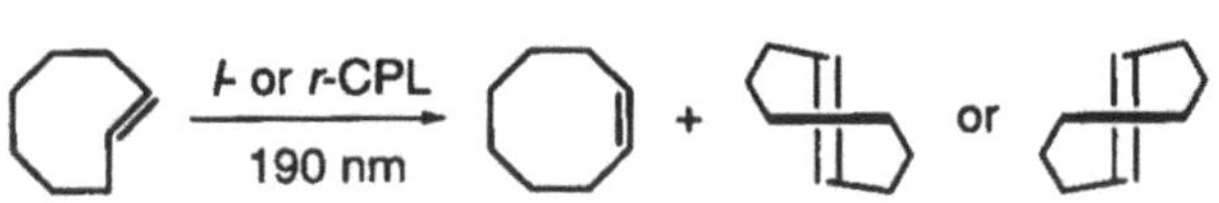

rac-9                          10        (*R*)-(−)-9      (*S*)-(+)-9
                                         favored         favored
                                         with *l*-CPL    with *r*-CPL

**Scheme 4**  *E-Z* photoisomerization of cyclooctene with CPL.

The use of a helical undulator enabling CPL excitation in the vacuum UV region should allow significant progress to be made using this versatile light source. Indeed, it is this technique which may reveal the cosmic roots of homochirality in the biosphere by giving us an understanding of the origin of optically active compounds, such as the L-enriched alanine found in the organic mantle of the Murchison meteorite [40–42].

## IV. ASYMMETRIC PHOTOSENSITIZATION

Asymmetric photosensitization offers an excellent opportunity to input a small amount of chiral information into a reaction, which can then proceed to give products with high op's. This type of reaction is a true example of photochemical asymmetric induction, as it is during the interaction of the photoexcited sensitizer with the prochiral substrate that chiral information is transferred, i.e., during the lifetime of the exciplex, which has recently been observed by fluorescence spectroscopy for some systems [43]. It is theoretically possible to carry out reactions using a catalytic amount of photosensitizer, allowing for extensive chirality amplification. At the time of our previous review [11], this area of photochemistry was able to report only low to moderate op's, but improvements in certain areas, particularly concerning photosensitized geometrical isomerization, have led to more respectable levels of optical activity in the products of these reactions [14].

It is possible for the chiral sensitizer to influence the photochemical processes that occur within the substrate after energy transfer through interactions in the exciplex. It is also important to note that the sensitization may proceed through a singlet, triplet or electron transfer mechanism, a feature of the reaction which is dependent on the energy relationship between the sensitizer and substrate. The mechanistic aspects of photosensitization are described briefly by Mattay and Müller [44], and in more depth by references contained therein.

### A. Photosensitized Destruction/Production

A stereoselective molecular recognition and subsequent destruction of enantiomeric [Co(acac)$_3$] (11) using Δ- or Λ-[Ru(menbpy)$_3$]$^{2+}$ (12) (menbpy = 4,4'-dimenthoxycarbonyl-2,2'-bipyridine) sensitizers has been reported by Ohkubo et al. [45]. Δ-12 was photoexcited in the presence of a racemic mixture of Δ- and Λ-11, to which it transferred an electron upon quenching. Different rates of destruction of Δ- and Λ-11 were observed, with Λ-11 quenching the sensitizer 1.28 times faster than Δ-11 (see Scheme 5). The stereoselectivity of the reaction was found to be affected by the solvent (EtOH/H$_2$O) [46]; during the initial stages the reactions of Δ- and Λ-[Co$^{III}$(acac)$_3$] followed psuedo-first-order kinetics, and at up to ~30% conversion, rate differences ($k_\Delta/k_\Lambda$) were found to be as high as

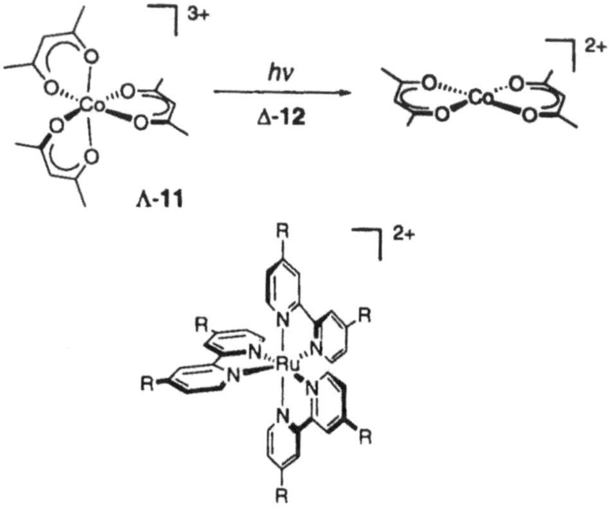

Scheme 5   Asymmetric photodestruction.

14.7 in 90% v/v EtOH/H$_2$O. This rate difference was found to be critically dependent on the solvent, falling to 8.65 in 80% v/v EtOH/H$_2$O [46].

## B.   Photosensitized Geometrical Isomerization

Since the ground-breaking discovery of enantiodifferentiating photosensitized geometrical isomerization by Hammond and Cole [47], extensive work has been carried out in this area. However, in spite of considerable efforts devoted to a variety of asymmetric photosensitizations using many different substrates and sensitizers, the op's reported have only recently begun to reach appreciable levels [11]. Indeed, in the recent literature, all attention in this area has been focused on the enantiodifferentiating Z-E photoisomerization of cyclooctene (**9** and **10**), which is sensitized by optically active (poly)alkyl benzene(poly)carboxylates (see Fig. 3) [43,48]. The mode of sensitization has been reported as following a singlet mechanism [43], which proceeded via an exciplex formed between the sensitizer and **9**. A rotational relaxation then occurred within the exciplex, producing a perpendicular cyclooctene singlet. This singlet state then decomplexed from the sensitizer, before relaxing either to **10** or the optically active (E)-cyclooctene ground states. It was shown that the rotational relaxation process within the exciplex was the only enantiodifferentiating step in this process ($k_S \neq k_R$), while the quenching of chiral sensitizer singlet by (R)- and (S)-**9** and the deactivation from

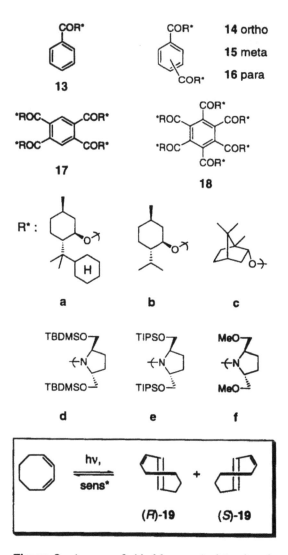

13

14 ortho

15 meta

16 para

17

18

R* :

a

b

c

d

e

f

(R)-19    (S)-19

**Figure 3**   A range of chiral benzene(poly)carboxylates and benzene(poly)amides used as photosensitizers for the photoisomerization of cyclooctene and 1,3-cyclooctadiene.

the free $^1p$ state were not enantiodifferentiating ($k_{qS} = k_{qR}$; $k_E(R) = k_E(S)$) (see Scheme 6). Interestingly, the temperature at which the sensitization process is carried out can have a very large influence on the outcome of the reaction, and simply by changing the temperature it is possible to alter the op's to such an extent that the chirality of the products is switched from one handedness to the other [49]. This phenomenon conflicts with the usually accepted belief that by lowering the temperature it is possible to attain higher op's (this is true if the entropy term is negligible), and it is thought that unequal entropies of activation for the rotational relaxation process are responsible for this unusual effect. Recent work by Inoue et al. [15] has shown that not only is the singlet-sensitized photo-isomerization of 9 and 10 controlled by changes in temperature, but also through

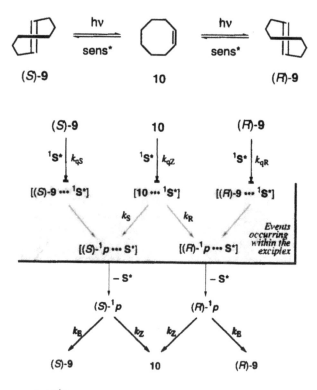

**Scheme 6** The mechanism of the enantiodifferentiating photosensitized isomerization of cyclooctene.

modification of the pressure at which the reaction is carried out, and this will be described below.

The singlet- versus triplet-sensitized photoisomerization of **9** and **10** has been investigated [50]. At low substrate concentrations triplet photoisomerization was observed, using sensitizers such as (−)-1,2-bis(menthyloxymethyl)benzene, which initially formed a singlet excited state before undergoing intersystem crossing to the triplet state. At low concentrations where the triplet mechanism operates, the $(E/Z)_{pss}$ ratios were relatively high (typically 0.22) but the ee's were extremely low (<0.5%). As the concentration of **10** increased, the mechanism was noted to change from triplet to singlet sensitization, and the short-lived sensitizer singlet was quenched during the lifetime to give low $(E/Z)_{pss}$ ratios (<0.02) and appreciably higher ee's of 1.3%. Such a concentration dependence was not observed for menthylbenzoate which functions as a singlet sensitizer; constant $(E/Z)_{pss}$ ratios of 0.24 and ee's of 2.6% were reported over the same concentration range [50].

The recent synthesis of optically active **9**, using highly congested polyalkyl and polyaralkyl benzenepolycarboxylate sensitizers afforded the best op to date; 64% at −89°C reported for benzenetetracarboxylate **17a** [14]. Regardless of the solvent employed, at ambient temperature no significant change in op was observed, and it was postulated that the photoexcited acceptor–donor sensitizer was quenched by cyclooctene, with which it formed an acceptor–donor–donor' triplex (see Fig. 4). Any charge that developed during the process was expected to delocalize extensively over the triplex, thus reducing ionic character, and thereby minimizing any solvent effects. In common with results previously reported [49], the temperature-controlled chirality inversion effect was observed for this type of sensitizer. Interesting results were also reported for the enantiodifferentiating photosensitized isomerization of 1-methylcylooctene with **13–18** [51]. In this work, a direct relationship was observed in the differential activation parameters

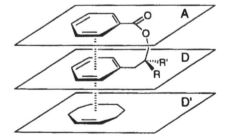

**Figure 4** The proposed acceptor–donor–donor (A-D-D') triplex in the asymmetric photoisomerization of cyclooctene sensitized by chiral paralkyl benzenecarboxylate.

for cyclooctene and 1-methylcyclooctene; as the steric hindrance was increased by introducing the methyl group, steady increments in $\Delta\Delta V^{\ddagger}$ and $\Delta\Delta S^{\ddagger}$ were observed for all chiral sensitizers employed. It was also found that as the temperature was reduced and/or the steric hindrance in sensitizer increased, the $(E/Z)_{pss}$ ratio fell dramatically, although op's reported were low (op <8% for **13b** at $-70°C$).

The photosensitized enantioselective synthesis and chiroptical properties of $(E,Z)$-1,3-cyclooctadiene (**19**) has also been described [52]. $(-)$-Menthyl benzenehexacarboxylate (**18b**) afforded the highest ee's in pentane of 10% and 18% at 25°C and $-40°C$, respectively. Polar solvents were found to diminish the ee, suggesting that in this case a radical ionic intermediate was formed in high dielectric solvents. However, in pentane, a new fluorescence peak (480 nm) was observed, and this was attributed to the exciplex. The low ee's obtained were attributed to a poorer interaction of the $(Z,Z)$-substrate with the benzenehexacarboxylate, which would be expected to exert lower control over the products as they formed. Changes in temperature were once again noted to cause a switching of the product chirality, demonstrating the role of the entropy term in the enantiodifferentiating process. The formation of an intimately interacting exciplex has been shown to be essential for good enantiodifferentiation in the photosensitized geometrical isomerization of cyclooctene and **19**. When (poly)alkyl aromatic(poly)amide sensitizers (**13d-f**, **15d-f**, **16d**, **17d** and **17e**) were used, the enantioselectivity was found to be sharply diminished, even over a wide range of temperatures, and the best ee of 14% was reported for the preparation of $(E,Z)$-cyclooctadiene at $-56°C$ in pentane, using **17d** [15]. In this case, small solvent polarity effects were noted, suggesting that more charge transfer character was present in the exciplex (perhaps a solvent separated ion pair) as compared to the aromatic ester–sensitized photoisomerizations.

Perhaps one of the most interesting recent developments in the field of asymmetric photosensitization is the observation that pressure, as well as temperature, can play a vital role in controlling the ee of the product. Increasing the pressure under which a reaction is carried out is known to affect the selectivity of the process as a result of more compact transition state preferences (for example, see the work of Buback et al. [53] and Chung et al. [54]). With this in mind, Inoue et al. have recently reported that they were able to control the ee of the product for the photoisomerization of **9** and **10** through the action of high pressure. In this example an analogous switching of the op was observed for reactions sensitized by menthyl and bornyl aromatic esters (**13–18a–c**). A plot of ln ($k_S/k_R$) versus pressure ($P$) shows the dramatic effect that pressure has on the enantioselectivity of the photoisomerization of **10** sensitized by **14a**, **17b**, and **18b**, with an equipodal pressure (i.e., the pressure at which the enantioselectivity changes from one to the other enantiomer) for two of the sensitizers studied (see Fig. 5) [15]. Depending on the chiral sensitizer employed, the differential activation vol-

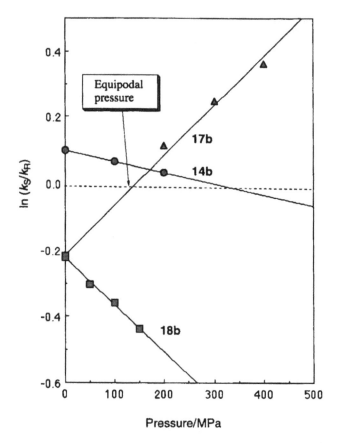

**Figure 5** The effect of pressure on the enantioselectivity of the asymmetric photoisom-erization of cyclooctene using several benzene(poly)carboxylate sensitizers (see Fig. 3 for sensitizer structure).

ume $\Delta\Delta V\ddagger$ was found to vary from $-3.7$ to $+5.6\ cm^3\ mol^{-1}$, which was unexpect-edly large for an enantiodifferentiation in the excited state (see Table 2). The best results were reported for 1,2,4,5-benzenetetracarboxylate and benzenehexa-carboxylate sensitizers (**17** and **18**) which were shown to cause an increase of $\sim100\%$ in the $(E/Z)_{pss}$ at 400 MPa ($E/Z$ ratios of up to 0.5 were reported) as compared to atmospheric pressure (typically the $E/Z$ ratio is 0.25–0.36 [55,56]. **17b** afforded $(R)$-$(-)$-9 (11% ee) at atmospheric pressure but gave the antipodal $(S)$-$(+)$-9 (18% ee) at 400 MPa. The observed ee's at 0.1-, 100-, 150-, and 200-MPa pressures (1 h irradiation) and kinetic parameters (at 25°C) for **17b**, **17c**, **18b**, and **18c** are given in Table 2. With these results in hand, the potential to

**Table 2** Pressure Effects on Photosensitization; Activation Enthalpies, Entropies, and Volumes for Four Crowded Sensitizers

| Sensitizer | Pressure MPa | 9/10 | Conversion % | % yield | % ee | $\Delta\Delta H^{\ddagger}_{S\text{-}R}/$ kcal mol$^{-1}$ | $\Delta\Delta S^{\ddagger}_{S\text{-}R}/$cal mol$^{-1}$ K$^{-1}$ | $\Delta\Delta V^{\ddagger}_{S\text{-}R}/$ cm$^3$ mol$^{-1}$ |
|---|---|---|---|---|---|---|---|---|
| 17b | 0.1 | 0.21 | 38 | 13 | −11.2 | −0.77 | −3.00 | −3.71 |
|     | 100 | 0.32 | 45 | 18 | −3.6 | | | |
|     | 200 | 0.36 | 41 | 21 | +5.7 | | | |
| 17c | 0.1 | 0.31 | 39 | 19 | +14.2 | −0.61 | −1.55 | +0.29 |
|     | 100 | 0.34 | 41 | 20 | +14.3 | | | |
|     | 200 | 0.43 | 46 | 23 | +13.0 | | | |
| 18b | 0.1 | 0.026 | 2.4 | — | −10.9 | −0.96 | −3.85 | +3.50 |
|     | 100 | 0.045 | 5.2 | 4.2 | −17.7 | | | |
|     | 150 | 0.056 | 9.3 | 5.1 | −21.4 | | | |
| 18c | 0.1 | 0.034 | 7.5 | 3.2 | +5.0 | −0.86 | −2.60 | −5.56 |
|     | 100 | 0.062 | 13 | 5.4 | +15.8 | | | |
|     | 200 | 0.073 | 15 | 6.2 | 26.8 | | | |

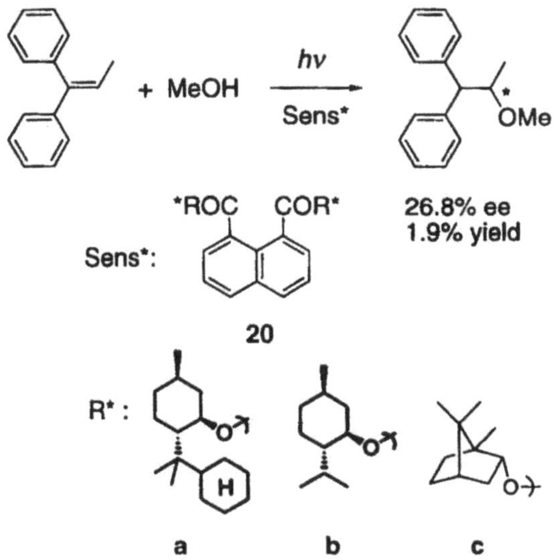

**Scheme 7** Photosensitized polar addition.

control the pressure and temperature of the photoreaction to give a product of known ee is not far from reach.

## C. Photosensitized Polar Addition

Asymmetric photosensitization techniques can also be used to facilitate the en-antiodifferentiating polar addition of methanol to 1,1-diphenylpropene [57]. A range of chiral naphthalenecarboxylate sensitizers bearing analogous auxiliaries to those employed in the photoisomerization of cyclooctene (e.g., **20a-c**) (see Scheme 7) were found to give the anti-Markovnikov product, 1,1-diphenyl-2-methoxypropane in optical yields of up to 27%.

The same authors have also reported on the enantiodifferentiating cis-trans photoisomerizations of 1,2-diarylcyclopropanes and 2,3-diphenyloxirane sensitized by the aromatic carboxylates **13-20** [58]. However, optical purities for these photoisomerizations were not high (op<10%).

## V. ASYMMETRIC INDUCTION BY CHIRAL COMPLEXING AGENT

An alternative approach to the use of chiral sensitizers is to form a complex of a prochiral photosubstrate or labile photoproduct with a chiral host, and then carry out the photoprocess. It is important to note that the asymmetry is induced in the thermal process that precedes or follows the photochemical reaction, separating the following examples from photosensitized processes, where the enantiodifferentiating interaction occurs in the excited state during the lifetime of the substrate–sensitizer exciplex.

Complexation techniques can also be very efficient in terms of the amount of chirality that must be added to a system before the photoprocess is carried out [59]. Asymmetric induction of the product within the complex is ideally followed by dissociation, allowing the formation of a new host–prochiral guest complex to occur. Catalytic turnover such as this allows for a very chiral source–efficient method for the preparation of optically enriched materials. It is clear that the host/photoactive guest molar ratio will be dependent on the strength of the binding interaction between the two species, and this can vary from catalytic quantities through to the other extreme, where the complexing agent is also the solvent or reaction matrix.

At this point it is appropriate to mention solid-state photochemistry, which is an example of complexation of either an achiral or prochiral photosubstrate in a chiral environment. It is important to note that the crystal structure can be inherently chiral without bearing any chirality in the molecules that have crystallized, and in certain circumstances there is no chiral information present in the molecular system prior to crystallization. In this respect, the generation of chirality can be wholly independent of the substrate. There have been extensive publications in this field, of which we have chosen a recent representative selection [60–82].

Recently, cyclodextrin has come forward as an excellent host molecule for asymmetric photochemistry, as it can complex a wide range of substrates [83], and many photoprocesses can be carried out in the chiral field of the host's bowl-shaped cavity. The nonasymmetric aspects of this work have been reviewed by Monti [84,85] covering unimolecular reactions (photorearrangements, intramolecular photoreactions, photoisomerizations) and bimolecular reactions (photosubstitutions, photocycloadditions, intermolecular hydrogen abstraction, photoinduced electron transfer, photoionization, and photocatalytic systems). It is interesting to note that the nonreactive pathways through which the photoexcited substrate can lose its energy are greatly reduced by the complexation of photosubstrate in cyclodextrin as a result of the removal of many of the standard modes of deactivation, resulting in longer lifetimes for the photoexcited species and higher quantum yields [86]. However, it should be mentioned that very few of these works explicitly describe asymmetric induction, which is the central theme of this chapter.

## A.  Geometrical Photoisomerization

The geometrical photoisomerization of (Z)-cyclooctene (10) to a pss enriched in either (R)-(−)-9 or (S)-(+)-9 is most commonly described for photosensitized processes (see Sec. IV). However, the first report of cyclooctene photoisomerization complexed by β-cyclodextrin (**21**) appeared recently [87], and in this instance no sensitizer was employed. This reaction, which was carried out in the solid state, involved the formation of a 1:1 complex of a sample of **10** in the cavity of **21**, which was subsequently isolated, dried, and photoirradiated at 185 nm, giving an $E/Z_{pss}$ with a ratio of 0.47. The enantioselectivity observed was low (0.24% ee), but this work paved the way for the solution state photosensitization of the same substrate using a benzoate sensitizer tethered to β-cyclodextrin, which gave ee's of ~10% [88].

We have already described the work of Zandomeneghi et al. [31], who reported the use of BSA to selectively bind the $(S)$-$(-)$ isomer of 1,1'-bi-2-naphthol (**3**), which was then excited with CPL, giving a selective photodestruction of this isomer to obtain 89% ee after 57% destruction. Using the same complex, but irradiating with a standard light source (Hg lamp, 100 W), the same author was able to demonstrate that **3** photoisomerized to the (R)-(+) enantiomer, **2**, when irradiated at 365 nm, with accompanying photodestruction of **3** [89]. Importantly, no isomerization was observed for the opposite process, i.e., the conversion of **2** to **3**. The enantioselective binding site was found to be basic in nature, and it was also shown that a similar photoisomerization could be carried out in basic MeCN, although to a much lesser extent. When the photoisomerization/photodestruction process using an equimolar ratio of **2** and **3** was carried out to 5% photodestruction (80 min irradiation), an enantiomeric excess of 98.4% **2** was observed, and after 26% photodestruction (280 min irradiation), 99.5% ee was reported.

## B. Photodestruction

By the enantiospecific complexation capabilities of BSA, Zandomeneghi and coworkers were also able to study the photodestruction of a racemic sample of 2-[(3-benzoyl)phenyl]propionic acid (**22**) in BSA [90]. Upon complexation, one of the two binding sites was observed to show a new absorption band at 365 nm, and because this complexation site was enantioselective, irradiation at 365 nm allowed enantioselective excitation of the complex such that the selective photodestruction of one of the enantiomers could be achieved. A modified anisotropy factor, $g'$, of 0.326 (365 nm) was calculated for this compound when complexed to BSA, and using this value it was calculated that an intolerably large photodestruction would be necessary to achieve high ee's (80% ee at 99.8% photodestruction). Interestingly, when human serum albumin was used as the complexing agent, no chirality recognition was noted.

## C. Photocyclization

Toda et al. [73] described photocyclization in the solid state, using various chiral lattices (e.g., **23**) as the host for the [2+2] photocycloaddition of an enone tethered to an olefin (24) (which bears no chiral auxiliaries). The photoreaction in a water suspension of the inclusion crystals prepared by cocrystallization was found to proceed with complete enantioselectivity (90% conversion), giving the cyclobutane derivative (25). When the photoreaction was carried out in acetonitrile, a racemic mixture was formed. Inclusion crystals of a similar composition were also formed by mixing in the solid state and, upon irradiation, the same product was formed with 99% ee (48% yield) (Scheme 8). The mixture was also analyzed by IR and shown to be identical to that formed from the cocrystallization procedure, and the authors suggest that in the suspension the same 2:1 complex is formed. The photocyclization of N,N-dialkylphenylglyoxamides (e.g., 26), prepared by cocrystallization and solid-state mixing with a similar chiral host compound, was also described. The results of the water suspension photocyclizations were impressive; both enantiomers of 27 were formed with 85 and 41% ee (48 and 39% yield) respectively from samples prepared by cocrystallization and solid-state mixing methods. Unfortunately, the stereochemistry of the products was not defined.

Prinsen and Laarhoven have reported work concerning the influence of the chiral environment in the photochemical synthesis of enantiomerically enriched hexahelicene (28) [74]. Because hexahelicene has an exceptionally large optical rotation ($[\alpha]_D^{25} = 3640°$) it is possible to detect even very small op's. A repeat

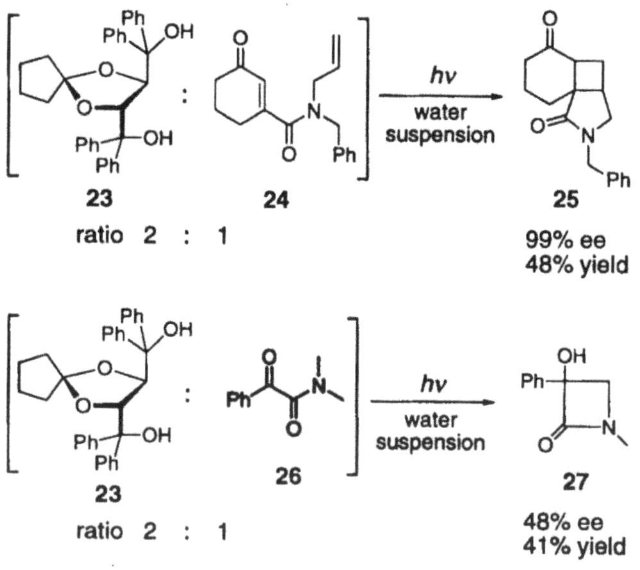

Scheme 8  Photocyclization in the presence of a chiral complexing agent.

of the work by Laarhoven and Cuppen [91] was carried out whereby a chiral isotropic solvent (diethyl (*R,R*)-(+)-tartrate {2.7% ee at −33°C}, diethyl (*R,R*)-(+)-*O,O′*-dibenzoyltartrate {1.1% ee at −25°C and 16°C}, ethyl (*S*)-(+)-*O*-(1-naphthoyl)lactate {1.3% ee at −33°C}, or ethyl (*S*)-(+)-*O*-(2-phenylbenzoyl)lactate {1.6% ee at −33°C}) was used for the photocyclizations of **29** (Scheme 9). In all cases, results were found to be consistent with the previous work. However, the paper then described the same reaction, carried out in the cholesteric phase of two liquid crystals (TM74 and TM75). Although ee's reported were ≤1%, (+)-**28** was found to form in slight excess (1.0% ee in a cholestric phase of 230-nm pitch length). This preference was thought to be the result of a partially enantioselective weak interaction with the cholestric phase, which forms a helical matrix. This was backed up by the knowledge that a chiral molecule can influence the phase transition from nematic to cholesteric phase (see Feringa et al. [33] for a recent account), indicating an interaction with the host lattice.

Although the control of the phase and pitch length of a liquid crystal using a fulgide dopant has also been reported [92], no asymmetric induction is described for this system.

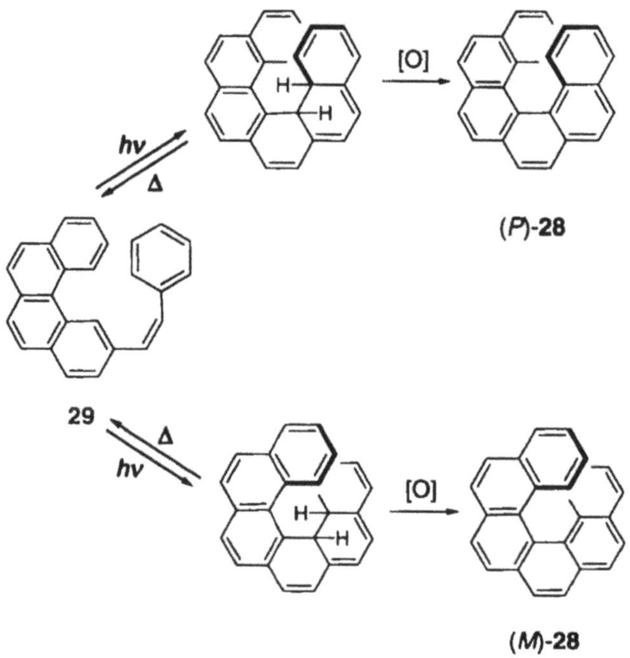

Scheme 9 Asymmetric photocyclization.

## D. Photooxidation

Hamada et al. have reported the enantioselective, photocatalytic oxidation of (R)-
and (S)-1,1′-bi-2-naphthol (2 and 3) with [Ru(menbpy)₃]²⁺ 12 a chiral ruthenium
complex [93]. Δ-12 was photoexcited using filtered visible light (λ > 400 nm),
and the excited complex oxidized by [Co(acac)₃], before returning to its ground
state oxidation level through reaction with the diol. The (S)-diol was found to
be the most reactive of the two enantiomers, and after 13.8% conversion, 2 was
present in 15.2% ee (Scheme 10). However, this value was observed to decrease
steadily as the percentage conversion increased.

The enantioselective oxidation of [Coᴵᴵ(acac)₂(H₂O)₂] in the presence of
acac and 12 has also been described [94]. In this case, atmospheric oxygen was
used to oxidize 12, which then reacted with the Co(II) complex. The formation
of both Δ- and Λ-[Co(acac)₃] occurred, and different rates, dependent on the
chirality of the ruthenium photocatalyst, were observed. It is noted that when Δ-12
was used, the predominant product was Λ-[Coᴵᴵᴵ(acac)₃], with 10% ee for ~20%

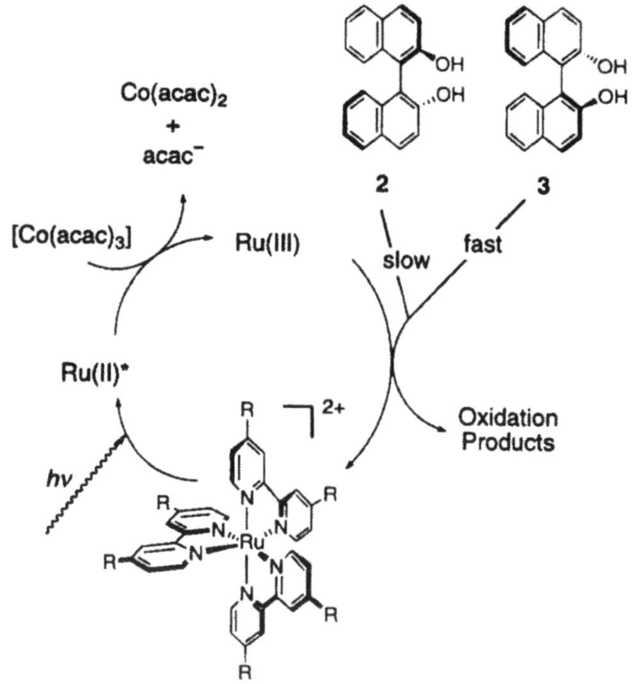

Scheme 10  Asymmetric photooxidation.

conversion. When a sample of racemic **12** was used, the reaction was found to proceed with no production of chirality. This group has also reported the enantioselective photoreduction of [Co(acac)₃] using achiral 1-benzyl-1,4-dihydronicotinamide (**30**) in the presence of molecular aggregates [95]. This work showed that micelles were able to concentrate the reactive species through hydrophobic interactions, but naturally, no induced enantioselectivity was observed in the photoreduction. However, if a carrier protein—in this case, BSA—was used, it was possible to concentrate the reactants in the protein host and to use its inherent chiral fields to induce an enantiodifferentiation. Thus, in BSA, Δ-[Co(acac)₃] was found to be photoreduced by **30** 1.22 times faster than the Λ isomer. Unfortunately, no op's were reported in this paper.

Another example of the cyclodextrin cavity's role in complexation photochemistry is given by Weber et al. [96] who reported the photosensitized enantiodifferentiating oxygenation of racemic α-pinene (**31**), using a porphyrin tethered to a β-cyclodextrin derivative (**32**) (Fig. 6). Upon irradiation with visible light ($\lambda > 350$ nm) in the presence of atmospheric oxygen for 8 h, several oxidation products were obtained. The ratios of these products were found to differ depending on the solvent employed, but it was noted that the (*S*) enantiomers were always formed in excess. For the reaction in the presence of 5 equivalents of 2-methylpyridine, excellent ee's of up to 67% for the (*S*)-1,2-epoxide (**33**) and 57%

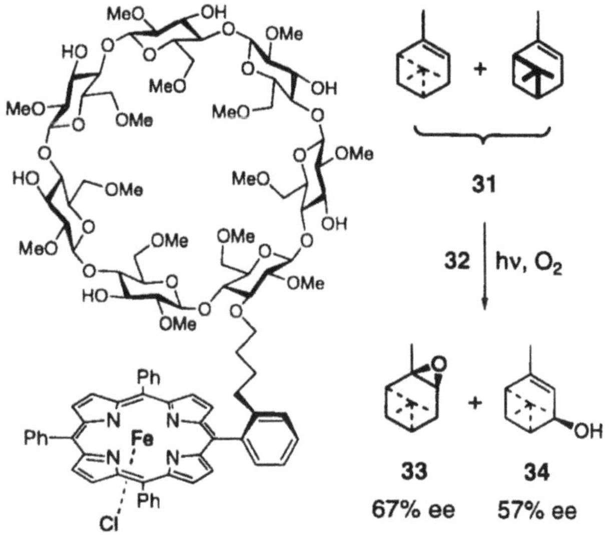

**Figure 6** Cyclodextrin derivatives act as chiral hosts for asymmetric photoprocesses.

for the 3-hydroxy compound (**34**) were observed, and this was presumed to be a result of axial complexation with the Fe by the base, thus preventing oxidation at the "achiral" face of the porphyrin. When the oxidation was carried out thermally in the presence of the untethered components, the yields were shown to increase, but only low ee's were reported (ee < 8%), pointing toward an oxygenation mechanism in which the cyclodextrin plays an important role.

## E.  Photosensitization

The photoisomerization process of (Z)-cyclooctene (**10**), complexed by a modified β-cyclodextrin tethered to a benzoate sensitizer (**35**), has been described [88]. Several solvents were considered, with the best results obtained in a 50:50 mixture of water and methanol, in which a catalytic amount of the host/sensitizer led to a pss with an $E/Z$ ratio of 0.8. This value far exceeded the usual pss ratio's reported for alkyl benzoate photoisomerizations of **10**, which are typically 0.25–0.36 [55,56], and is almost comparable to the pss ratio reported for direct photoisomerization [97]. The water content was presumed to account for the increased inclusion of cyclooctene in the cyclodextrin cavity; in MeOH the $K$, was believed to be small (<100 M$^{-1}$), but when the solvent was 50:50 water/methanol this value rose dramatically to 20100 M$^{-1}$, and it was calculated that in this solvent, the host was 73% occupied by cyclooctene (see Scheme 11).

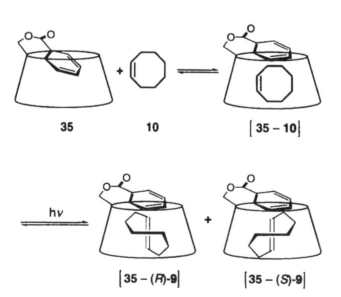

**Scheme 11**   Competitive complexation, and photosensitized geometrical isomerization of cyclooctene within a β-cyclodextrin cavity.

## VI. ASYMMETRIC INDUCTION BY CHIRAL SUBSTITUENT

This section of our review has recently received the most extensive literature attention, with most coverage given to the area of the [2+2] photocycloaddition [98]. It is perhaps a reflection of the ease with which asymmetric induction with chiral substituents can be carried out that this is so. Indeed, very little in the way of specialist equipment is required, and the complexity of products that can be obtained with exceptionally high de's in relatively few synthetic steps can be quite astounding. This, coupled with the fact that a very large number of substrates are amenable to these types of photoreactions, leads to a good level of acceptance of some of these processes as "standard" synthetic routes, e.g., the preparation of a four-membered ring, which is difficult to achieve through non-photochemical methods. It is clear that the use of a chiral auxiliary or building block must constitute at least a molar ratio of chiral information (usually taken in pure form from the chiral pool) necessitating the additional steps of substrate preparation and auxiliary recovery, but the control and versatility of these processes make this "expense" worthwhile, and it is in this area that asymmetric photochemistry has had the largest impact on the more general field of chemistry.

Several reviews concentrating on specific areas within the field of asymmetric induction by chiral substituents have been written, including an excellent review of the asymmetric photoreactions of conjugated enones and esters, which has been written by Pete [98]. This work gives an extensive description of the theoretical basis behind several of the photoreactions that we will discuss, and it should be considered as essential reading for anyone interested in this area. The reactions falling into the general category of asymmetric induction by a chiral substituent that have received attention in the recent literature are 1.) photocyclization, 2.) [2+2] photocycloaddition (inter- and intramolecular), 3.) the Paternò-Büchi reaction, 4.) the Schenck reaction, 5.) [4+2] photocycloaddition, 6.) photodeconjugation, and 7.) geometrical isomerization.

### A. Photocyclization

Photocyclization is usually preceded by a hydrogen abstraction process, or an electron transfer–proton transfer sequence, which results in the formation of a biradical species. This reactive intermediate may then rearrange, collapse back to starting materials, or react with its surroundings, either inter- or, more usually, intramolecularly. The stereoselective photocyclization of β-aryl amines (**36**) to 2-aminocyclopropanols in polar and nonpolar solvents is reported by Weigel et al. [99,100]. When photoirradiation was carried out in diethyl ether, a β-hydrogen abstraction, or more probably an electron transfer from the amino group to the carbonyl followed by a β-proton transfer, led to the production of a 1,3-biradical (**37**), which then collapsed to the cyclopropane (**38** or **39**). The proposed mecha-

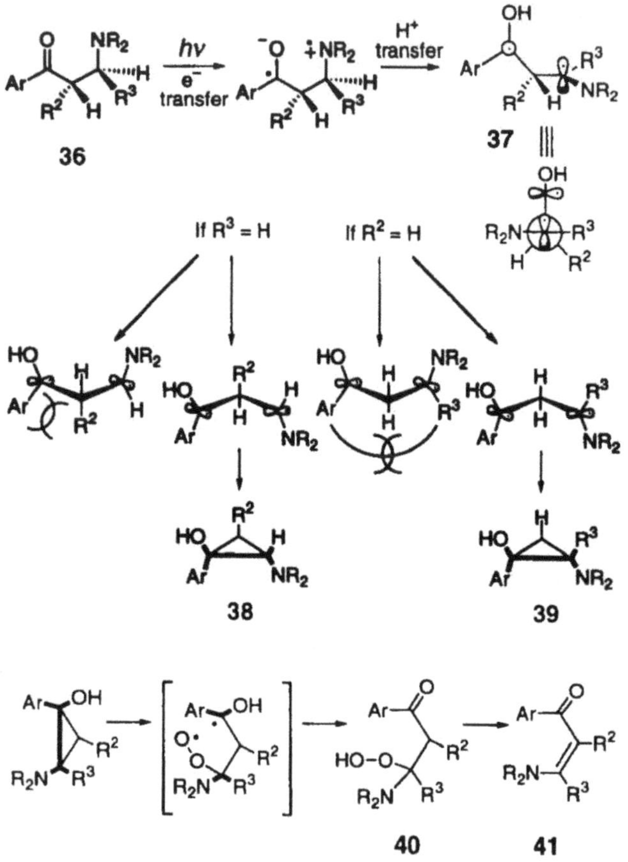

**Scheme 12**  Asymmetric induction by chiral substituent in the photocyclization process.

nism for the selectivity observed is shown in Scheme 12. Nonpolar solvents such as hexane were expected to promote the back reaction, but no recemization was observed, and it was assumed either that the biradical collapsed before bond rotation could occur or that no hydrogen transfer initiated by an electron transfer occurred in the nonpolar solvent. When the reaction was carried out in methanol, the cyclopropane ring opened, then reacted with atmospheric oxygen to form a hydroperoxide (**40**). Hydrogen peroxide was then eliminated to give an $\alpha,\beta$-unsaturated enone (**41**).

Steiner et al. reported that the enantioselective cyclization of Cl′ substituted N-(2-benzoylethyl)glycine esters (**42**) gave 3-hydroxyprolines (**43**) [101]. Substituent-controlled asymmetric induction of the stereogenic centers of C2 and C3 during the photocyclization of n,π* excited derivatives was discussed (see Scheme 13). After hydrogen abstraction, a 1-hydroxy-1,5-biradical was formed which was stabilized by an intramolecular hydrogen bond, adopting a preferred conformation, which then reacted to give a single diastereomer (**43**). When the second ester group was exchanged for a *t*-butyl group, the opposite conformational selectivity was observed, affording **44** with complete diastereoselectivity. The Norrish type II process was also observed when a γ proton was available (e.g., R = CH₂OTDS, **45**), giving the 1,4 biradical which resulted in the formation of the cyclobutane derivative (**46**), but this occurred in a relatively unselective fashion, accompanied with the formation of several byproducts.

The photocyclization of chiral *N*-(2-benzoylethyl)glycinamides (**47** and **48**; see Scheme 14) leads to the formation of 3-hydroxyprolines (**49, 50,** and **51**), as reported by Wessig et al. [102]. The scheme shows two reactions, the first of which proceeded with 7.3:1 selectivity, and the second of which was completely diastereoselective, probably as a result of the steric influence of the auxiliary on

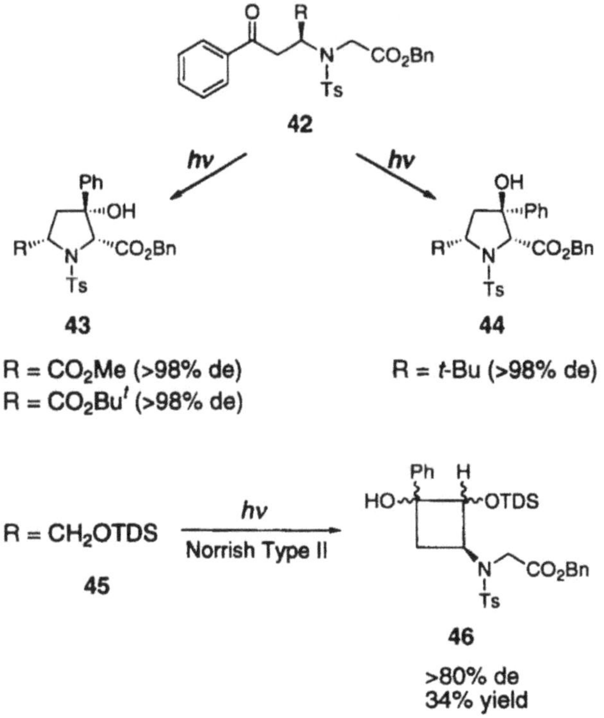

**Scheme 13** Asymmetric photocyclization.

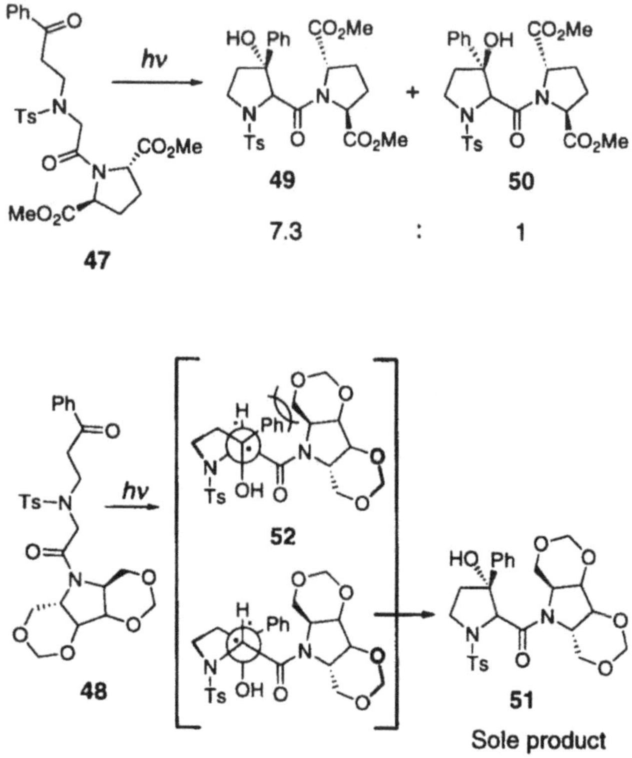

**Scheme 14**   Asymmetric photocyclization.

the 1,5-biradical intermediate (**52**) formed after hydrogen abstraction. A temperature effect was also noted for the reactions, with the diastereoselectivity increasing as the temperature fell. This work allowed the $\Delta\Delta H^{\ddagger}$ (0.79 kcal mol$^{-1}$) and $\Delta\Delta S^{\ddagger}$ (1.2 cal mol$^{-1}$) to be calculated for the first reaction in the scheme.

Fulgides (e.g., **53**, see Scheme 15) are capable of photochromic reactions. Yokoyama et al. reported the preparation and subsequent photochromic isomerization of a fulgide which was designed to resist enantiotopomerization [103]. A sample of the resolved fulgide was irradiated (405 nm light in toluene solution) and was observed to come to a pss with a ratio of 19:81. Irradiation with visible light ($\lambda > 580$ nm) led to the complete recovery of the initial conformation. An advance on this system was made by the same authors, who described the process of diastereomeric photochromism, in which a fulgide derivatized with a binaphthyl auxiliary was allowed to thermally equilibrate, and a photocyclization process carried out [104]. As a result of the relative populations of photoreactive

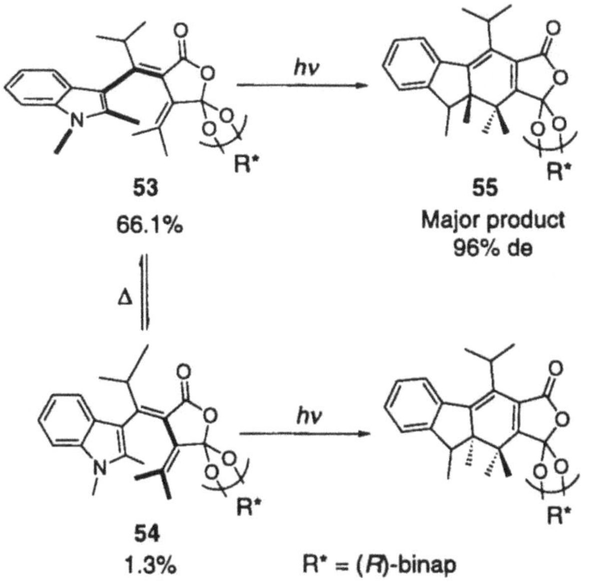

**Scheme 15** Diastereoselective photocyclization of a fulgide.

conformers [66.1% (**53**) and 1.3% (**54**), respectively], the (S) product (**55**) was found to predominate, with 96% de. When the cyclized product was then exposed to visible light, decomposition back to the fulgide was observed.

A large number of photocyclization reactions have also been described in the solid state [60,71,79], generally giving products with high op's and in good yield.

## B. [2+2] Photocycloaddition

Intra- and intermolecular [2+2] photocycloaddition has received a great deal of interest, possibly because it can be easily applied to the single-step synthesis of complex ring systems which contain diverse functionality [105]. Moreover, the (partial) control over the chirality at four new stereogenic centers has rendered this reaction a very attractive choice to the synthetic chemist, for both the formation of natural and unnatural target molecules [106,107]. The mechanism of [2+2] photocycloaddition, first proposed by Corey et al. [108], can be considered as a two-step process [98,109–111] initially involving the formation of a 1,4-biradical, formed by the interaction of a $^3\pi,\pi^*$ enone with a ground state alkene, through attack at the $\beta$ carbon [112]. The triplet biradical then undergoes an intersystem crossing process, affording the singlet biradical [109], which can then undergo a retrocleavage, reforming the ground state starting materials, or cyclize, giving the cyclobutane derivatives. The mode of relaxation of the $^3\pi,\pi^*$ enone is thought to involve a rotation around the ethylenic linkage, which may result in highly twisted states which are biradical in nature, especially where small cyclic enones are involved. The literature in this field through 1993 has been reviewed by Schuster et al. [113].

The [2+2] photocycloaddition process can be split into two sections, these being the inter- and intramolecular reactions. Although a good degree of stereocontrol is observed for intermolecular examples, it is unusual to achieve complete regio- and stereocontrol, whereas the intramolecular process with short tethers possesses no problems concerning regioselectivity, resulting in much fewer products which have higher ee's. Clever synthetic strategies that allow for tether cleavage after photocycloaddition have made the intramolecular [2+2] photocycloaddition the route of choice in most instances.

Although [2+2] photocycloaddition is not limited to reactions of olefins with enone derivatives (styrene and other aromatic alkenes can also undergo this type of photoreaction [11]), the majority of the recent literature has focused on the enone systems, and this will make up the bulk of our discussion.

## 1. Intermolecular [2+2] Photocycloadditions

If the starting alkene and/or enone contains stereogenic centers, the faces of the substrate are rendered diastereotopic, resulting in some degree of facial selectivity in the [2+2] photocycloaddition reaction, and the products of this reaction are rarely evenly distributed through all of the possible outcomes. However, stereo- and regioselectivity are not always high, as we shall see from examples given below. Approach from the less hindered face leads to the 1,4 biradical, and the stability of this intermediate often governs the stereoselectivity of the reaction. The high energy of the biradical should induce an early transition state for the cyclization process. As steric demands of the groups on the biradical increase, so too does the chance of regeneration to the starting materials, a factor which allows for effective competition between the transition states that lead to the diastereomeric cyclobutanes [98]. High facial selectivities are generally observed for photocycloadditions involving C6-substituted cyclohexenes, or C5-substituted cyclopentenones, with the best results obtained when the asymmetric carbon is $\alpha$ to the carbonyl group.

Mechanistic details for the attack of olefins on C5-substituted cyclopentenones are discussed at length by Scharf et al. [114]. All steps of the photocycloaddition are reversible except for the final cyclization, and diastereoselectivity is dominated by only one mechanistic step, that being the approach of the olefin to the vibrationally relaxed $^3\pi,\pi^*$ furanone (56, see Fig. 7). The $\beta$ carbon structure is governed by the substituent R'. For small R' (e.g., H, Me) lower face attack is preferred. As the size of R' increases, the upper face attack becomes increasingly

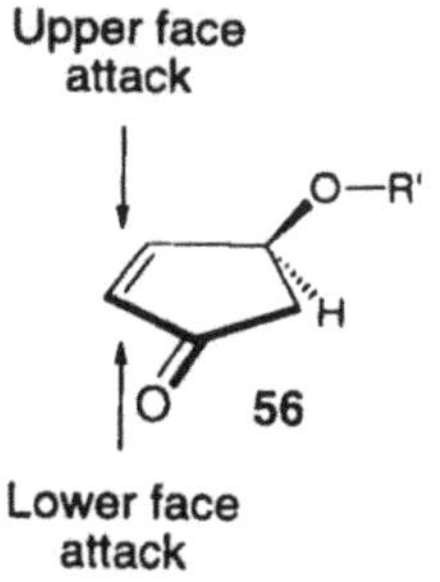

Upper face
attack

56

Lower face
attack

**Figure 7**  Attack of an olefin on the diastereotopic faces of a chiral enone in the [2+2] photocycloaddition.

favored because the radical becomes diffused, and the β carbon becomes more planar (the homoanomeric effect). The steric bulk plays little role in the attack of the olefin when R' = H, Me, and Et, but the cone of rotation blocks the upper face when R' = i−Pr. It is concluded that the first bond formation occurs at the β carbon.

Intermolecular [2+2] photocycloaddition has been applied to the synthesis of a wide variety of natural products and their analogs, which frequently contain four-membered rings in highly ordered chiral environments, or a structure which can be readily derived from the [2+2] cycloadduct. The marine natural product Spatol (57) contains a three fused ring system (see Scheme 16). This has been

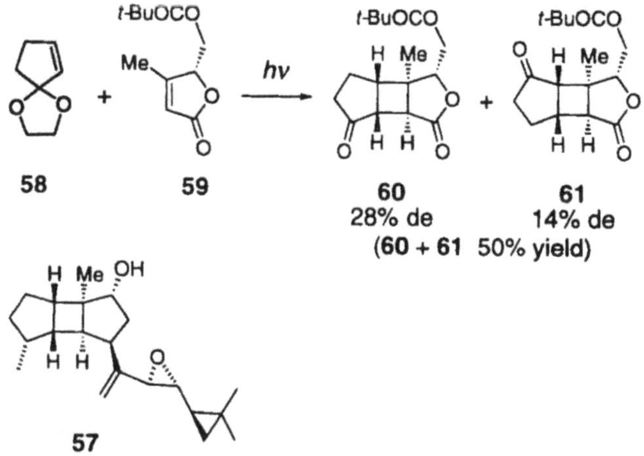

**Scheme 16**  Intermolecular [2+2] photocycloaddition: Spatol precursors.

probed through the reaction of olefin (58) with the α,β-unsaturated lactone derivative (59), which afforded the two regioisomers (60 and 61) in an approximately
1:2 ratio, with moderate selectivity (~50% combined yield, 28% de and 14%
de, respectively) [115]. The low regioselectivity was acceptable to the authors,
who devised a method to convert one of the unwanted isomers to the desired
product, thus increasing the efficiency of the total synthesis.

The reaction of an olefin with a 1,3-diketone enol, known as the de Mayo
reaction [116], is an important member of the [2+2] photocycloaddition reaction
family. This and related processes were discussed by Sato et al. [117]. 1,3-Dioxi-
nones (62) react with ethylene to give cyclobutane products. (Kaneko et al. [118]
and Demuth et al. [119] have written reviews on this subject.) The intermolecular
reactions of olefins with enones, carried out by Organ et al. [120], are complementary to the work on spirodioxinone derivatives (e.g., 63) by Sato et al. [121].
Reaction of enone 64 with cyclohexene led to a mixture of seven products, with
the all-cis isomer formed in 32% yield. However, higher selectivity was seen for
the reaction with a protected cyclohexenone (65), which afforded the all-cis isomer (66) in 54% yield, and reaction with cyclopentene (67), which gave 68 as
a single product in 90% yield, as shown in Scheme 17 (also see Fig. 8).

The reactions of 1,4-dihydropyridines with acrylonitrile has also been examined [122]. When a tetraacetylglucose chiral auxiliary was used, moderate
regio- and enantioselectivity was observed [with the best result of 45% diastereomeric excess (de) for the reaction of acrylonitrile with 69]. Interestingly, the 5,6

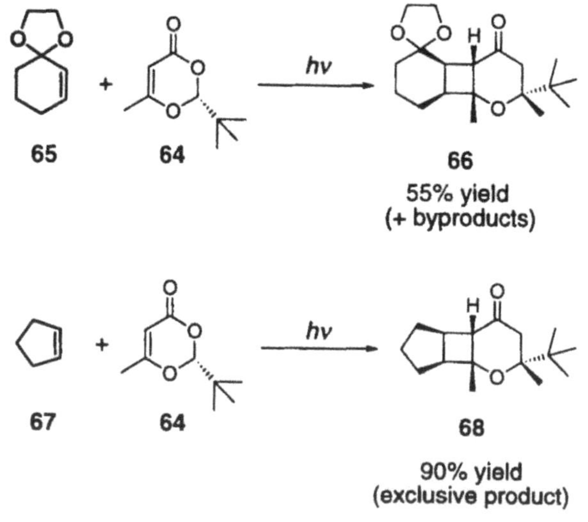

65          64                          66
                                     55% yield
                                  (+ byproducts)

67          64                          68
                                     90% yield
                                 (exclusive product)

Scheme 17   Intermolecular [2+2] photocycloaddition using a 1,3-dioxinone.

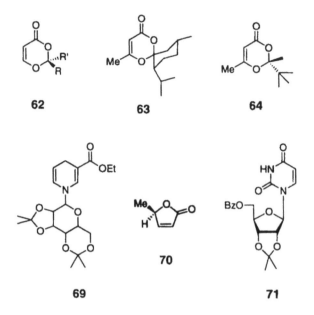

**Figure 8**   Chiral enones used as substrates in intermolecular [2+2] photocycloaddition.

double bond was shown to play an important role in the stereochemistry observed for the cyclobutane, as demonstrated by the reaction of the 5,6-saturated system, which afforded a complex mixture of all possible regio- and stereoisomers.

The reaction of several α,β-butenolides (**70**) with two simple, fully symmetric olefins [ethylene and tetramethylethylene (TME)] revealed that the exo product was always favored (73:27 ratio exo/endo), a finding consistent with approach of the olefin from the less hindered side of the enone [123]. The size of the substituent at C5 also played an important role in controlling the de of the reaction, and changing the substituent from methyl to *t*-butoxycarbonyl increased the de from 18% to 56% for reactions with ethylene. TME has also been reacted with uridine and related compounds (**71**) [124]. Endo attack was greatly favored over exo attack (9:1 in the best case). As the bulk of the sugar moiety increased, the amount of exo product increased slightly, but endo was always found to be the predominant mode of attack.

## 2.   Intramolecular [2+2] Photocycloaddition

It is noted from the above reactions that the stereo- and regioselectivity of intermolecular [2+2] photocycloaddition are not completely satisfying. However, the strategy of using a tether to bring the olefin and the enone substrates together has led to a much greater degree of control than observed for the intermolecular examples [125]. If the tether chosen is long enough to allow reaction to occur but short enough to restrict the approach of the olefin to only one regioisomeric position, good control can be applied (low regioselectivity can be a serious problem for the intermolecular [2+2] photocycloaddition) [126]. Because the tethered olefin is frequently in close proximity to the chiral centers, it is reasonable to assume that a higher degree of asymmetric induction will take place. Several workers have used the tether as part of the framework of the target molecule, but increasingly, tethers which can be readily cleaved are employed [127]. Cleavable tethers allow for simple modification of the structure, which is diminished when the tether is included in the final product, while offering the control that is required for a good asymmetric synthetic procedure. It should be noted that high levels of asymmetric induction are achieved when the chiral centers are included in the tether unit. Pete's review [98] is the source of much inspiration in this area, and many interesting points are discussed, as well as in the review of Winkler et al. [128]. In the subsequent sections, we will first consider all-carbon tethers and then heteroatom-containing tethers.

*All-Carbon Tethers.*    One of the simplest natural products to be prepared using the tethered [2+2] photocycloaddition reaction is grandisol (**72**) [129–131]. When the intramolecular synthesis was carried out in the presence of a Cu(I) complex (which was shown to coordinate to the two olefinic groups), the desired product was formed with 60% ee. The level of chelate control in this reaction is quite high, and structures for this complex were proposed by the authors (Fig. 9).

In the report by Resek et al. [132], the tether was joined at a chiral center in the C4 position of the cyclohexenone ring, and only two carbons separated the reacting centers. If a substituent, R, was added to the olefin terminus, reasonable stereoselectivity in this position was observed. When R = H, 95% yield conversion to a single product was reported (**73**). An example of three-carbon alkyl tether with a chiral side chain adjacent to the enone has been described by Crimmins et al. [133]. This system afforded products with a high diastereoselectivity which could be switched through the choice of solvent. In high-polarity solvents, the selectivity was 5:1 for the all-cis diastereomer (**74**, R' = OEt). This selectivity fell to 1:1 in dichloromethane or hexanes. When the amide substrate (**74**, R' = NMe₂) was used, a 1:3 ratio was found in dichloromethane, which was the opposite selectivity to that normally observed. The same authors have reported a very similar system for the synthesis of a (±)-bilobalide precursor (**75**) via the cyclobutane derivative (single product, 69% de) [134] and have examined other all-alkyl tethers [135,136].

*Heteroatom-Containing Tethers.*    Several chiral esters and amides (**76**) derived from menthol (15% de), phenylethylamine (35% de), 2,5-dimethylpyrrolidine (90% de), and camphorsultam (90% de) have been used to direct the [2+2]

(a) all carbon tethers

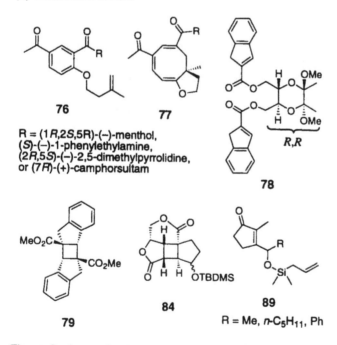

72

73

74

75

R = H or Ac, R' = OEt
R = H, R' = NMe₂

(b) heteroatom tethers

76

77

R = (1R,2S,5R)-(−)-menthol,
(S)-(−)-1-phenylethylamine,
(2R,5S)-(−)-2,5-dimethylpyrrolidine,
or (7R)-(+)-camphorsultam

78

R,R

79

84

89

R = Me, n-C₅H₁₁, Ph

**Figure 9**  Intramolecular [2+2] photocycloaddition.

photocycloaddition of the homoallyl group to a benzene ring [137]. Interestingly, the product formed passed through a cyclooctatriene (77) and then rearranged via another photochemical [2+2] process to give a tricyclic species, which slowly rearranged back to 77.

Fleming has reported the use of several cinnamic acid derivatives tethered through a siloxane moiety in the preparation of a variety of different truxinates [138,139]. This reaction was further developed by Scharf [140], who included the cinnamic acid in a chiral framework, such as that derived from erythritol (78). This compound preferentially adopted a chair conformation with minimal axial interactions, and the cinnamates aligned in a head-to-head fashion, reacting to give the major product in an ~5:1 ratio with all other products at −60°C, and ~2:1 ratio at +8°C. The introduction of a bulkier aryl group rendered the head-to-head conformation unfavorable, and the cis isomer became the prominent product (79). These tethers were readily cleaved.

Since the first bond in the cyclization process is thought to form at the β carbon, it is reasonable to expect chiral auxiliaries in this region of the molecular framework to have a high degree of influence on the stereochemical outcome of the reaction. Lactic acid derivatives (80 and 81) were investigated as conveniently prepared chiral substrates for the intramolecular [2+2] photocycloaddition. Although the monolactate derivative 80 gave intermolecular [2+2] cycloadduct, high selectivity for specific intramolecular process was observed in the case of dilactate derivative 81 [only two products observed (82 and 83)] [141], but there was only low control over the regioselectivity, probably because the tether was too long to restrict the reaction to only one approach for the olefinic moiety (see Scheme 18). The esters were then readily cleaved after the reaction, giving products that were easily modified.

High levels of selectivity have been observed for the reaction of an α,β-unsaturated lactone tethered to the enone via a shorter ester linkage [142]. The main carbon skeleton resulting from the [2+2] photocycloaddition was formed as a single enantiomer. The reaction of tethered allenylsilanes (84) afforded cyclobutane derivatives with an appended double bond (85) which offered a new point of modification [143]. De's as high as 99% (90% yield) were observed. When the tether was changed to a thioether (86), the yield dropped to 80%, with a de of 89%. This work was based on the *t*-butylallenyl analogs (e.g., 88), prepared by the same group [144], which also underwent photocycloaddition with 99% stereoselectivity. Esters, diisopropysiloxanes, and ethers have also been used as tethers, undergoing intramolecular [2+2] photocycloaddition in a highly diastereoselective fashion, with up to 77% de [145]. High optical purity products were observed, suggesting that the mechanism involved the establishment of stereochemistry through kinetic processes upon the addition of the enone excited state to the least hindered face of the allene. Interestingly, the lifetime of the 1,4 biradical intermediate is sufficient for a rotation to occur, giving both the *E* and *Z* products (Scheme 19).

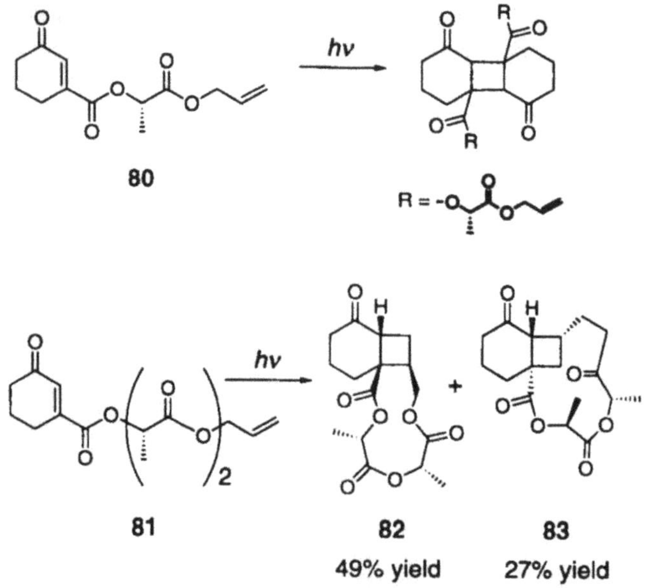
hν
R = -O...O...
80
81
82
83
49% yield
27% yield


**Scheme 18** Intramolecular [2+2] photocycloaddition of a substrate with a lactate-derived tether.

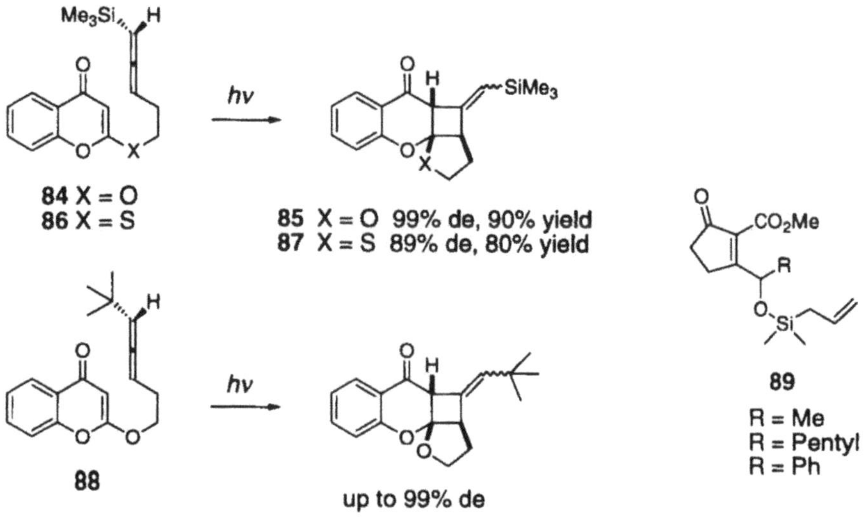
Me3Si
SiMe3
hν
84 X = O
86 X = S
85 X = O 99% de, 90% yield
87 X = S 89% de, 80% yield
88
up to 99% de
CO2Me
89
R = Me
R = Pentyl
R = Ph


**Scheme 19** Intramolecular [2+2] photocycloaddition of allenyl derivatives.

Crimmins also reported on the use of the siloxane group as part of a chiral, four-atom tether [127]. For **89**, R = Me, a de of 74% was observed, and this was increased to 70% for R = pentyl, and 90% for R = phenyl.

It is clear that [2+2] photocycloaddition had dominated the recent literature of the asymmetric photochemical process in solution, and we have highlighted some of the more interesting examples that have appeared over the last 6 years. However, this chapter does not seek to describe this area in depth, and the reader's attention should be directed to the reviews mentioned to obtain a more specialized view of this area of photochemistry.

## C. The Paternò-Büchi Reaction

The Paternò-Büchi reaction, which bears many similarities to the [2+2] photo-cycloaddition of two-olefin moieties, consists of the reaction of a ketone or thio-ketone with an olefin, leading to the formation of the four-membered oxetane or thiooxetane ring system, usually in good yield. Chiral groups can be attached to either of the reacting species, and high selectivities are observed. Several possible Paternò-Büchi reactions are shown in Scheme 20 [11]. However, even though it is similar to the all-carbon reaction, there have been only a few reported Paternò-Büchi reactions in the recent literature. The Norrish type I photoinduced cleavage of the C12—C13 bond, followed by a hydrogen shift of a hecogenine precursor

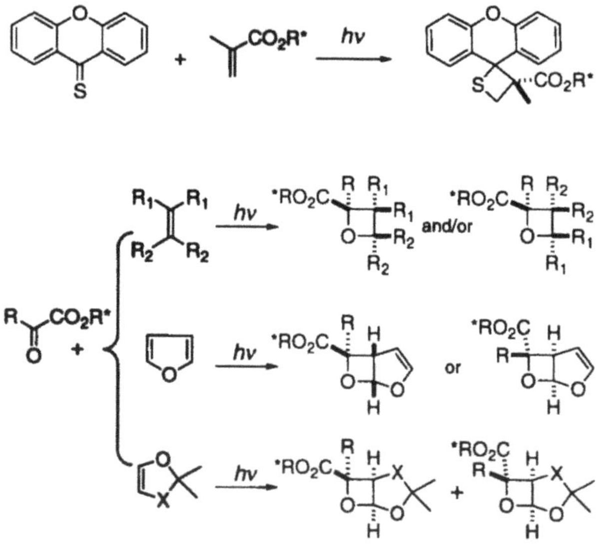

**Scheme 20**  Examples of the Paternò-Büchi reaction.

(**90**), led to the formation of a 1,4 biradical, which then recombined to form an oxetane (**92**) [146]. The reaction was found to pass through an aldehyde intermediate (**91**), which then underwent an intramolecular Paternò-Büchi reaction (see Scheme 21).

The photocycloaddition of L-ascorbic acid derivatives (e.g., **93**) with 4-chlorobenzaldehyde (**94**) and benzyl methyl ketone led to preferential attack on the less hindered α-face of the enone with approximately 2:1 regioselectivity (33% de for **96**) (see Scheme 22) [147]. When the substrate was changed to benzophenone, the regioselectivity was reversed, even though the facial selectivity remained the same (35% de). This was proposed to be the result of a mechanistic switchover, from a 1,4 diradical process for benzophenone to a photoinduced electron transfer process for the other substrates.

Buschmann et al. have described a paper concerning the linear free energy relationships in the Paternò-Büchi reaction [148]. In this paper, an attempt was made to correlate $\Delta G^{\ddagger}$ with the free energy of the ground state conformers for cyclohexanol derivatives. No Hammett relationship for different aromatic substituents was observed, as the effect that this group has on the reaction is predominantly steric in nature, meaning that any effects on the selectivity of the reaction are derived purely from the auxiliary, Aux*. For the photoreaction between **97**

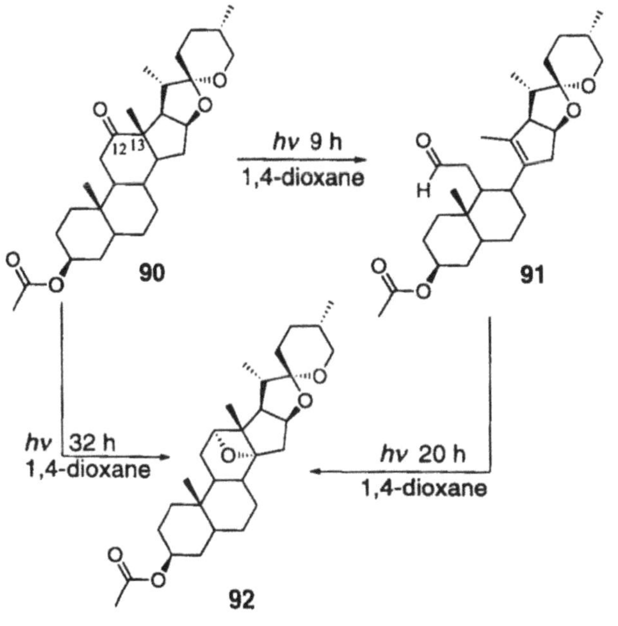

**Scheme 21**  Hecogenine precursors in the Paternò-Büchi reaction.

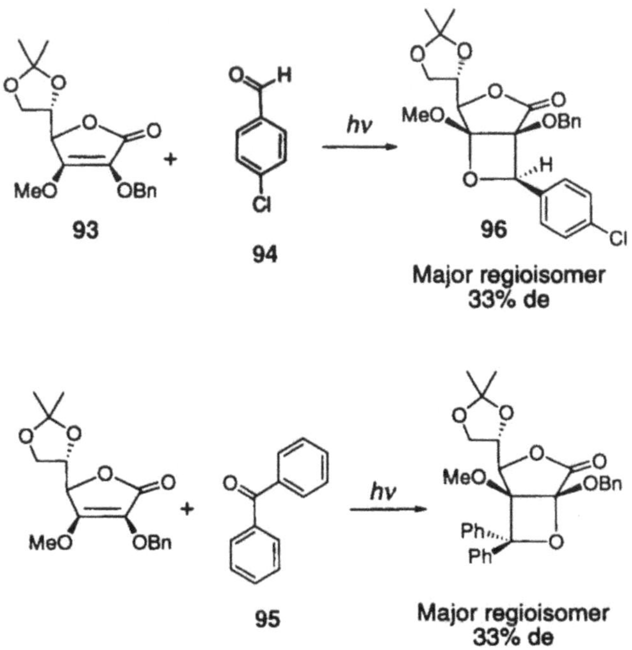

**Scheme 22**   Regioselectivity/diastereoselectivity in the Paternò-Büchi reaction.

and unsaturated cyclic ethers there are two extreme cases: 1.) the conformational energy barrier to ring flipping of the cyclohexane is *lower* than the activation energy relationship for the formation of X or Y ($k_{AB•} \gg k_A, k_B$) (here the Curtain-Hammett principle applies, so the ratio X/Y is determined by $\Delta G_X^{\ddagger}$ and $\Delta G_Y^{\ddagger}$), or 2.) the energy barrier for ring flipping is *higher* than the energy of activation of the formation of the products ($k_{AB•} \ll k_A, k_B$), and the X/Y ratio is determined by [A*]/[B*] (see Fig. 10). The selectivity may also occur as a result of the influence of the auxiliary on the 1,4 biradical, making the bis-equatorial form most readily revert to starting materials.

## D.   The Schenck Reaction

The photooxygenation of olefins with singlet molecular oxygen ($^1O_2$) (known as the Schenck reaction, or the "ene" reaction) offers the synthetic chemist a convenient and effective route to allylic hydroperoxides. Adam has contributed largely to knowledge concerning the Schenck reaction and has written an extensive review covering a large amount of allylic photooxygenations [13]. Cyclic olefins, bi- and polycyclic hydrocarbons, cyclic olefins with a heteroatom substituent,

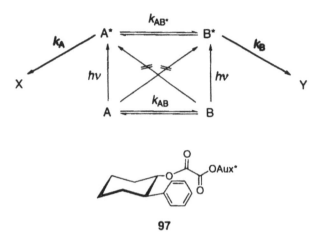

Figure 10   Allowed processes in the Paternò-Büchi reaction.

acyclic systems (with or without a chiral auxiliary), and allylic alcohols/amines are reviewed. It is noted that effective control has not yet been achieved for cyclic systems, and polycyclic systems may give high levels of control, but only if the substrate itself bears certain characteristics. However, for acyclic systems, especially allylic derivatives, the selectivity is both high and controllable.

Linker and Frölich have described the effects of the 6-position substituent on 1,4-cyclohexadiene on the photooxygenation process [149]. Dienes **99**, prepared from the Birch reduction of suitable benzene derivatives **98**, provided a substrate with diastereotopic faces, and the stereoselectivity was derived from an interaction between the incoming $^1O_2$ and the group in the 6 position, which may either be attractive (OH) or repulsive ($CO_2R$), affording the trans (**100**) or cis (**101**) products, respectively (see Scheme 23). Interestingly, although the dominant steric factor in the reaction is initially the ester group, when an ethyl group is also bound to the 6 position this becomes the dominant steric factor, making the cis diastereomers the major products of the reaction. High cis selectivity is also seen for the alcohol derivatives **102**, which react with $^1O_2$ to give **103** almost exclusively.

Brünker and Adam have described the photooxygenation of allylic amines (**104**) and their acyl derivatives (**105**) (Scheme 24) [150]. This paper details a new mechanism proposed by the author in an attempt to explain why the traditionally used mechanism, based on the "cis effect" [151,152], is not valid in all cases. It was concluded that only the 1,3-allylic strain would lead to diastereoselectivity and that amines, ammonium chlorides, and alcohols would lead to threo selectivity. For *threo*-selective reactions, the best results were obtained in

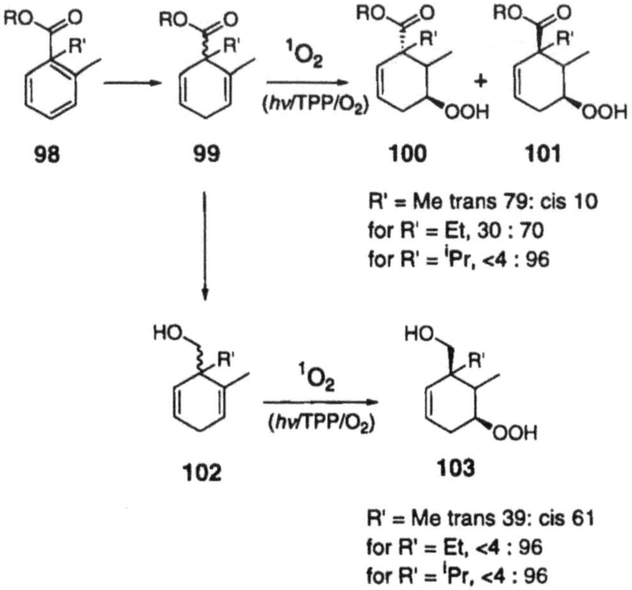
RO
O
R'
98
99
$^1O_2$
(hv/TPP/$O_2$)
OOH
100
101
R' = Me trans 79: cis 10
for R' = Et, 30 : 70
for R' = $^i$Pr, <4 : 96
HO
102
103
R' = Me trans 39: cis 61
for R' = Et, <4 : 96
for R' = $^i$Pr, <4 : 96

**Scheme 23** The Schenck reaction.

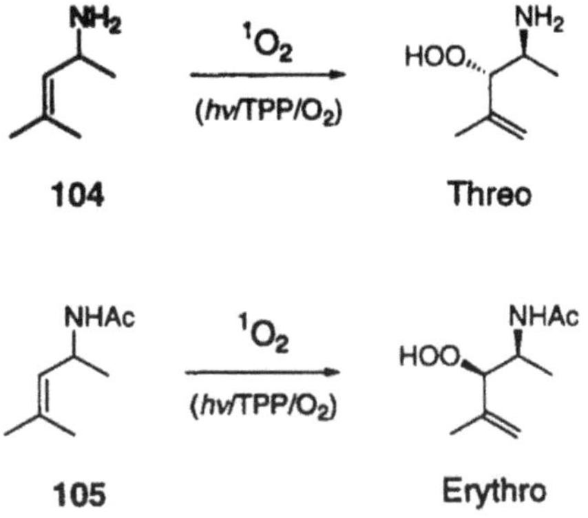
$NH_2$
$^1O_2$
(hv/TPP/$O_2$)
HOO
104
Threo
NHAc
105
Erythro

**Scheme 24** Erythro and threo selectivity in the Schenck reaction.

nonpolar aprotic media, whereas for *erythro*-selective reactions, no solvent effects were noted. It was also noted that all substrates react regioselectively, with a preference for the (Z) isomer in the minor product. The "cis effect" mechanism does not take into account any possible interactions between the substrate and the singlet oxygen, and predicts the same selectivity for a methyl or amine derivative, which is incorrect. It was reasoned that the reaction could not be subject to steric control only because *t*-butyl derivatives gave the erythro product, also not predicted by the cis effect. The new mechanism proposed that as the $^1O_2$ approaches, it feels either a repulsive interaction [ester, amide, NHAc, etc. (**106**)] or an attractive interaction (NH$_2$, NH$_3^+$, OH) with the allylic substrate (**107**) through hydrogen bonding. Scheme 25 shows how these interactions can lead to the products **108–111** that are indeed obtained. This mechanism is further discussed in a later paper by Adam et al. [153], which compares the photooxygenation with epoxidation using perbenzoic acid, which shows similar selectivities.

Dussault and Woller have also reported on the enantioselective photooxygenation of allylic alcohols (e.g., **112**) with the aim of synthesizing 1,4-dioxygenated peroxides (**113**) [154]. The photooxygenation illustrated in Scheme 26 gave **113** in 58% yield, with good stereoselectivity.

## F.  [4+2] Photocycloaddition

Very few asymmetric [4+2] photocycloadditions have been described in the literature [11]. However, reported examples are usually found to proceed with good levels of selectivity. The intramolecular [4+2] photocycloaddition of the diarylethene derivative (**114**) was found to proceed smoothly with good diastereoselectivity (de = 86.6% in toluene at −40°C) (Scheme 27) [155]. When the dielectric constant was increased slightly, the de was found to increase; this was also the case as the temperature was lowered. The mechanism of the asymmetric induction is described in detail.

The [4+2] photooxygenation of benzene rings lead to endoperoxides, and these are the subject of the paper by Linker et al. [156]. When **115** was reacted with $^1O_2$, two products, derived from the single (**116**) and double (**117**) addition, were formed with complete diastereoselectivity. The researchers considered R as OH, ester, and acid, and found that for carbonyl derivatives the reaction proceeded slowly, affording more of the doubly substituted product, whereas the alcohol reacted quickly, giving the mono derivative in excess (Scheme 28). The selection of the transition state to give either **116** or **117** may be governed by the interaction between the terminal oxygen in the transition state.

Prein et al. have also described the [4+2] photocyclization of singlet oxygen to 3-(1-hydroxyethyl)-4-methylphenol (**118**) (see Scheme 29) [157]. The selectivity for **119** and **120** reached a value of 85:15 at 95% conversion in CDCl$_3$. When the isomeric 4-(1-hydroxyethyl)-3-methylphenol (not shown) was sub-

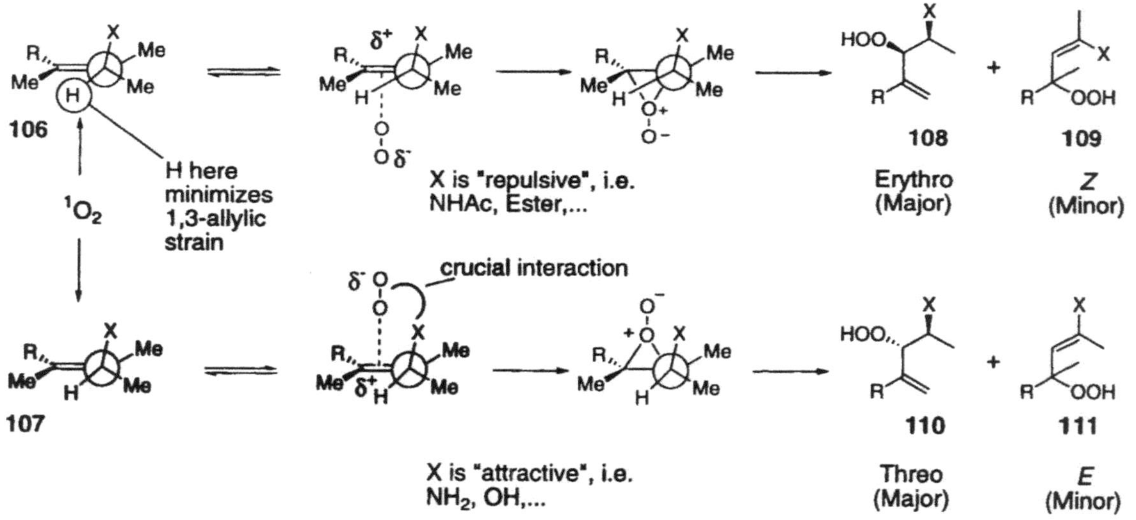

**Scheme 25** Mechanism of the Schenck reaction.

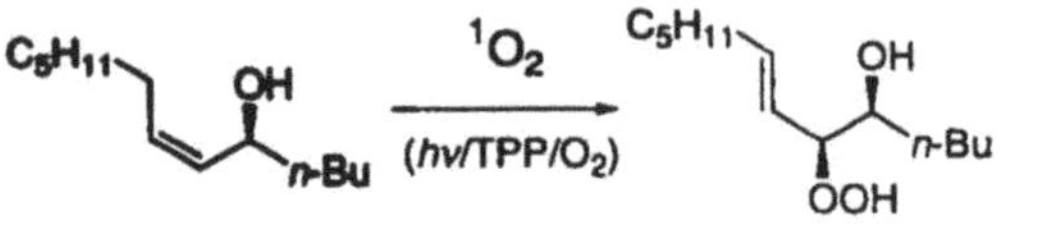

112                                    113

58% (9:1 syn : anti)

**Scheme 26**  Preparation of a natural product using the Schenck reaction.

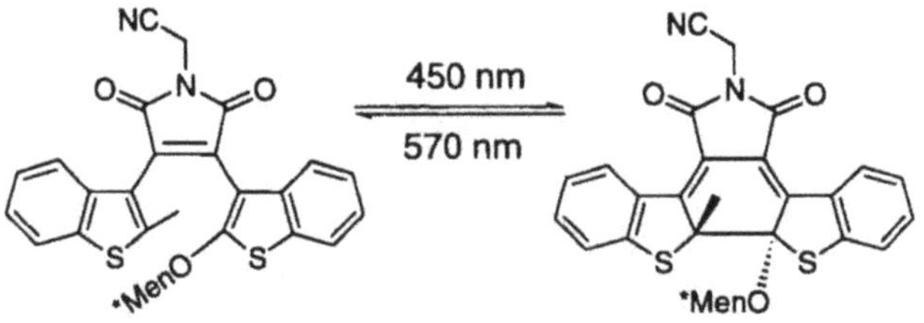

114                         de$_{pss}$ = 86.6 % (450 nm)

**Scheme 27**  [4+2] Photocycloaddition.

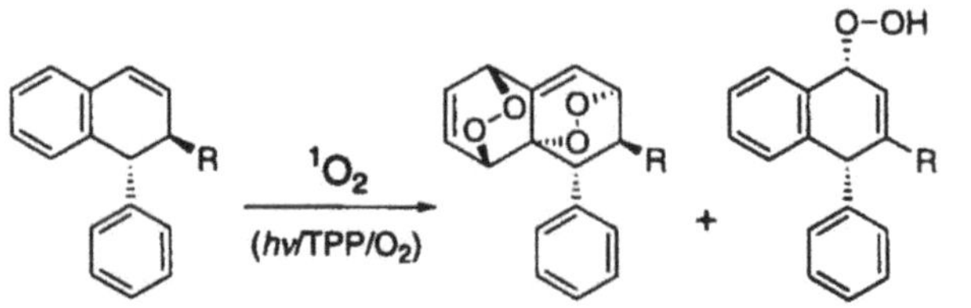

115                         117            116

81 : 19 (R = CO$_2$H)
78 : 22 (R = CO$_2$Me)
37 : 63 (R = CO$_2$Me)

**Scheme 28**  [4+2] Photocycloaddition for the oxygenation of aromatic compounds.

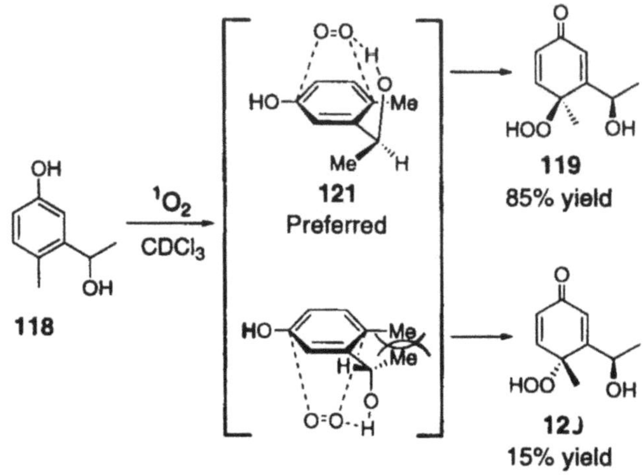

**Scheme 29** Intramolecular hydrogen bonding for preferential transition state conformation in [4+2] photocycloaddition.

jected to the same oxygenation, the much higher de of >90% was obtained although the conversion was kept deliberately low (40%). The best de values were reported in nonpolar solvents, and this was suggested to be due to lower interference with the intra- and intermolecular hydrogen bonds that control the conformation of the transition state like **121** during the reaction. Hydrogen bonding or polar solvents interfere with this process, thus reducing the selectivity.

## F.  Photodeconjugation

Irradiation of enones in the $n,\pi^*$ or $\pi,\pi^*$ absorption bands leads to the efficient Z-E isomerization, and a pss is reached. At longer irradiation times it is possible for a second photoprocess to occur. If there are hydrogen atoms in the allylic position, $\gamma$-hydrogen abstraction can occur, leading to a photodienol, which rapidly collapses back to the starting material or undergoes photodeconjugation [98]. At low temperature, the thermally driven back-reaction, a [1,5] sigmatropic shift, will not normally occur. Only a small amount of recent material has been added to this field by Pete et al., and this work has been reviewed by themselves [98].

The irradiation of $\alpha,\beta$-unsaturated esters (**122**; see Scheme 30) derived from diacetone glucose in the presence of an achiral proton source, such as alcohol, amine, or aminoalcohol, gave the $\beta,\gamma$-unsaturated product (**123**) in moderate to high yield, with good de (for *N,N*-dimethylaminoethanol additive, de $\geq$ 98%) [158,159]. Interestingly, when the proton source was changed from an amine to an alcohol, the opposite facial selectivity was observed. The products were

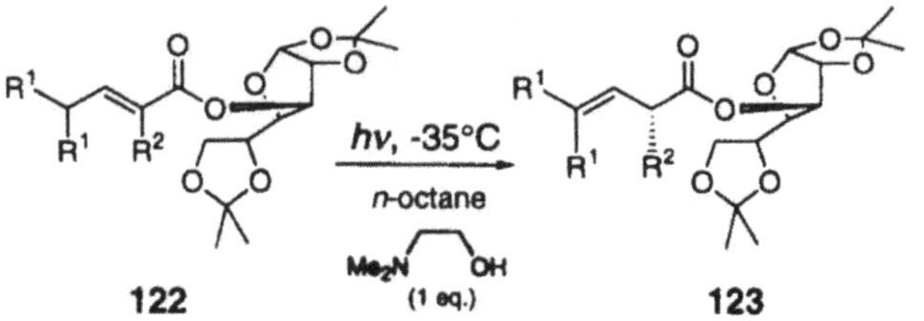

**Scheme 30** Photodeconjugation.

subsequently reduced using achiral hydrogenation techniques to give highly en-
antiopure products. A simple natural product synthesis is also reported using the
diacetone D-glucose auxiliary, affording products with up to 98% ee [159]. In
this case, the methodology is used without explanation.

## G.  Geometrical Isomerization

Harada and Feringa have reported on the role of chiral substituents in the optically
controlled photoisomerization of sterically overcrowded olefins [160]. The en-
antiopure substrate (**124**; see Scheme 31) was photoisomerized from trans to cis
(a pss was attained) using a high-pressure Hg lamp. The product conformation
was determined by the chiral centers on the substrate, and if optically pure starting
material was used, the product's helicity was found to be predictable, with **124**
giving the (*P,P*)-helical product (**125**) in 55% yield.

In an different approach to that normally employed for the photosensitized
geometrical isomerization procedure, Sugimura et al. [161] reported on the teth-

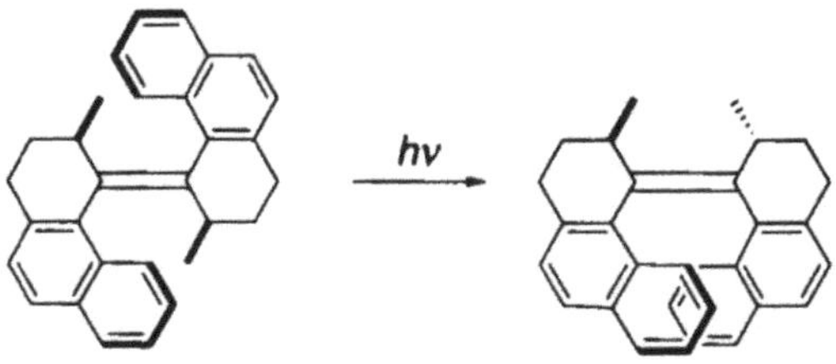

**Scheme 31** Stereoselective geometrical isomerization of an overcrowded olefin.

ering of the sensitizer to the substrate, and chirality included in the tether unit led to diastereoselectivity in the products which were formed. Thus, a chiral tether was used to attach a benzoate sensitizer to the 1-position of cyclooctene (126), allowing high de's to be achieved for the intramolecular photosensitized isomerization (Scheme 32). When the system was irradiated using a low-pressure Hg lamp, irrespective of the solvent employed, a $(Z/E)_{pss}$ of 0.8 was rapidly reached, ruling out the possibility of a radical ion intermediate, and supporting the formation of an exciplex in all cases [14]. Constant de values throughout the irradiation period showed that the only diastereoselective process was the Z-to-E photoisomerization, while the back reaction was found not to be at all diastereoselective. The de of the product is known to be dependent on the reaction temperature [49] and the best de reported for the benzoate sensitized process was 33% at $-65°C$. Similar results were obtained for the tere- and isophthalate derivatives (126, R = $p$-CO$_2$Me and $m$-CO$_2$Me, respectively), and the highest de was obtained for the terephthalate (44% at $-65°C$). The differential enthalpy ($\Delta\Delta H^{\ddagger}$) and entropy ($\Delta\Delta S^{\ddagger}$) values were calculated from Eyring treatments of the data, again highlighting that changes in the temperature of the reaction could lead to an inversion of the product's de, and this inversion was observed over the temperature range studied for the terephthalate derivative. When the same chiral tether/sensitizer unit was used in the intermolecular photosensitized isomerization of 10, only low ee's of up to 7% were observed.

## H.  Meta Cycloaddition

Sugimura et al. have also used tethers comprising chiral 2,4-pentanediol derivatives (127) to control the intramolecular photocycloaddition of a styrene unit to

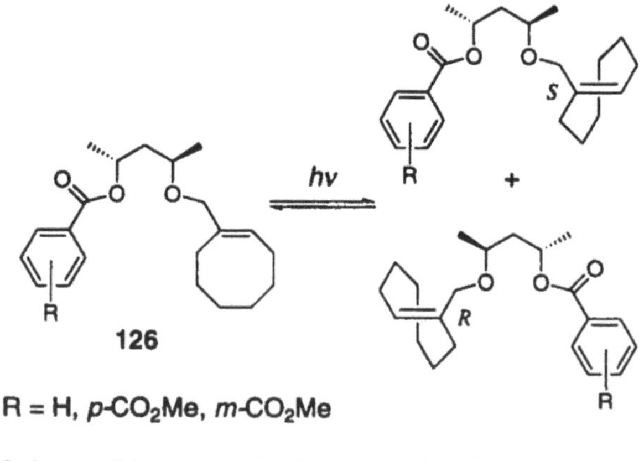

R = H, $p$-CO$_2$Me, $m$-CO$_2$Me

**Scheme 32** Stereoselective geometrical isomerization of cyclooctene tethered to the sensitizer.

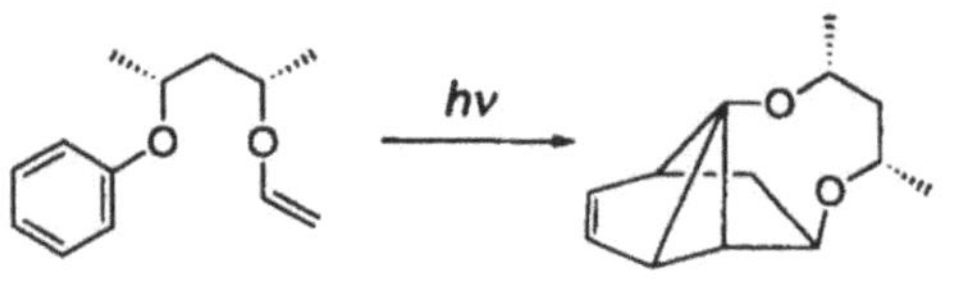

127 128

100% de
70% yield

**Scheme 33**   Meta arene-alkene photocycloaddition.

a phenol ring [162]. This resulted in the diastereoselective *meta*-arene-alkene photocycloaddition to give 3-tricyclo[3.2.1.0]octenes (e.g., **128**) with complete diastereoselectivity, in 70% yield (see Scheme 33).

## VII.   DISCUSSION/CONCLUSIONS

Asymmetric photochemistry in solution is still a growing field, and we have endeavored to highlight the recent developments that have taken place. The advantages of photochemical processes over conventional ground state reactions include the possibility of accessing electronically excited states at low temperatures, enabling selective excitation of specific components in the system. Furthermore, the products generated are frequently unique to photoreactions. These reasons have encouraged photochemists to attempt to use light to carry out many interesting transformations and syntheses, which are complementary in terms of the products generated to more traditional, ground state reactions. The role of CPL in direct asymmetric photochemistry has increased in importance, from being simply a tool to evaluate the molar ellipticity ($g$) factor, to carrying out processes where moderate op's can be obtained. This work, particularly in the vacuum UV region, is beginning to shed some light on the potential role of CPL in the generation of optical activity in the prebiotic state. Work elsewhere in this field has shown that through the use of complexing agents, particularly biological carrier proteins, it is possible to greatly enhance effective $g$ factors, allowing for good levels of enantioselectivity in photodestruction processes, which had hitherto required the almost complete destruction of the sample before an appreciable op was attained.

Developments in the understanding of the interaction between a chiral sensitizer and substrate in the exciplex have also led to increased control over the op's that can be obtained from a reaction, and it has recently become possible to use both temperature and pressure to govern which of the two enantiomers in a geometrical photoisomerization process is prepared. Temperature and pressure switching of optical activity is likely to enable complete control in this reaction. It is this work which has shown us a unique possibility for photochemical processes in the investigation of the role of entropy in the transition state of the reaction. This necessitates modifying the reaction temperature over a wide range and such changes cannot be made for ground state (i.e., thermal) reactions without changing the reaction mechanism. Advances in the photochemistry of complexations have allowed the use of imaginative supramolecular systems to control the thermal processes that occur close in time to photoexcitation, and the transfer of chiral information from the excited host to the guest enables efficient chirality amplification to occur. However, it is the field of chiral substituent–induced asymmetric photochemistry that has seen the most recent interest, and several methods, including the chiral tethering of substrates to one another, offer high levels of diastereoselectivity with accompanying near-quantitative yields, for a wide range of substrates. Although we have described this process as being the least efficient in terms of the amount of chiral information that must be added to the system in the form of the chiral auxiliary/building block prior to the photoprocess, it is satisfying to see that these photoreactions now feature highly in the current literature.

The search for new photoprocesses must be encouraged if an expansion of the scope of stereodifferetiating photoreactions is to occur. It is important to elucidate the (external) factors and mechanisms governing photochirogenesis procedures from a more global point of view and to utilize them more efficiently. It will then be possible to utilize and discuss thermal and photochemical asymmetric reactions on a common ground. However, none of the areas described above have fully succumbed to the skills of the modern photochemist, and it is clear that further understanding of the mechanisms of all of the photoprocesses described will be needed if the techniques are to become fully integrated into the synthetic chemist's armory.

## ACKNOWLEDGMENTS

We would like to thank Prof. J.-P. Pete of the University of Reims for stimulating discussion, in particular concerning the topics described in Sections V and VI and Ms. Makiko Niki for her extensive assistance in collecting and sorting the necessary references.

## REFERENCES

1. Collins, A.N.; Sheldrake, G.N.; Crosby, J. *Chirality in Industry*; Wiley: Chichester, 1992; Vol. 1.
2. Collins, A.N.; Sheldrake, G.N.; Crosby, J. *Chirality in Industry*; Wiley: Chichester, 1997; Vol. 2.

3. Gawley, R.E.; Aube, J. *Principles of Asymmetric Synthesis*; Pergamon: Oxford, 1996.

4. Hassner, A. *Advances in Asymmetric Synthesis*; Hassner, A., Ed.; JAI Press: Greenwich, CT, 1995; Vol. 1.

5. Noyori, R. *Asymmetric Catalysis in Organic Synthesis*; Wiley: New York, 1994.

6. Nógrádi, M. *Stereoselective Synthesis*; 2nd ed.; VCH: Weinheim, 1995.

7. Procter, G. *Asymmetric Synthesis*; Oxford University Press: New York, 1996.

8. Seyden-Penne, J. *Chiral Auxiliaries and Ligands in Asymmetric Synthesis*; Wiley: New York, 1995.

9. Sheldon, R.A. *Chirotechnology—Industrial Synthesis of Optically Active Compounds*; Marcel Dekker: New York, 1993.

10. Wong, C.H.; Whitesides, G.M. *Enzymes in Synthetic Organic Chemistry*; Pergamon: Oxford, 1994.

11. Inoue, Y. *Chem. Rev.* **1992**, *92*, 741–770.

12. Henderson, G.L. *Trans. Ill. State Acad. Sci* **1968**, *61*, 360–366. Kagan, H.B.; Fiaud, J.C. *Top. Stereochem.* **1978**, *10*, 175–285. Rau, H. *Chem. Rev.* **1983**, *83*, 535–547. Jarosz, S. *Wiad. Chem.* **1983**, *37*, 167–192. Griesbeck, A.G. *EPA Newslett.* **1986**, *28*, 13–20. Bonner, W.A. *Top. Stereochem.* **1988**, *18*, 1–96.

13. Prein, M.; Adam, W. *Angew. Chem., Int. Ed. Engl.* **1996**, *35*, 477–494.

14. Inoue, Y.; Yamasaki, N.; Yokoyama, T.; Tai, A. *J. Org. Chem.* **1993**, *58*, 1011–1018.

15. Inoue, Y.; Matsushima, E.; Wada, T. *J. Am. Chem. Soc.* **1998**, *120*, 10687–10696.

16. Joy, A.; Robbins, R.J.; Pitchumani, K.; Ramamurthy, V. *Tetrahedron Lett.* **1997**, *38*, 8825–8828.

17. Leibovitch, M.; Olovsson, G.; Sundarababu, G.; Ramamurthy, V.; Scheffer, J.R.; Trotter, J. *J. Am. Chem. Soc.* **1996**, *118*, 1219–1220.

18. Sundarababu, G.; Leibovitch, M.; Corbin, D.R.; Scheffer, J.R.; Ramamurthy, V. *Chem. Commun.* **1996**, 2159–2160.

19. Inoue, Y.; Tsuneishi, H.; Hakushi, T.; Yagi, K.; Awazu, K.; Onuki, H. *Chem. Commun.* **1996**, 2627–2628.

20. le Bel, J.-A. *Bull. Soc. Chim. (Paris)* **1874**, *22*, 337–347. English translation: le Bel, J.A. In *Classics in the Theory of Chemical Combination*; O.T. Benfey, Ed.; Dover: New York, 1963; pp 161–171.

21. van't Hoff, J.H. *Die Lagerung der Atome im Raume*, 2nd ed.; Vieweg: Braunschweig, 1894; p 30.

22. Kuhn, W.; Knopf, E. *Z. Phys. Chem., Abt. B* **1930**, *7*, 292–310.

23. Kuhn, W. *Trans. Faraday Soc.* **1930**, *26*, 293–309.

24. Nikogosyan, D.N.; Khoroshilova, E.V. *Dokl. Akad. Nauk SSSR* **1988**, *300*, 1172–1177.

25. Nikogosyan, D.N.; Repevev, Y.A.; Khoroshilova, E.V.; Kryukov, I.V.; Khoroshilov, E.V.; Sharkov, A.V. *Chem. Phys.* **1990**, *147*, 437–445.

26. Kagan, H.B.; Fiaud, J.C. *Top. Stereochem.* **1988**, *18*, 249–331. Nicoud, J.F.; Eskenazi, C.; Kagan, H.B. *J. Org. Chem.* **1977**, *42*, 4270–4272.

27. Rau, H. *Chem. Rev.* **1983**, *83*, 535–547.

28. Balavoine, G.; Moradpour, A.; Kagan, H.B. *J. Am. Chem. Soc.* **1974**, *96*, 5152–5158.

29.  Kagan, H.B.; Balavoine, G.; Moradpour, A. *J. Mol. Evol.* **1974**, *4*, 41–48.
30.  Shimizu, Y.; Kawanishi, S. *Chem. Commun.* **1996**, 819–820.
31.  Zandomeneghi, M. *J. Am. Chem. Soc.* **1991**, *113*, 7774–7775.
32.  Stevenson, K.L.; Verdieck, J.F. *Mol. Photochem.* **1969**, *1*, 271–288.
33.  Huck, N.P.M.; Jager, W.F.; Delange, B.; Feringa, B.L. *Science* **1996**, *273*, 1686–1688.
34.  Feringa, B.L.; Jager, W.F.; De Lange, B. *Tetrahedron* **1993**, *49*, 8267–8310.
35.  Feringa, B.L.; Schoevaars, A.M.; Jager, W.F.; De Lange, B.; Huck, N.P.M. *Enantiomer* **1996**, *1*, 325–335.
36.  , Huck, N.P.M.; Feringa, B.L. *J. Chem. Soc. Chem. Commun.* **1995**, 1095–1096.
37.  Feringa, B.L.; Huck, N.P.M.; van Doren, H.A. *J. Am. Chem. Soc.* **1995**, *117*, 9929–9930.
38.  Feringa, B.L.; Jager, W.F.; De Lange, B. *J. Chem. Soc., Chem. Commun.* **1993**, 288–290.
39.  Zhang, M.; Schuster, G.B. *J. Org. Chem.* **1995**, *60*, 7192–7197.
40.  Engel, M.H.; Macko, S.A.; Silfer, J.A. *Nature (London)* **1990**, *348*, 47–49.
41.  Engel, M.H.; Macko, S.A. *Nature (London)* **1997**, *389*, 265–268.
42.  Pizzarello, S.; Cronin, J.R. *Nature (London)* **1998**, *394*, 236.
43.  Inoue, Y.; Yamasaki, N.; Yokoyama, T.; Tai, A. *J. Org. Chem.* **1992**, *57*, 1332.
44.  Müller, F.; Mattay, J. *Chem. Rev.* **1993**, *93*, 99–117.
45.  Ohkubo, K.; Hamada, T.; Ishida, H.; Fukushima, M.; Watanabe, M. *J. Mol. Catal.* **1994**, *89*, L5–L10.
46.  Ohkubo, K.; Hamada, T.; Ishida, H. *J. Chem. Soc., Chem. Commun.* **1993**, 1423–1425.
47.  Hammond, G.S.; Cole, R.S. *J. Am. Chem. Soc.* **1965**, *87*, 3256–3257.
48.  Inoue, Y.; Yokoyama, T.; Yamasaki, N.; Tai, A. *J. Am. Chem. Soc.* **1989**, *111*, 6480–6482.
49.  Tsuneishi, H.; Inoue, Y.; Hakushi, T.; Tai, A. *J. Chem. Soc. Perkin Trans. 2* **1993**, 457–462.
50.  Tsuneishi, H.; Hakushi, T.; Inoue, Y. *J. Chem. Soc., Perkin Trans. 2* **1996**, 1601–1605.
51.  Tsuneishi, H.; Hakushi, T.; Tai, A.; Inoue, Y. *J. Chem. Soc., Perkin Trans. 2* **1995**, 2057–2062.
52.  Inoue, Y.; Tsuneishi, H.; Hakushi, T.; Tai, A. *J. Am. Chem. Soc.* **1997**, *119*, 472–478.
53.  Buback, M.; Bünger, J.; Tietze, L.F. *Chem. Ber.* **1992**, *125*, 2577–2582.
54.  Chung, W.S.; Turro, N.J.; Mertes, J.; Mattay, J. *J. Org. Chem.* **1989**, *54*, 4881–4887.
55.  Inoue, Y.; Takamuku, S.; Kunitomi, Y.; Sakurai, H. *J. Chem. Soc., Perkin Trans. 2* **1980**, 1672–1677.
56.  Yamasaki, N.; Yokoyama, T.; Inoue, Y.; Tai, A. *J. Photochem. Photobiol., A: Chem.* **1989**, *48*, 465–467.
57.  Inoue, Y.; Okano, T.; Yamasaki, N.; Tai, A. *J. Chem. Soc., Chem. Commun.* **1993**, 718–720.
58.  Inoue, Y.; Yamasaki, N.; Shimoyama, H.; Tai, A. *J. Org. Chem.* **1993**, *58*, 1785–1793.

59. Lehn, J.-M. *Supramolecular Chemistry*; VCH: Weinheim, 1995.
60. Akutsu, S.; Miyahara, I.; Hirotsu, K.; Miyamoto, H.; Maruyama, N.; Kikuchi, S.; Fumio, T. *Mol. Cryst. Liq. Cryst. Sci. Technol., Sect. A* **1996**, *277*, 447–453.
61. Asahi, T.; Nakamura, M.; Kobayashi, J.; Toda, F.; Miyamoto, H. *J. Am. Chem. Soc.* **1997**, *119*, 3665–3669.
62. Borecka, B.; Fu, T. Y.; Gudmundsdottir, A.D.; Jones, R.; Liu, Z.; Scheffer, J.R.; Trotter, J. *Chem. Mater.* **1994**, *6*, 1094–1095.
63. Fu, T.Y.; Liu, Z.; Olovsson, G.; Scheffer, J.R.; Trotter, J. *Acta Crystallogr., Sect. B: Struct. Sci.* **1997**, *B53*, 293–299.
64. Gudmundsdottir, A.D.; Scheffer, J.R. *Photochem. Photobiol.* **1991**, *54*, 535–538.
65. Gudmundsdottir, A.D.; Scheffer, J.R.; Trotter, J. *Tetrahedron Lett.* **1994**, *35*, 1397–1400.
66. Gudmundsdottir, A.D.; Lewis, T.J.; Randall, L.H.; Scheffer, J.R.; Rettig, S.J.; Trotter, J.; Wu, C.-H. *J. Am. Chem. Soc.* **1996**, *118*, 6167–6184.
67. Hirotsu, K.; Okada, K.; Mizutani, H.; Koshima, H.; Matsuura, T. *Mol. Cryst. Liq. Cryst. Sci. Technol., Sect. A* **1996**, *277*, 459–466.
68. Jones, R.; Scheffer, J.R.; Trotter, J.; Yang, J. *Tetrahedron Lett.* **1992**, *33*, 5481–5484.
69. Koshima, H.; Ding, K.L.; Chisaka, Y.; Matsuura, T. *Tetrahedron: Asymmetry* **1995**, *6*, 101–104.
70. Leibovitch, M.; Olovsson, G.; Sundarababu, G.; Ramamurthy, V.; Scheffer, J.R.; Trotter, J. *J. Am. Chem. Soc.* **1996**, *118*, 1219–1220.
71. Leibovitch, M.; Olovsson, G.; Scheffer, J.R.; Trotter, J. *J. Am. Chem. Soc.* **1997**, *119*, 1462–1463.
72. Leibovitch, M.; Olovsson, G.; Scheffer, J.R.; Trotter, J. *Pure Appl. Chem.* **1997**, *69*, 815–823.
73. Toda, F.; Miyamoto, H.; Kikuchi, S. *J. Chem. Soc., Chem. Commun.* **1995**, 621. Miyamoto, H.; Kikuchi, S.; Oki, Y.; Inoue, M.; Kanemoto, K.; Toda, F. *Mol. Cryst. Liq. Cryst. Sci. Technol., Sect. A* **1996**, *277*, 433–438.
74. Prinsen, W.J.C.; Laarhoven, W.H. *Recl. Trav. Chim. Pays-Bas* **1995**, *114*, 470–475.
75. Roughton, A.L.; Pieper, W.; Muneer, M.; Demuth, M.; Krueger, C. *Mol. Cryst. Liq. Cryst. Sci. Technol., Sect. A* **1994**, *248*, 651–5 CODEN: MCLCE9; ISSN: 1058–725X;.
76. Sakamoto, M.; Takahashi, M.; Fujita, T.; Nishio, T.; Iida, I.; Watanabe, S. *J. Org. Chem.* **1995**, *60*, 4682–4683.
77. Sakamoto, M.; Takahashi, M.; Shimizu, M.; Fujita, T.; Nishio, T.; Iida, I.; Yamaguchi, K.; Watanabe, S. *J. Org. Chem.* **1995**, *60*, 7088–7089.
78. Sakamoto, M.; Takahashi, M.; Kamiya, K.; Yamaguchi, K.; Fujita, T.; Watanabe, S. *J. Am. Chem. Soc.* **1996**, *118*, 10664–10665.
79. Sakamoto, M.;Takahashi, M.; Moriizumi, S.; Yamaguchi, K.; Fujita, T.; Watanabe, S. *J. Am. Chem. Soc.* **1996**, *118*, 8138–8139.
80. Sakamoto, M.; Takahashi, M.; Fujita, T.; Watanabe, S.; Nishio, T.; Iida, I.; Aoyama, H. *J. Org. Chem.* **1997**, *62*, 6298–6308.
81. Suzuki, T. *Pure Appl. Chem.* **1996**, *68*, 281–284.

82. Toda, F.; Miyamoto, H.; Koshima, H.; UrbanczykLipkowska, Z. *J. Org. Chem.* **1997**, *62*, 9261–9266.
83. Rekharsky, M.V.; Inoue, Y. *Chem. Rev.* **1998**, *98*, 1875–1917.
84. Bortolus, P.; Monti, S. *Adv. Photochem.* **1996**, *21*, 1–133.
85. Bortolus, P.; Grabner, G.; Köhler, G.; Monti, S. *Coord. Chem. Rev.* **1993**, *125*, 261–268.
86. Grabner, G.; Köhler, G.; Monti, S. *J. Phys. Chem.* **1993**, *97*, 13011–13016.
87. Inoue, Y.; Kosaka, S.; Matsumoto, K.; Tsuneishi, H.; Hakushi, T.; Tai, A.; Nakagawa, K.; Tong, L.-H. *J. Photochem. Photobiol. A: Chem.* **1993**, *71*, 61–64.
88. Inoue, Y.; Dong, F.; Yamamoto, K.; Tong, L.-H.; Tsuneishi, T.; Tai, A. *J. Am. Chem. Soc.* **1995**, *117*, 11033–11034.
89. Levi-Minzi, N.; Zandomeneghi, M. *J. Am. Chem. Soc.* **1992**, *114*, 9300–9304.
90. Festa, C.; Levi-Minzi, N.; Zandomeneghi, M. *Gazz. Chim. Ital.* **1996**, *126*, 599–603.
91. Laarhoven, W.H.; Cuppen, T.J.H.M. *J. Chem. Soc., Perkin Trans.* 2 **1978**, 315–318.
92. Janiki, S.Z.; Schuster, G.B. *J. Am. Chem. Soc.* **1995**, *117*, 8524–8527.
93. Hamada, T.; Ishida, H.; Usui, S.; Tsumura, K.; Ohkubo, K. *J. Mol. Catal.* **1994**, *88*, L1–L5.
94. Ohkubo, K.; Hamada, T.; Watanabe, M.; Fukushima, M. *Chem. Lett.* **1993**, 1651–1654.
95. Ohkubo, K.; Yamashita, K.; Ishida, H.; Haramaki, H.; Sakamoto, Y. *J. Chem. Soc., Perkin Trans.* 2 **1991**, 1833–1838.
96. Weber, L.; Imiolczyk, I.; Haufe, G.; Rehored, D.; Hennig, H. *J. Chem. Soc., Chem. Commun.* **1992**, 301–303.
97. Inoue, Y.; Takamuku, S.; Sakurai, H. *J. Phys. Chem.* **1977**, *81*, 7–11.
98. Pete, J.-P. *Adv. Photochem* **1996**, *21*, 135–216.
99. Weigel, W.; Schiller, S.; Henning, H.-G. *Tetrahedron* **1997**, *53*, 7855–7866.
100. Weigel, W.; Schiller, S.; Reck, G.; Henning, H.-G. *Tetrahedron Lett.* **1993**, *34*, 6737–6740.
101. Steiner, A.; Wessig, P.; Polborn, K. *Helv. Chim. Acta* **1996**, *79*, 1843–1862.
102. Wessig, P.; Wettstein, P.; Giese, B.; Neuburger, M.; Zehnder, M. *Helv. Chim. Acta* **1994**, *77*, 829–837.
103. Yokoyama, Y.; Uchida, S.; Shimizu, Y.; Yokoyama, Y. *Mol. Cryst. Liq. Cryst. Sci. Technol., Sect. A* **1997**, *297*, 85–91.
104. Yokoyama, Y.; Uchida, S.; Yokoyama, Y.; Sugawara, Y.; Kurita, Y. *J. Am. Chem. Soc.* **1996**, *118*, 3100–3107.
105. Winkler, J.D.; Hong, B.-C.; Bahdor, A.; Katanietz, M.G.; Blumberg, P.M. *J. Org. Chem.* **1995**, *60*, 1381–1390.
106. Blechert, S.; Jansen, R.; Velder *Tetrahedron* **1994**, *50*, 9649–9656.
107. Benchikh-le-Hocine, M.; Do Khac, D.; Fétizon, M.; Guir, F.; Guo, Y.; Prangé, T. *Tetrahedron Lett.* **1992**, *33*, 1443–1446.
108. Corey, E.J.; Bass, J.D.; LeMahieu, R.; Mitra, R.B. *J. Am. Chem. Soc.* **1964**, *86*, 5570–5583.
109. Becker, D.; Denekamp, C.; Haddad, N. *Tetrahedron Lett.* **1992**, *33*, 827–830.
110. Andrew, D.; Hastings, D.J.; Weedon, A.C. *J. Am. Chem. Soc.* **1994**, *116*, 10870–10882.

111. Andrew, D.; Weedon, A.C. *J. Am. Chem. Soc.* **1995**, *117*, 5647–5663.
112. Winkler, J.D.; Shao, B. *Tetrahedron Lett.* **1993**, *34*, 3355–3358.
113. Schuster, D.I.; Lem, G.; Kaprindis, N.A. *Chem. Rev.* **1993**, *93*, 3–22.
114. Hoffmann, N.; Buschmann, H.; Raabe, G.; Scharf, H.-D. *Tetrahedron* **1994**, *50*, 11167–11186.
115. Tanaka, M.; Tomioka, K.; Koga, K. *Tetrahedron* **1994**, *50*, 12843–12852.
116. de Mayo, P. *Acc. Chem. Res.* **1971**, *4*, 41–47.
117. Sato, M.; Ohuchi, H.; Abe, Y.; Kaneko, C. *Tetrahedron: Asymmetry* **1992**, *3*, 313–328.
118. Kaneko, C.; Sato, M.; Sakaki, J.; Abe, Y. *J. Heterocycl. Chem.* **1990**, *27*, 25–30.
119. Demuth, M.; Palomer, A.; Sluma, H.-D.; Dey, A.K.; Kruger, C.; Tsay, Y.-H. *Angew. Chem.* **1986**, *98*, 1093–1095; *Angew. Chem. Int. Ed. Eng.* **1986**, *25*, 1117.
120. Organ, M.G.; Froese, R.D.J.; Goddard, J.D.; Taylor, N.J.; Lange, G.L. *J. Am. Chem. Soc.* **1994**, *116*, 3312–3323.
121. Sato, M.; Takayama, K.; Furuya, T.; Inukai, N.; Kaneko, C. *Chem. Pharm. Bull.* **1987**, *35*, 3971–3974.
122. Adembri, G.; Donati, D.; Fusi, S.; Ponticelli, F. *J. Chem. Soc., Perkin Trans. 1* **1992**, 2033–2038.
123. Alibes, R.; Bourdelande, J.L.; Font, J. *Tetrahedron: Asymmetry* **1991**, 2, 1391–1402.
124. Ishikawa, I.; Itoh, T.; Takayanagi, H.; Oshima, J.; Kawahara, N.; Mizuno, Y.; Ogura, H. *Chem. Pharm. Bull.* **1991**, *39*, 1922–1930.
125. Amougay, A.; Pete, J.-P.; Piva, O. *Bull. Soc. Chim. Fr.* **1996**, *133*, 625–635.
126. Gleiter, R.; Fischer, E. *Chem. Ber.* **1992**, *125*, 1899–1911.
127. Crimmins, M.T.; Guise, L.E. *Tetrahedron Lett.* **1994**, *35*, 1657–1660.
128. Winkler, J.D.; Mazur Bowen, C.; Liotta, F. *Chem. Rev.* **1995**, *95*, 2003–2020.
129. Alibes, R.; Bourdelande, J.L.; Font, J.; Gregori, A.; Parella, T. *Tetrahedron* **1996**, *52*, 1267–1278.
130. Langer, K.; Mattay, J.; Heidbreder, A.; Möller, M. *Leibigs Ann. Chem.* **1992**, 257–260.
131. Langer, K.; Mattay, J. *J. Org. Chem.* **1995**, *60*, 7256–66.
132. Resek, J.E.; Meyers, A.I. *Synlett* **1995**, 145–146.
133. Crimmins, M.T.; Choy, A.L. *J. Am. Chem. Soc.* **1997**, *119*, 10237–10238.
134. Crimmins, M.T.; Kung, D.K.; Gray, J.L. *J. Am. Chem. Soc.* **1993**, *115*, 3146–3155.
135. Crimmins, M.T.; King, B.W.; Watson, P.S.; Guise, L.E. *Tetrahedron* **1997**, *53*, 8963–8974.
136. Crimmins, M.T.; Watson, P.S. *Tetrahedron Lett.* **1993**, *34*, 199–202.
137. Wagner, P.J.; McMahon, K. *J. Am. Chem. Soc.* **1994**, *116*, 10827–10828.
138. Ward, S.C.; Fleming, S.A. *J. Org. Chem.* **1994**, *59*, 6476–6479.
139. Ward, S.C.; Fleming, S.A. *Tetrahedron Lett.* **1992**, *33*, 1013–1016.
140. Haag, D.; Scharf, H.-D. *J. Org. Chem.* **1996**, *61*, 6127–6135.
141. Faure, S.; Blanc, S.P.-L.; Piva, O.; Pete, J.-P. *Tetrahedron Lett.* **1997**, *38*, 1045–1048.
142. Tanaka, M.; Tomioka, K.; Koga, K. *Tetrahedron* **1994**, *50*, 12829–12842.
143. Shepard, M.S.; Carreira, E.M. *J. Am. Chem. Soc.* **1997**, *119*, 2597–2605.

144. Carreira, E.M.; Hastings, C.A.; Shepard, M.S.; Yerkley, L.A.; Millward, D.B. *J. Am. Chem. Soc.* **1994**, *116*, 6622–6630.
145. Booker-Milburn, K.I.; Gulten, S.; Sharpe, A. *Chem. Commun.* **1997**, 1385–1386.
146. Jautelat, R.; Winterfeldt, E.; Müller-Fahrnow, A. *J. Prakt. Chem./Chem.-Ztg.* **1996**, *338*, 695–701.
147. Thopate, S.R.; Kulkarni, M.G.; Puranik, V.G. *Angew. Chem. Int. Ed. Eng.* **1998**, *37*, 1110–1112.
148. Buschmann, H.; Hoffmann, N.; Scharf, H.D. *Tetrahedron: Asymmetry* **1991**, *2*, 1429–1444.
149. Linker, T.; Frölich, L. *J. Am. Chem. Soc.* **1995**, *117*, 2694–2697.
150. Brünker, H.-G.; Adam, W. *J. Am. Chem. Soc.* **1995**, *117*, 3976–3982.
151. Orfanopoulos, M.; Gardina, M.B.; Stephenson, L.M. *J. Am. Chem. Soc.* **1979**, *101*, 275–276.
152. Adam, W.; Catalani, L.H.; Griesbeck, A. *J. Org. Chem.* **1986**, *51*, 5494–5496.
153. Adam, W.; Brünker, H.-G.; Kumar, A.S.; Peters, E.-M.; Peters, K.; Schneider, U.; von Schnering, H.G. *J. Am. Chem. Soc.* **1996**, *118*, 1899–1905.
154. Dussault, P.H.; Woller, K.R. *J. Am. Chem. Soc.* **1997**, *119*, 3824–3825.
155. Yamaguchi, T.; Uchida, K.; Irie, M. *J. Am. Chem. Soc.* **1997**, *119*, 6066–6071.
156. Linker, T.; Rebien, F.; Tóth, G. *Chem. Commun.* **1996**, 2585–2586.
157. Prein, M.; Mauer, M.; Peters, E.M.; Peters, K.; von Schnering, H.G.; Adam, W. *Chem. Eur. J.* **1995**, *1*, 89–94.
158. Piva, O.; Caramelle, D. *Tetrahedron: Asymmetry* **1995**, *6*, 831–832.
159. Piva, O.; Pete, J.-P. *Tetrahedron: Asymmetry* **1992**, *3*, 759–768.
160. Harada, N.; Koumura, N.; Feringa, B.L. *J. Am. Chem. Soc.* **1997**, *119*, 7256–7264.
161. Sugimura, T.; Shimizu, H.; Umemoto, S.; Tsuneishi, H.; Hakushi, T.; Inoue, Y.; Tai, A. *Chem. Lett.* **1998**, 233–234.
162. Sugimura, T.; Nishiyama, N.; Tai, A.; Hakushi, T. *Tetrahedron: Asymmetry* **1994**, *5*, 1163–1166.

# 3

# Photochemical cis-trans Isomerization in the Triplet State

**Tatsuo Arai**
University of Tsukuba, Tsukuba, Ibaraki, Japan

## I. INTRODUCTION

Since the one-way cis → trans isomerization was discovered in several types of arylethenes, interest in the photochemical cis-trans isomerization has continued to grow in recent years [1–11]. Conventional two-way isomerization and highly selective photochemical cis-trans isomerization can take place on the singlet excited state as well as in the triplet excited state [11–15]. In this chapter, a concept for various types of cis-trans isomerization in the triplet state is described, mainly from the viewpoints of the modes of isomerization and the difference in potential energy surfaces of cis-trans isomerization.

Stilbene **1** undergoes isomerization in the singlet excited state on direct irradiation [13,16–23]. The mechanism was established by J. Saltiel et al. from the experiments of the azulene effect on the photostationary state isomer composition [13,24,25]. Thus, usually, quenching experiments as well as observation of the transient spectra are necessary to distinguish the singlet and triplet mechanisms of isomerization of olefin on direct irradiation. Furthermore, triplet sensitization is necessary to study the isomerization in the triplet manifold.

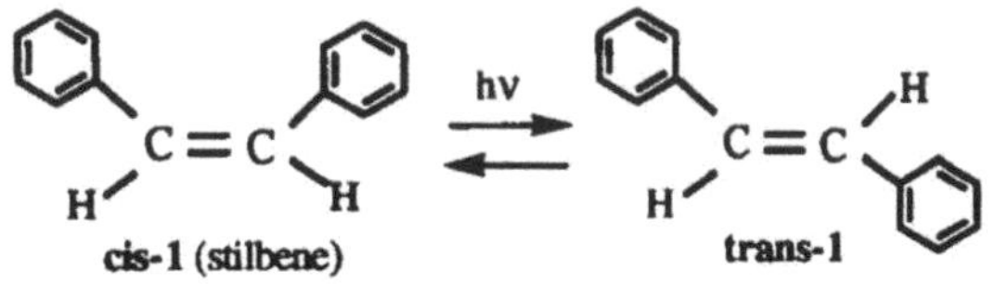

cis-1 (stilbene)                              trans-1

On sensitized irradiation with carbonyl compounds such as benzophenone, it is expected that the triplet state of olefins will be produced. However, a radical chain mechanism is sometimes discussed when benzil or biacetyl is used as a sensitizer. Thus, some other experimental evidence such as direct observation of the triplet state by laser photolysis is needed to establish the triplet mechanism even on triplet sensitization.

## II. ONE-WAY AND TWO-WAY ISOMERIZATION

Stilbene 1 undergoes cis-trans mutual isomerization (two-way isomerization) in the singlet as well as in the triplet state [13–15,24–32]. The potential energy surfaces are shown in Fig. 1.

On triplet sensitization the resulting cis ($^3c^*$) and trans triplet excited state ($^3t^*$) convert to the perpendicular excited triplet state ($^3p^*$); $^3p^*$ is equilibrated with $^3t^*$ by an equilibrium constant of $K_{tp} = [^3p^*]/[^3t^*]$ and deactivated to the ground state $^1p$, giving $^1c$ and $^1t$ [6,13–15,24,33–41]. Thus, the isomerization of stilbene takes place as a diabatic process.

The typical one-way isomerization was reported in 1983 for 2-anthrylethenes 2. In the first report, an adiabatic isomerization from $^3c^*$ to $^3t^*$ was proposed from the quantum yield of isomerization ($\Phi_{c \to t} \gg 1$ and $\Phi_{t \to c} = 0$) and the photostationary state isomer composition ([c]/[t]$_{pss}$ = 0/100). By the aid of transient spectroscopy more detailed potential energy surfaces were proposed (Fig. 2a, b) [1,2,42–50].

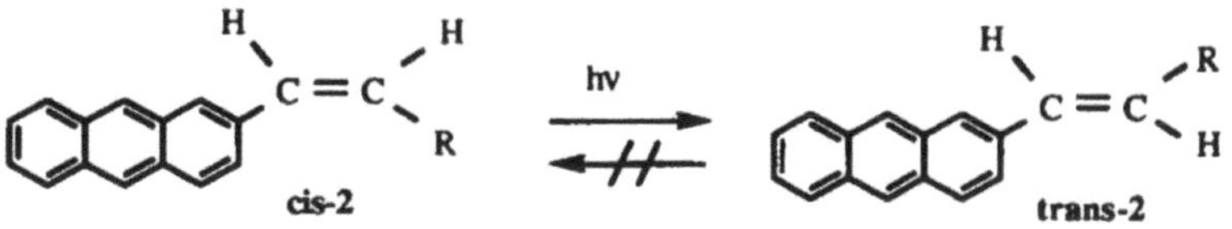

cis-2                                                trans-2

R=Me (a), $^t$Bu (b), Ph (c), 2-naphthyl (d)

The name of one-way isomerization was used by Ramamurthy and Liu in 1976 in the triplet-sensitized isomerization of an olefin related to vitamin A [51]. β-Ionol (3) underwent cis-trans isomerization on triplet sensitization with high-

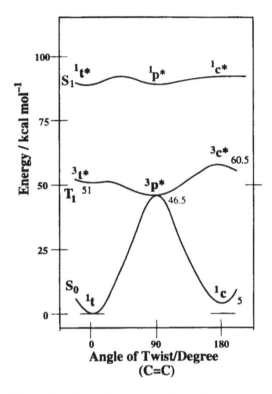

**Figure 1**   Potential energy surfaces of cis-trans isomerization of stilbene (1).

energy sensitizers such as acetophenone, but the photostationary state cis-to-trans isomer composition ($[c]/[t]$)$_{\mathrm{pss}}$ increased with decreasing triplet energy of the sensitizer, finally giving the 100% cis at the photostationary state on 2-acetylnaphthalene sensitization (Fig. 3).

In this case, due to differences in triplet energy between the cis ($E_T \approx 74$ kcal mol$^{-1}$) and trans isomers of **3** ($E_T \approx 59$ kcal mol$^{-1}$), the rate of the triplet

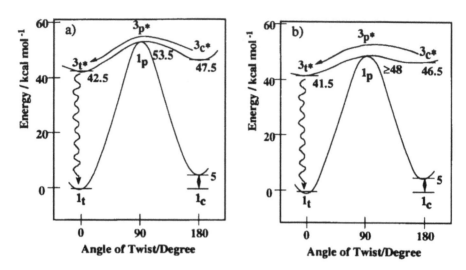

**Figure 2**  Potential energy surfaces of one-way cis-trans isomerization of **2b** (a) and **2c** (b).

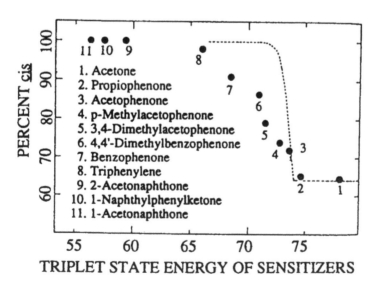

**Figure 3**  Photostationary state composition of β-ionol (**3**) as a function of sensitizer excitation energy. (Adapted from Ref. 51; copyright 1998 American Chemical Society.)

energy transfer is much faster for the trans isomer ($k_{st}$) than the cis isomer ($k_{sc}$), with sensitizers having the triplet energy close to that of the trans isomer.

Thus, the concept of one-way isomerization on sensitized isomerization of β-ionol 3 appeared several years before the finding of the typical one-way isomerization in 2-anthrylethenes (2).

It should be noted that by proper choice of the sensitizer and/or excitation wavelength one could observe the isomerization undergoing in a seemingly one-way manner for both cis-to-trans and trans-to-cis directions.

This chapter discusses the typical one-way isomerization taking place as a quantum chain process as observed in 2-anthrylethenes.

## III. FEATURES OF THE TYPICAL TWO-WAY AND ONE-WAY ISOMERIZATION

Table 1 compares features of photoisomerization of 2-styrylanthracene (2c) and stilbene (1) as typical examples for the one-way and two-way isomerizations. First, on triplet sensitization, 2c undergoes isomerization only from the cis isomer to the trans isomer, whereas stilbene undergoes isomerization for both cis-to-trans and trans-to-cis directions.

Second, the quantum yield of cis → trans isomerization increases with increasing cis isomer concentration and exceeds unity for 2-styrylanthracene (Fig. 4), but the value for stilbene remains ~0.5 under usual conditions. An exceptional case for the quantum chain process will be discussed later. The mechanisms of typical two-way and one-way isomerizations and the quantum yields of isomerization are shown in Schemes 1 and 2.

Third, the triplet lifetime of the 2-styrylanthracene is in the order of 100 μs, whereas the triplet lifetime of stilbene is 60 ns.

**Table 1** Typical Features of Isomerization of Stilbene and 2-Styrylanthracene in the Triplet State

| Feature | Stilbene | 2-Styrylanthracene |
|---|---|---|
| Mode of isomerization | Two-way | One-way |
| $([c]/[t])_{pss}$ | 60/40 | 0/100 |
| Quantum yield of isomerization | $\phi_{c \to t} = {\sim}0.5$ | $\phi_{c \to t} \gg 1$ |
|  | $\phi_{t \to c} = {\sim}0.5$ | $\phi_{t \to c} = 0$ |
| Triplet lifetime, $\tau_T$ | 60 ns | ~100 μs |
| Intermediate for deactivation | Twisted triplet ($^3p^*$) | Trans triplet ($^3t^*$) |

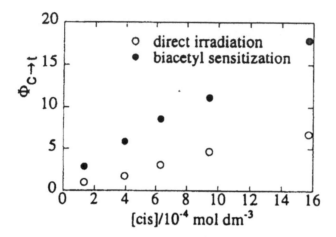

**Figure 4**   Effect of isomer concentration on the quantum yield of cis-trans isomerization of **2c**. (Reprinted from Ref. 4; copyright 1998 Chemical Society of Japan.)

$$^1S \xrightarrow{\ h\nu\ } {}^1S^* \qquad\qquad {}^3c^* \xrightarrow{\quad\quad} {}^3p^*$$

$$^1S^* \xrightarrow{\ \Phi_T^s\ } {}^3S^* \qquad\qquad {}^3t^* \underset{\longleftarrow}{\xrightarrow{\ K_{tp}\ }} {}^3p^*$$

$$^3S^* + {}^1c \xrightarrow{\qquad} {}^1S + {}^3c^* \qquad {}^3p^* \xrightarrow{\ k_{pd}\ } (1\text{-}\alpha)\,{}^1c \ + \ \alpha\,{}^1t$$

$$^3S^* + {}^1t \xrightarrow{\qquad} {}^1S + {}^3t^*$$

$$\Phi_{c\rightarrow t} = \alpha \qquad\qquad \Phi_{t\rightarrow c} = 1\text{-}\alpha$$

**Scheme 1**   Mechanism for the two-way isomerization in the triplet state.

$$^1S \xrightarrow{\ h\nu\ } {}^1S^* \qquad\qquad {}^3c^* \xrightarrow{\quad\quad} {}^3t^*$$

$$^1S^* \xrightarrow{\ \Phi_T^s\ } {}^3S^* \qquad\qquad {}^3t^* \xrightarrow{\ k_{td}\ } {}^1t$$

$$^3S^* + {}^1c \xrightarrow{\qquad} {}^1S + {}^3c^* \qquad {}^3t^* + {}^1c \xrightarrow{\ k_{tc}\ } {}^1t + {}^3c^*$$

$$^3S^* + {}^1t \xrightarrow{\qquad} {}^1S + {}^3t^*$$

$$\Phi_{c\rightarrow t} = 1 + k_{tc}/k_{td} \qquad \Phi_{t\rightarrow c} = 0$$

**Scheme 2**   Mechanism for the one-way isomerization in the triplet state.

## IV.  EFFECTS OF ARYL GROUP ON THE MODE OF ISOMERIZATION AND THE POTENTIAL ENERGY SURFACE

Table 2 summarizes the substituent effects on the mode of cis-trans isomerization in the triplet state [33,41,52–74]. Roughly speaking, the triplet energy of the aryl substituent on the ethylenic carbon controls the mode of isomerization. Thus, olefin substituted by an aryl group with a very low triplet energy tends to undergo one-way cis → trans isomerization in the excited triplet state with a quantum chain process.

Further lowering the triplet energy of the aryl substituent reduces the efficiency of cis → trans isomerization as observed in 3-(1-propenyl)perylene (11) [74].

**Table 2**  Modes of Isomerization, Triplet Lifetimes $\tau_T$, and Equilibrium Constant $K_{tp}$ of ArCH=CHR

| Ar | Triplet energy of ArH (kcal mol$^{-1}$) | $\tau_T^a$ ($K_{tp}$) R = $^t$Bu | R = Ph | Refs. |
|---|---|---|---|---|
| Phenyl | 84.3 | Two-way 46$^b$ (10$^{-3}$) | Two-way 0.063 (~8) | 33,41,52–59 |
| 2-Naphthyl | 60.9 | Two-way 0.13 (~5) | Two-way 0.14 (~2) | 60,61,62 |
| 3-Chrysenyl | 56.6 | Two-way 0.36 (~0.16) | Two-way 0.14 (~1) | 63 |
| 8-Fluoranthenyl | 54.2 | One-way 25 (≪10$^{-3}$) | Two-way 0.50 (~0.11) | 64,65 |
| 1-Pyrenyl | 48.2 | One-way 54 (≪10$^{-3}$) | Two-way 27 (~10$^3$) | 66 |
| 2-Anthryl | 42 | One-way 280 (≪10$^{-3}$) | One-way 190 (≪10$^{-3}$) | 42–47,67–69 |
| Ferrocenyl | 40 | | Inefficient one-way | 70,71 |
| 3-Perylenyl | 35 | Two-way$^b$ ~50 (≪10$^{-3}$) | One-way ~100 (≪10$^3$) | 72–74 |

$^a$ In μs.
$^b$ R = Me.

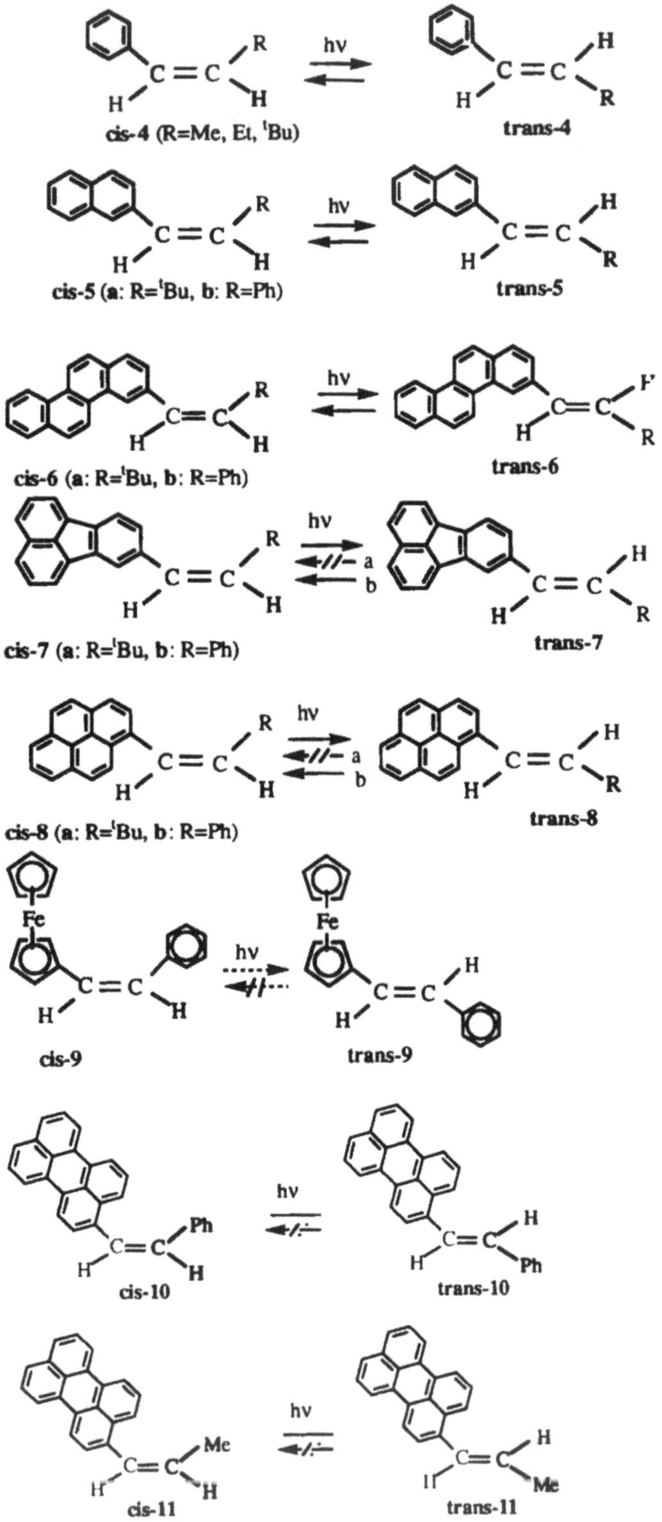

cis-4 (R=Me, Et, $^t$Bu)     trans-4

cis-5 (a: R=$^t$Bu, b: R=Ph)     trans-5

cis-6 (a: R=$^t$Bu, b: R=Ph)     trans-6

cis-7 (a: R=$^t$Bu, b: R=Ph)     trans-7

cis-8 (a: R=$^t$Bu, b: R=Ph)     trans-8

cis-9     trans-9

cis-10     trans-10

cis-11     trans-11

In order to discuss the potential energy surfaces of isomerization one can assume the following points. On the triplet potential energy surface potential energy minima could exist at only three conformations: cis ($^3c*$), perpendicular ($^3p*$), and trans ($^3t*$). In addition, the deactivation can take place only from $^3c*$, $^3p*$, and $^3t*$ with different contributions depending on the shape of the potential energy surface.

Some of the triplet energies of $^3c*$, $^3p*$, and $^3t*$ estimated from the experimental results are summarized in Table 3 [40,44–46,58–61,65,66,73,74]. The energies of $^3t*$ over ground state $^1t$ and $^3c*$ over $^1c$ decrease by introduction of an aryl group with a low triplet energy, and finally in the case of an aryl group with a very low triplet energy the energies of $^3t*$ and $^3c*$ become almost the same values as those of the parent aryl group. However, the energy of $^3p*$ over $^1t$ is estimated to be nearly constant as ~53 and ~46 kcal mol$^{-1}$ for the series of 1-arylalkenes and diarylethylenes, respectively.

Due to the steric hindrance, $^3c*$ is less stable by ~5 kcal mol$^{-1}$ than $^3t*$ in olefins such as stilbene [39,40] and 2-styrylanthracene, and therefore, it is enough to take only the deactivation from $^3t*$ and $^3p*$ into account.

The potential energy minima can be estimated from quenching experiments by azulene and oxygen as well as determination of the triplet lifetime. Oxygen can quench both $^3t*$ and $^3p*$, whereas azulene quenches only $^3t*$ [41,58,78]. Since neither azulene nor oxygen affects ([t]/[c])$_{pss}$ in β-alkylstyrenes (**4**), their energy minimum exists only at $^3p*$. In stilbene (**1**), azulene increases the ([t]/[c])$_{pss}$ value on triplet sensitization, whereas oxygen does not affect it [13–15,24,26]. These results indicate that $^3p*$ and $^3t*$ are both populated as a mixture of $K_{tp} = [^3p*]/[^3t*]$, but the equilibrium is shifted to $^3p*$ as estimated from $K_{tp} = 8$ [13–15,24,41]. In triplet-sensitized isomerization of a naphthylethylene, **5a**, both azulene and oxygen increase the ([t]/[c])$_{pss}$ value, indicating that $^3p*$ and $^3t*$ are

**Table 3** Triplet Energies of ArCH=CHR

| Ar | Triplet energy of ArH (kcal mol$^{-1}$) | Triplet energies, kcal mol$^{-1}$ | | | | | | |
|---|---|---|---|---|---|---|---|---|
| | | R = $^t$Bu | | | R = Ph | | | |
| | | $^3t*$ | $^3p*$ | $^3c*$ | $^3t*$ | $^3p*$ | $^3c*$ | Refs. |
| Phenyl | 84.3 | 60ᵃ | 53ᵃ | | 51 | 46.5 | 55.5 | 40 |
| 2-Naphthyl | 60.9 | ~53 | ~53 | | ~49 | ~49 | | 58–61 |
| 8-Fluoranthenyl | 54.2 | 49 | ≥56 | ~52 | ~44 | ~45.5 | ~46 | 65 |
| 1-Pyrenyl | 48.2 | 44 | ≥51 | ~46 | 41 | ~46 | ~42 | 66 |
| 2-Anthracenyl | 42 | 42.5 | ~53.5 | 42.5 | 41.5 | ≥48 | 41.5 | 44–46 |
| 3-Perylenyl | 35 | 35ᵃ | 50ᵃ | 35ᵃ | 35 | ~46.5 | 35 | 73,74 |

ᵃ R = Me.

equally populated ($K_{tp} = 1$) [60]. The mechanisms of isomerization of β-alkyl-styrene (**4**), stilbene (**1**), and **5a** in the triplet state are summarized in Schemes 3–5.

The equilibrium constant $K_{tp}$ is also estimated from the triplet lifetime. The lifetime of the triplet states, $\tau_T$, is expressed by Eq. (1), where $k_{pd}$ and $k_{td}$ stand for the rate constants for deactivation of $^3p^*$ and $^3t^*$, and are assumed to be similar to those for stilbene ($k_{pd} \approx 2 \times 10^7 \, s^{-1}$) [41] and one-way isomerizing ethylenes ($k_{td} \leq 2 \times 10^4 \, s^{-1}$) [1,2], respectively. With decreasing $K_{tp}$, $\tau_T$ should increase. In the limiting cases of typical two-way and one-way isomerizations, $\tau_T$ values are described by Eqs. (2) and (3) for **4** and **2**, respectively. The estimated $K_{tp}$ values for **6a**, **7b**, and **8b** from the lifetime determination [63–66] are also listed

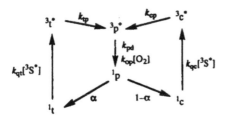

**Scheme 3**   Mechanism for the isomerization of **4** in the triplet state.

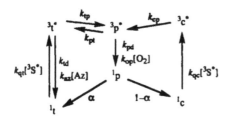

**Scheme 4**   Mechanism for the isomerization of **1** in the triplet state.

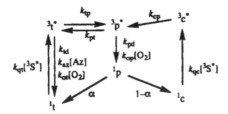

**Scheme 5**   Mechanism for the isomerization of **5b** in the triplet state.

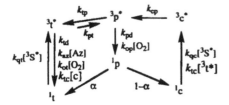

**Scheme 6**  Mechanism for the isomerization of **8b** in the triplet state.

in Table 2. The mechanisms of isomerization of **8b** and **2** are shown in Schemes 6 and 7, respectively.

$$\tau_T = (1 + K_{tp})/(K_{tp}k_{pd} + k_{td}) \tag{1}$$

$$\tau_T = 1/k_{pd} \tag{2}$$

$$\tau_T = 1/k_{td} \tag{3}$$

The deactivation from the triplet state does not necessarily take place only from the stable conformation because the deactivation rate constants from $^3t^*$ and $^3p^*$ are different by more than three orders of magnitude as mentioned above. When the free energy change $\Delta G_{tp}$ from $^3t^*$ to $^3p^*$ is $-3$ kcal mol$^{-1}$, i.e., $^3p^*$ is more stable than $^3t^*$ by 3 kcal mol$^{-1}$, most of the triplet state exists as $^3p^*$ ($K_{tp} = 150$) and the deactivation takes place from this conformation as in β-alkylstyrene **4**. When $\Delta G_{tp} \approx 0$, $^3t^*$ and $^3p^*$ are equilibrated with each other ($K_{tp} = 1$); in this case, unimolecular deactivation takes place from $^3p^*$, giving the cis and trans isomers, while in the presence of an appropriate quencher $^3t^*$ is exclusively quenched and the trans isomer composition at the photostationary state increases (**1,5b**, etc).

However, if $\Delta G_{tp} \approx 4$ kcal mol$^{-1}$ ($K_{tp} \approx 10^{-3}$), the deactivation from $^3t^*[(1/(1 + K_{tp})] \times k_{td} \approx 2 \times 10^4$ s$^{-1}$) and $^3p^*[(K_{tp}/(1 + K_{tp})) \times k_{pd} \approx 2 \times 10^4$ s$^{-1}$] is estimated to be 1:1 (**8a**). When $\Delta G_{tp} > 7$ kcal mol$^{-1}$ ($K_{tp} < 10^{-5}$),

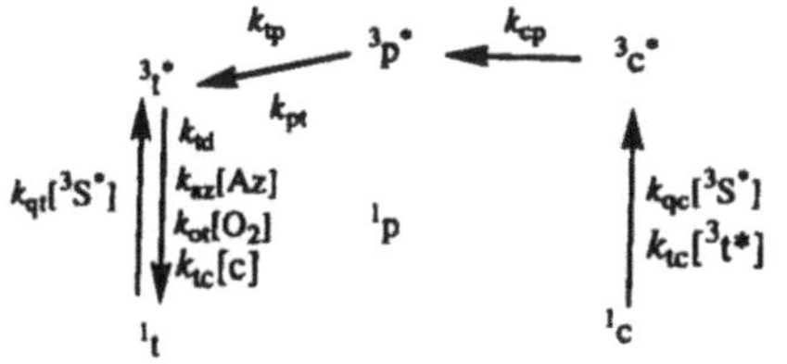

**Scheme 7**  Mechanism for the one-way isomerization of **2** in the triplet state.

the deactivation of the triplet state takes place only from $^3t^*$ and the typical one-way isomerization is to be observed.

## V. ISOMERIZATION OF SUBSTITUTED STILBENES AND RELATED COMPOUNDS

Table 4 lists the quantum yields of cis-trans isomerization determined on triplet sensitization as well as the T-T absorption spectra and the triplet lifetimes for stilbene and related compounds [33,77–92].

Although *trans*-stilbene exhibits T-T absorption spectra on direct irradiation at −196°C in ether–isopentane–ethanol (5:5:2), no triplet could be detected

**Table 4**  T-T Absorption Maxima, Triplet Lifetimes, Quantum Yields of Isomerization, and Photostationary State Isomer Ratio

| | $\lambda_{max}$(T-T), nm | $\tau_T$, ns (Solv) | $K_{tp}$ | $\Phi_{c \to t}$ | $\Phi_{t \to c}$ | $([c]/[t])_{pss}$ | Refs. |
|---|---|---|---|---|---|---|---|
| 1 | <360 | 62 (PhH)[a] | ≈5–10 | 0.38 | 0.55 | 1.5 | 41,77–79 |
| 12a | | | | 0.45 | 0.55 | 1.2 | 33,77 |
| 12b | 355,400 | 60 | ≈5–10 | | | | 80 |
| 12c | | | | 0.40 | 0.60 | | 81 |
| 12d | 370 | 50 (PhH) | | | | | 82 |
| 12e | 400,520 | 60 ($C_6H_{12}$)[b] | ≈5–10 | | | | 83 |
| 12f | 430,530 | 50 (PhH) | | | | | 82 |
| 12g | 440 | 55 ($CH_3CN$)[c] | | | | | 84 |
| 12h | 500 | 70 ($C_6H_{12}$) | ≈3 | | | | 83 |
| 12i | 410,540 | 85 (PhH) | ≈2 | | | | 82 |
| 12j | 435,600 | 150 (PhH) | ≈0.5 | 0.50 | 0.40 | 1.3 | 83,85 |
| 12k | 450,550,790 | 200 (PhH) | ≈0.3 | | | | 82 |
| 12l | 480,680 | 300 (PhH) | ≈0.2 | | | | 82 |
| 12m | 790 | 1000,1400 (PhH) | ≈0.05 | | | | 82,86 |
| 12n | 380 | 60 ($C_6H_{12}$) | ≈5–10 | | | | 87 |
| 12o | 380 | 85 (MTHF)[d] | ≈2 | | | | 87 |
| 13a | 490 | 300 (MTHF) | ≈0.2 | 0.47 | 0.41 | | 88,89 |
| 13b | 495 | 500 (MTHF) | ≈0.1 | 0.46 | 0.45 | | 88,89 |
| 13c | 490 | 40 (MTHF) | | 0.42 | 0.47 | | 88,89 |
| 13d | 480,565 | 500 (PhH) | ≈0.1 | | | | 90 |
| 14a | | | | 0.18 | 0.50 | | 91,92 |
| 14b | | | | 0.22 | 0.51 | 1.78 | 91,92 |
| 4c | | | | 0.21 | 0.44 | 1.4 | 91,92 |

[a] In benzene.
[b] In cyclohexane.
[c] In acetonitrile.
[d] In 2-methyltetrahydrofuran.

under direct irradiation at room temperature [93]. On triplet sensitization *cis*- and *trans*-stilbene give the T-T absorption with a lifetime of 62 ns and $\lambda_{max} < 360$ nm [41]. Since nitro and carbonyl groups enhance the intersystem crossing from the singlet excited state to the triplet excited state, the T-T absorption has been observed for nitro- or carbonyl-substituted arylethenes even on direct irradiation [82–90]. At room temperature $\tau_T$ is of the order of 100 ns for stilbene (**1**), 4-phenylstilbene (**12b**), 4-acetylstilbene (**12n**), 4-benzoylstilbene (**12o**), and nitro-stilbenes (**12d–i**), except methoxy- or amino-substituted nitrostilbenes (**12j–m**); 4-nitro-4'-dimethylaminostilbene (**12m**) has a $\tau_T$ of $\sim 1$ μs. Thus, the equilibrium between $^3t^*$ and $^3p^*$ shifted to the $^3t^*$ side with the introduction of the push-and-pull substituents on the para positions of stilbene. 1-Naphthylethenes (**13**) also have a longer triplet lifetime of 300–500 ns except **13b**. The equilibrium constant $K_{tp}$ estimated from Eq. (1) is also listed in Table 4.

On benzopheneone sensitization the quantum yields of cis-trans isomerization of stilbene, $\Phi_{c \to t}$ and $\Phi_{t \to c}$, were 0.55 and 0.38, respectively, and the value $\Phi_{t \to c}/\Phi_{c \to t}$ (1.45) $\approx$ ([c]/[t])$_{pss}$ ($\approx 1.48$) [77,78]. Detailed experiments revealed that with sensitizers having excitation energies much higher than those of *trans*- and *cis*-stilbene there is a variation in the photostationary state isomer composition as well as in the quantum yields of isomerization; ([c]/[t])$_{pss}$ varies from 1.27 with acetophenone ($E_T = 73.6$ kcal mol$^{-1}$) to 1.83 with 2-acetylfluorene ($E_T = 62$ kcal mol$^{-1}$) [77]. In most cases, the sums of $\Phi_{c \to t}$ and $\Phi_{t \to c}$ are almost 1, but in dipyridylethenes (**14**) the values are considerably lower than 1 with relatively low $\Phi_{c \to t}$ values [91,92].

cis-12                                trans-12

a: $R_1$=2-Me, $R_2$=H, b: $R_1$=2-Ph, $R_2$=H, c: $R_1$=4-Cl, $R_2$=H, d: $R_1$=3-NO$_2$, $R_2$=H, e: $R_1$=4-NO$_2$, $R_2$=H, f: $R_1$=2,4-di-NO$_2$, $R_2$= H, g: $R_1$=$R_2$=2-NO$_2$, h: $R_1$=$R_2$=4-NO$_2$, i: $R_1$=4-NO$_2$, $R_2$=3-OMe, j: $R_1$=4-NO$_2$, $R_2$=4-OMe, k: $R_1$=4-NO$_2$, $R_2$=2,5-di-OMe, l: $R_1$=4-NO$_2$, $R_2$=4-NH$_2$, m: $R_1$=4-NO$_2$, $R_2$=4-N(CH$_3$)$_2$, n: $R_1$= 4-COCH$_3$, $R_2$=H, o: $R_1$=4-COPh, $R_2$=H

cis-13                                trans-13

a: R=H, b: R=Cl, c: R=Br, d: R=NO$_2$

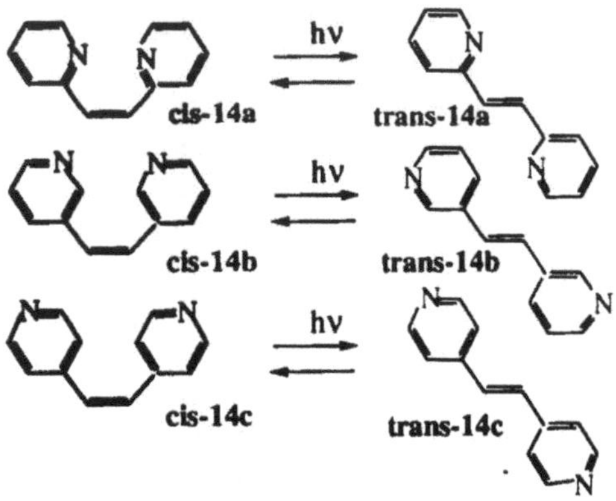

## VI.  TWO-WAY ISOMERIZATION BY THE QUANTUM CHAIN PROCESS

The quantum chain process is not limited to the one-way isomerizing olefins but could be observed in olefins two-way isomerizing in the triplet state where the energy minimum exists at $^3t^*$. **6a**, **7b**, and **8b** undergo two-way isomerization in the triplet state [63–66]. However, the quantum yields of cis → trans isomerization of **6a**, **7b**, and **8b** increase with increasing cis isomer concentration and exceed unity; the $\Phi_{c \rightarrow t}$ value of **8b** was reported to be 42 on benzil sensitization at [cis-**8b**] = $2.7 \times 10^{-3}$ M [66]. Here are mentioned 1.) porphyrin-sensitized isomerization of stilbene and 2.) effects of additives on the quantum chain process of styrylstilbene.

### A.  Porphyrin-Sensitized Isomerization of Stilbene and Its Analogs

Quantum chain reactions were already reported 20 years ago on the metalloporphyrin sensitized isomerization of stilbene (**1**) and 1-(1-naphthyl)-2-(4-pyridyl)-ethene (**15**) [94,95]. In these cases the metalloporphyrin acts not only as a sensitizer but as a mediator of excitation energy.

Irradiation with visible light of 400–600 nm in the presence of zinc or magnesium etioporphyrins (**16**; $E_T$ = 40–42 kcal mol$^{-1}$) **1** and **15** underwent cis-trans isomerization [94]. While the quantum yields of isomerization for stilbene ($\Phi_{c \rightarrow t}$ = 0.01 and $\Phi_{t \rightarrow c}$ = 0.001) are low for both directions, the quantum yield of cis → trans isomerization of **15** exceeded unity ($\Phi_{c \rightarrow t}$ = 6.6 and $\Phi_{t \rightarrow c}$ = 0.2).

The mechanism involving a complex was proposed in the isomerization of **15**, where the light is absorbed by the ground state complex between the **16** and **15**. The triplet lifetimes of metalloporphyrins **16** are long enough to undergo energy transfer to produce the triplet state of **15** ($^3$**15**\*), which undergoes deactivation to the ground state cis and trans isomers or undergoes deactivation to give exclusively the trans isomer. Since the triplet energies of metalloporphyrins were in the range of 40–42 kcal mol$^{-1}$ and those of cis- and trans-**15** were estimated to be higher than these values; the energy transfers from the triplet porphyrin to the cis- and trans-**15** are endoergonic. The considerably high quantum yield of cis → trans isomerization indicates a nonvertical energy transfer from the porphyrin triplet to cis-**15**.

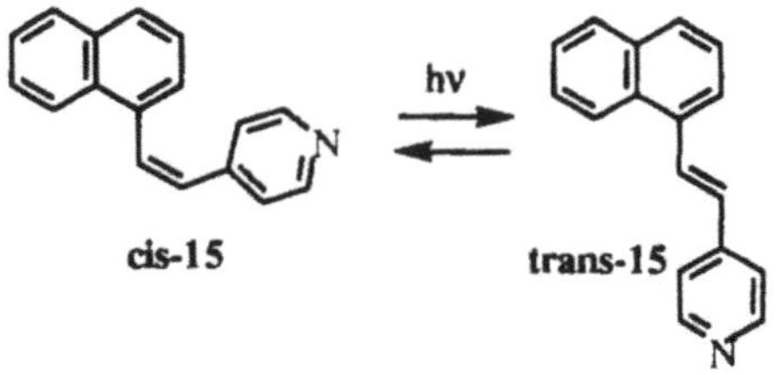

cis-**15**                    trans-**15**

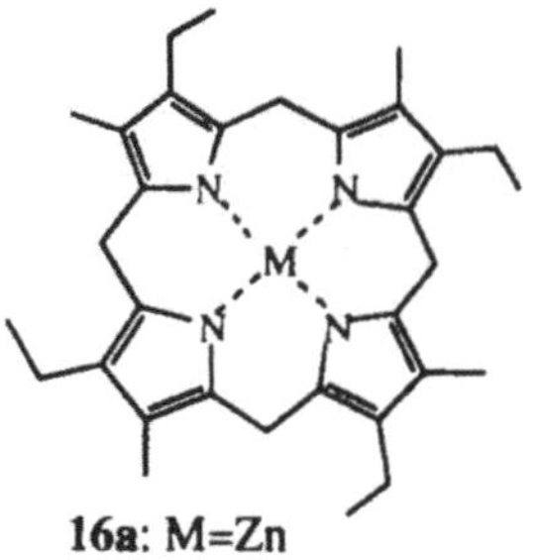

**16a**: M=Zn          **17a**: M=Pd
**16b**: M=Mg          **17b**: M=Pt

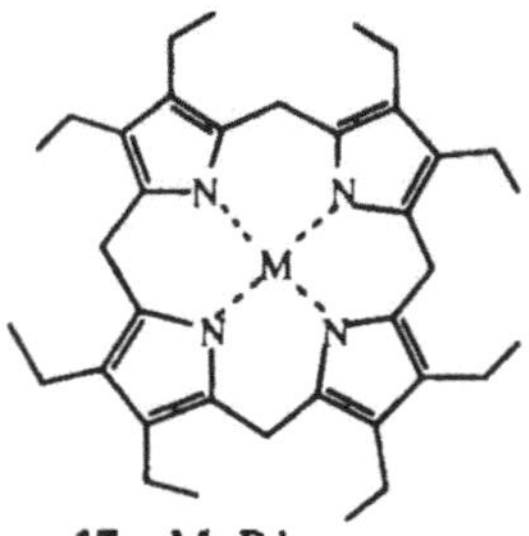

When porphyrins with much higher triplet energies such as palladium octaethylporphyrin (**17**; $E_T$ = 44.8 kcal mol$^{-1}$) were used as sensitizers, even the cis → trans isomerization of stilbene took place as a quantum chain process ($\Phi_{c \to t}$ = 1.6) [95]. The high quantum efficiencies were explained by a quantum chain process in which the metalloporphyrin serves as both an energy donor and an acceptor. Since the quantum yield of cis → trans isomerization of 1,2-diphenylpropene ($\Phi_{c \to t}$ = 0.37) remained as a normal value under the same experimental conditions as those of stilbene, the potential energy surface of the triplet state is an important factor for occurrence of the quantum chain cis-trans isomerization. That is, in 1,2-diphenylpropene the triplet state exists exclusively as a perpendicular conformation, where the triplet state and the ground state lay very close in energy and the deactivation can only take place thermally.

In the case of stilbene, the trans triplet and the perpendicular triplet state are equilibrated with each other and the triplet state exists as a mixture of them. The transoid triplet state ($E_T$ = 51 kcal mol$^{-1}$) [40] has enough energy to reexcite the porphyrin to the triplet state by deactivation to the ground state trans isomer. However, if an energy transfer occurred from the porphyrin triplet to stilbene to produce stilbene triplet (Scheme 8, b), there would be no possibility of observing a quantum chain process under the actual experimental condition of low porphyrin concentrations and the short triplet lifetime of stilbene. Therefore, formation of an exciplex, where the porphyrin and stilbene are held together long enough, permits the isomerization to occur. In such a complex, stilbene preferentially exists as a nearly transoid conformation, resulting in production of the porphyrin triplet and *trans*-stilbene (Scheme 8, f).

## B. Catalytic Behavior of Additives

In order to increase the quantum yield of cis → trans isomerization, it is proposed to use aromatic hydrocarbons such as anthracene as a carrier of the quantum chain process. (Z,E)-(cis-**18**) and (E,E)-1,4-di(3,5-di-*tert*-butylstyryl) benzene (trans-**18**) underwent mutual isomerization on biacetyl sensitization with a quantum chain process to afford a $\Phi_{c \to t}$ value of 1.2–1.3 in a cis isomer concentration of ~1 × 10$^{-2}$ M [96]. The quantum chain process proceeds through the energy transfer from the trans triplet ($^3t^*$) to the ground state cis isomer ($^1c$). On addition of anthracene (AN) as a quencher or a carrier of the excited state, the quantum

$$^1S \xrightarrow{\;h\nu\;} {}^1S^* \xrightarrow{\;isc\;} {}^3S^* \qquad\qquad (a)$$

$$^3S^* + cis \longrightarrow {}^3St^* + {}^1S \qquad\qquad (b)$$

$$^3St^* + {}^1S \longrightarrow trans + {}^3S^* \qquad\qquad (c)$$

$$^3S^* + cis \longrightarrow {}^3[S, St]^* \qquad\qquad (d)$$

$$^3S^* + trans \longrightarrow {}^3[S, St]^* \qquad\qquad (e)$$

$$^3[S, St]^* \longrightarrow {}^3S^* + trans \qquad\qquad (f)$$

$$^3[S, St]^* \longrightarrow {}^1S + (1 - \beta)\,cis + \beta\,trans \qquad (g)$$

$$^3[S, St]^* \longrightarrow {}^1S + {}^3St^* \qquad\qquad (h)$$

$$^3St^* \longrightarrow (1 - \alpha)\,cis + \alpha\,trans \qquad\qquad (i)$$

**Scheme 8**  Mechanism of porphyrin-sensitized isomerization of stilbene.

yield of cis-to-trans isomerization increased more than 50 times [97]. The mechanism of the enormous increase of the quantum yield of isomerization is illustrated in Fig. 5.

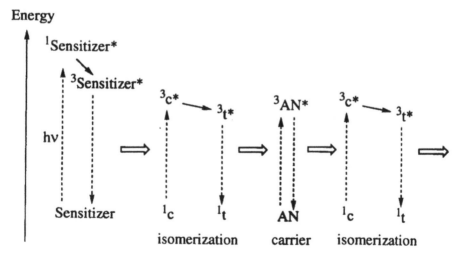

The biacetyl triplet state produced on irradiation with 436-nm light transfers its energy to the ground state cis isomer to produce the triplet state cis isomer ($^3c*$), which undergoes twisting around the double bond to give the $^3p*$ and $^3t*$. The $^3p*$ and $^3t*$ are equilibrated with each other. Deactivation from the former gave either the cis or trans isomer, where as that from the latter gave only the trans isomer.

The quantum chain process can also take place by quenching of the $^3t*$ by anthracene to produce the ground state trans and the triplet state anthracene

**Figure 5** Effect of additive on a quantum chain process. (Adapted from Ref. 97.)

($^3$AN*). The $^3$AN* thus produced can transfer its energy to the ground state cis to regenerate $^3$c*, which undergoes isomerization to $^3$t*.

Requirements of such a catalyst are 1.) a long triplet lifetime and 2.) an appropriate triplet energy. Anthracene has a long triplet lifetime in the range of 100 μs at ambient temperature under degassed conditions and its triplet energy (42.5 kcal mol$^{-1}$) is located between the reactant cis (43 kcal mol$^{-1}$) and the product trans (41 kcal mol$^{-1}$).

Another example of additives to increase the quantum yield has been reported in biacetyl-sensitized isomerization of 8-styrylfluoranthene in the presence of acridine has been reported [98].

## VII.  QUANTUM CHAIN PROCESS IN THE SOLID STATE

The conditions necessary to observe a quantum chain process (a quantum gain) in solid matrices using the domino mechanism were reported by Ebbesen et al. to find new application of the amplification of the effect of photon in molecular optics [99]. The quantum yield of reaction from A to B ($\Phi_G$) in solution is described as Eq. (4), where $\Phi$* is the quantum yield of production of the reactive excited state and $k_q$ and $k_d$ are the rate constants of energy transfer and unimolecular deactivation, respectively. Since the molecular diffusion is slow in the solid matrices, the equation is modified by using Perrin's equation as shown in Eq. (5), where $N' = 6.02 \times 10^{20}$ molecules and $v$ (cm$^3$) is the volume of the quenching spheres.

$$\Phi_G = \Phi^*(k_q[A]/k_d + 1) \tag{4}$$
$$\Phi_G = \Phi^* \exp(vxN'x[A]) \tag{5}$$

The $\Phi_G$ value depends on the quenching sphere radii $R$ and can reach an extremely high value ($>10^{10}$) with appropriate molecules and conditions.

## VIII.  TRANS → CIS ISOMERIZATION
##          OF ANTHRYLETHENES

We should again discuss the requirements of the one-way isomerization of arylethenes: 1.) the substitution with an aryl group having a low triplet energy and 2.) the difference of the steric hindrance to change the energy of the cis and trans isomers. Namely, the cis isomer is located at ~5 kcal mol$^{-1}$ higher energy than the trans isomer in the ground state for stilbene [39,40]. A similar value can be estimated for the series of diarylethenes such as 2-styrylanthracene [1,2].

If one can prepare a compound whose trans isomer has more steric hindrance than the cis isomer, then the trans isomer must locate at a higher energy than the cis isomer and the one-way trans → cis isomerization could proceed.

Furthermore, when the cis and trans isomers are not very different with respect to the steric effect, and the cis isomer is almost the same as, or slightly higher in energy than, the trans isomer in the ground state as well as in the triplet state, $^3c^*$ and $^3t^*$ are in equilibrium and the deactivation may take place from both $^3c^*$ and $^3t^*$ (Fig. 6). Thus, the two-way isomerization takes place as an adiabatic process as described later.

As to the possibility of the reverse process in one-way isomerizing olefins, Fischer et al. argued that the excited state behavior of anthryl analogs is essentially similar to that of stilbenes and that the anthrylethenes should undergo mutual isomerization [100]. They studied the trans → cis isomerization of several anthrylethenes.

On irradiation with 405 nm light at the longer wavelength edge of the absorption spectra where most of the photon are absorbed by the trans isomers of

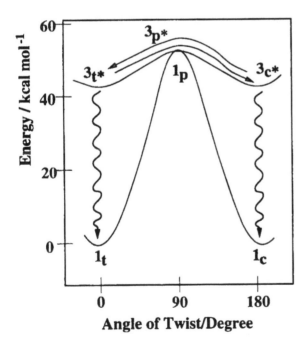

**Figure 6**   Potential energy surfaces of adiabatic two-way cis-trans isomerization.

**19b** and **19c**, they underwent isomerization to give the cis isomers; the photostationary states with up to 50% cis content were obtained. However, **19a** did not
give any cis isomer on irradiation even at the edge of the absorption spectrum.
Since the above results are obtained by using only the trans isomers, these substituent effects on the reverse isomerization of the trans isomers are not clear. However, the small change of the substituent may change the mode of isomerization
as well as the potential energy surfaces of isomerization even for the typical one-
way isomerizing anthrylethenes.

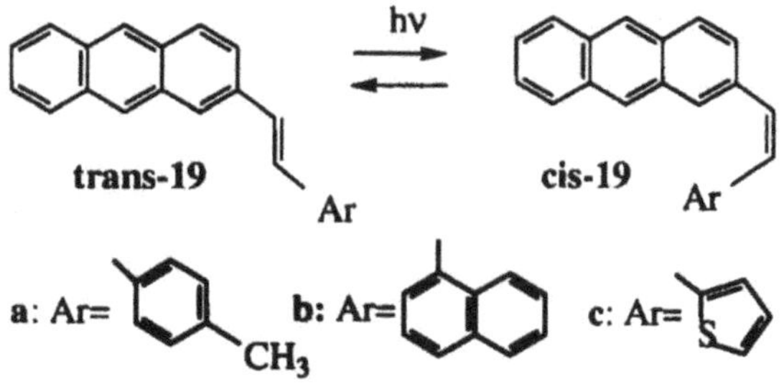

## IX.  INEFFICIENT ISOMERIZATION

As already mentioned, the introduction of an aryl group with a low triplet energy
changed the potential energy surface of isomerization to put the ${}^3t^*$ and ${}^3c^*$ as
energy minima and ${}^3p^*$ as an energy barrier. Therefore, substitution by an aryl
group with a very low triplet energy was studied to increase the energy barrier
of isomerization and suppress the efficiency of isomerization in the triplet state.
In addition, if the deactivation of ${}^3c^*$ and ${}^3t^*$ is accelerated by the heavy atom
effect, the deactivation prevails over the cis-trans isomerization. In this respect,
one-way isomerization of styrylferrocene [71] and one-way and two-way adiabatic isomerization of perylenylethenes [72–74] are discussed.

## A.  Styrylferrocene

Since the triplet energy of ferrocene is similar to that of anthracene the potential
energy surface of isomerization of styrylferrocene in the triplet state is considered
as similar to that of one-way isomerizing anthrylethenes. However, the substitution of a phenyl group in stilbene by a ferrocenyl group suppressed the efficiency
of isomerization due to the acceleration of deactivation of the cis and trans triplet
state before undergoing isomerization around the double bond; a very short life-

time of ferrocene triplet assumed to be 0.6 ns is responsible for the quick deactivation of the triplet styrylferrocene (**9**) [71,102].

## B.  Perylenylethenes

3-Styrylperylene (**10**) underwent one-way cis → trans isomerization in the triplet excited state [72,73]. The cis ($^3c*$) and trans isomers ($^3t*$) exhibited considerably different absorption spectra and the $^3c* \rightarrow {}^3t*$ conversion was followed by laser photolysis (Fig. 7). The activation energy and the frequency factor of $^3c* \rightarrow {}^3t*$ conversion was determined as 6.6 kcal mol$^{-1}$ and $2.1 \times 10^{12}$ s$^{-1}$, respectively [73]. Thus, the introduction of the very low triplet energy substituent in ArCH=CHPh series lowered the planar triplet energy in order to observe the $^3c* \rightarrow {}^3t*$ isomerization.

In the ArCH=CHAlkyl series, the $^3p*$ over $^1t$ is estimated to be ~7 kcal mol$^{-1}$ higher than that in ArCH=CHPh series; thus, photochemical behavior of 3-(1-propenyl)perylene **11** was examined.

Figure 8 shows the T-T absorption spectra of **11** observed on Michler's ketone sensitization. cis- and trans-**11** exhibited different T-T absorption spectra which decayed without changing the spectral profile, indicating that the conversion between $^3c*$ and $^3t*$ is inefficient even in the 10 μs time range [74].

On steady-state irradiation the isomerization between cis- and trans-**11** could take place with a very small quantum efficiency, $([c]/[t])_{pss} = 15.5/84.5$

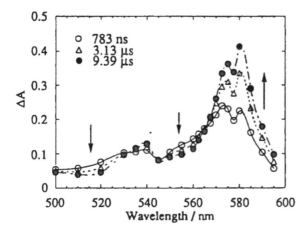

**Figure 7**  T-T absorption spectra determined on excitation of cis-**10** on 308-nm laser in methylcyclohexane at 205.4 K showing $^3c* \rightarrow {}^3t*$ conversion. (Reprinted from Ref. 73; copyright 1998 Chemical Society of Japan.)

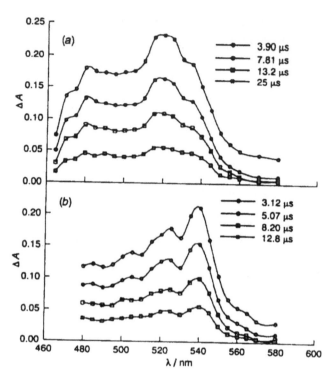

**Figure 8** T-T absorption spectra of *cis*- (a) and *trans*-11 (b) in benzene under an argon atmosphere. (Reprinted from Ref. 74; copyright 1998 the Royal Society of Chemistry.)

and the quantum yield of isomerization is $\sim 10^{-3}$ for both directions. These values indicate that deactivation to the ground state precedes the cis-trans isomerization in the triplet state and the potential energy surface of isomerization is proposed as shown in Fig. 9 [74]. The energy barrier of 13–16 kcal mol$^{-1}$ must be the highest value in the cis-trans isomerization in the excited state.

## X.  SINGLE BOND ROTATION IN THE EXCITED STATE

By careful examination of the fluorescence spectra Saltiel et al. reported that the cis → trans one-way isomerization of 2-styrylanthracene (**2c**) partly takes place in the excited singlet state as an adiabatic process [103,104]. In this case the isomerization is rotamer-specific, taking place only from the s-cis conformer.

Usually rotational isomerization cannot take place in the excited state as proposed by Havinga (NEER principle) [105,106]. The excited singlet state of **2c** follows this principle. 'ever, in 2-vinylanthracene and its alkyl derivatives,

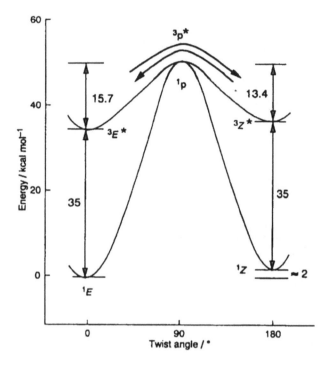

**Figure 9**   Potential energy surfaces of cis-trans isomerization of **11**. (Reprinted from ref. 74; copyright 1998 the Royal Society of Chemistry.)

s-trans → s-cis isomerization takes place in the excited singlet state [107, 108].

In 2-styrylanthracene (**2c**) the rotational isomerization does not occur in the singlet excited state, but occurs in the triplet state in a one-way manner from the s-trans isomer to the s-cis isomer as revealed by the observation of the change of the T-T absorption spectra [109]. Thus, the potential energy surface is proposed as shown in Fig. 10. Furthermore, in *N*-methoxy-1-(2-anthryl)ethanimine (**20**) the rotational isomerization took place in the excited singlet state as well as in the excited triplet state [110]. The former behavior is followed by the change of the fluorescence spectra from the s-trans to the s-cis isomer, whereas the latter is studied by the spectral change of the T-T absorption spectra.

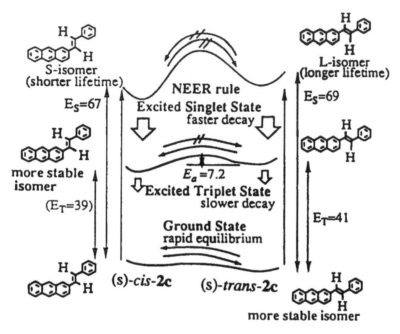

**Figure 10** Potential energy surfaces of rotational isomerization of **2c**. (Reprinted from Ref. 109; copyright 1998 Chemical Society of Japan.)

The low energy barrier of rotational isomerization in 2-anthryl-substituted C=C or C=N compounds could be attributed to the localization of the excitation energy at the anthryl group.

## XI. STRUCTURAL CHANGE ALONG THE ISOMERIZATION

In stilbene (**1**), the trans isomer takes an almost planar conformation around the single bond connecting the phenyl and styryl groups. However, the torsion angle around the single bond is estimated to be 30–40° in cis-stilbene (cis-**1**) [111–113]. Therefore, in the course of cis-trans isomerization around the double bond, the aromatic nucleus should rotate around the single bond connected to the ethylenic carbon to increase the conjugation between the two chromophores.

The above discussion is applied to most of the arylethenes. The importance of the single-bond rotation was argued from the view point of nonvertical excitation transfer from the low-energy sensitizer to the cis-stilbene [33–36,39, 112,113]. In the case of **8a**, even for the trans isomer the dihedral angle between the pyrenyl plane and the double bond is estimated to be about 30° due to the steric effect of the hydrogen atom at the peri position. In contrast, the dihedral

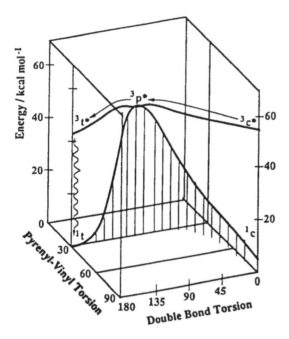

**Figure 11** Potential energy surafes of cis-trans isomerization of **8b**. (Reprinted from ref. 114; Potential energy surfaces of a one-way photoisomerizing olefin studied by photoacoustic calorimetry and X-ray crystallography, T. Arai, H. Okamoto, K. Tokumaru, T. Ni, R. A. Caldwell, and K. Ueno, *J. Photochem. Photobiol. A: Chem.* **75**, 85–90 (1993): copyright (1998), with permission from Elsevier Science.)

angle between the pyrenyl plane and the double bond in the cis isomer was determined 90° [114]. This means that the pyrenyl ring has a completely perpendicular arrangement from the double-bond plane. However, in the excited triplet state the one-way cis → trans isomerization took place very efficiently. The excitation is to be localized in the pyrene nucleus in the cis isomer, and for the isomerization to take place, the conjugation between the pyrenyl group and the adjacent etheylenyl carbon is required. The potential energy surface of one-way isomerization including the single-bond rotation is depicted in Fig. 11.

## XII. EFFECT OF SUBSTITUTION POSITION ON THE POTENTIAL ENERGY SURFACE OF BUTENYLANTHRACENE

**21** and **22** undergo one-way cis → trans isomerization in the triplet state in a similar way to that of **2b** [8–11,69,115,116]. In **2b**, **21a**, and **22a** the conversion

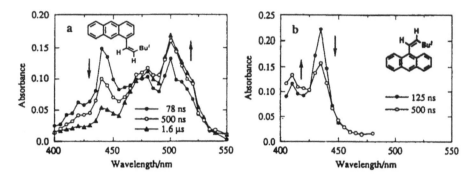

**Figure 12** Transient absorption spectra observed on pulsed laser excitation of *cis*-21a at 24°C (a) and *cis*-22a at 25°C (b) in deaerated toluene. (Reprinted from Ref. 116; copyright 1998 Chemical Society of Japan.)

from $^3c^*$ to $^3t^*$ can be detected by laser transient spectroscopy as the change of the T-T absorption spectra. Typical examples for **21a** and **22a** are shown in Fig. 12 [116]. The activation energy of the adiabatic $^3c^* \rightarrow {}^3t^*$ isomerization is determined to be 6.0, 4.6, and 3.1 kcal mol$^{-1}$ with the frequency factors of $5 \times 10^{10}$, $5 \times 10^9$, and $4.0 \times 10^8$ s$^{-1}$. Thus, the activation energy decreases in the order of **2b**, **21a**, and **22a**, indicating that the energy barrier around $^3p^*$ conformation is affected by the position of anthracene nucleus at which the ethenyl moiety is substituted. Similar results were obtained for the cis-trans isomerization of deuterated vinylanthracene, 1-, 2-, and 9-(ethenyl-2-d)anthracene [117]. The potential energy surfaces of the triplet isomerization of **2b**, **21**, and **22** are depicted in Fig. 13 [116].

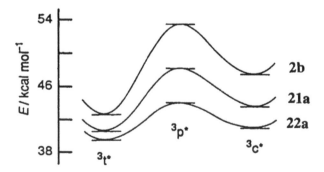

**Figure 13** The estimated triplet energy surfaces for **2b**, **21a**, and **22a**. (Adapted from Ref. 116; copyright 1998 Chemical Society of Japan.)

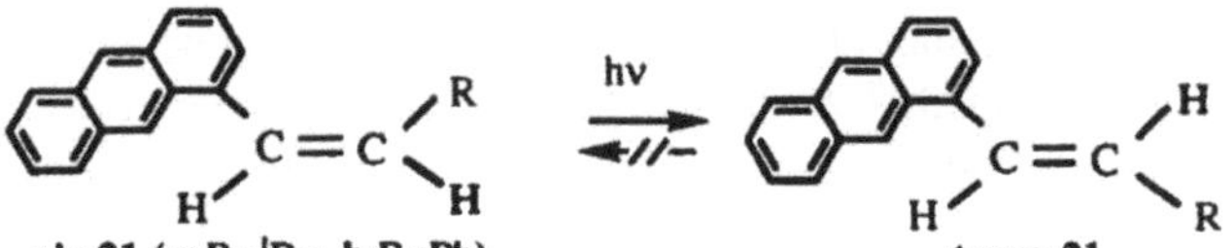

cis-21 (a: R=$^t$Bu, b:R=Ph)  hv $\rightleftharpoons$ trans-21

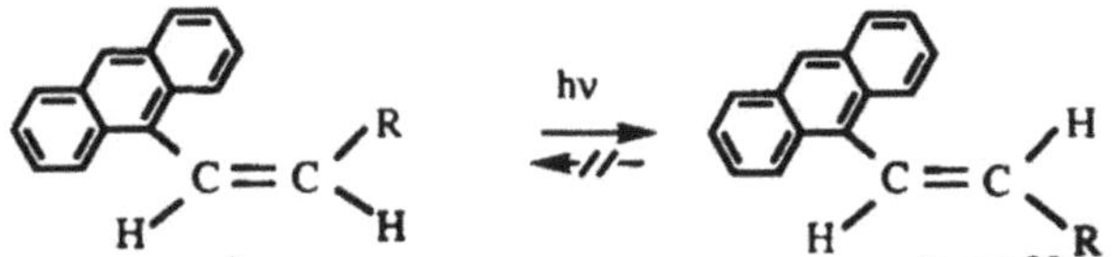

cis-22 (a: R=$^t$Bu, b:R=Ph)  hv $\rightleftharpoons$ trans-22

## XIII. ISOMERIZATION OF POLYENES WITH A QUANTUM CHAIN PROCESS FOR SEVERAL DIRECTIONS

### A. Dienes, Trienes, and Tetraenes

The effect of concentration on the quantum yield of isomerization as well as the photostationary state isomer composition has also been reported in the isomerization of polyenes [118–120]. The quantum yield of cis-trans isomerization increased with increasing concentration of the cis isomer on benzophenone sensitization of 1,3-pentadiene (23) from 0.55 ([23] = 0.2 M) to 0.84 ([23] = 9.8 M) [119]. Furthermore, in the isomerization of 2,4-hexadienes (24), not only cc → ct isomerization but also cc → tt, tt → cc, and tt → ct isomerizations proceed through the quantum chain process [119]. Although the s-cis rotamers are less populated than the s-trans totamers in the ground state due to the steric effect, the s-cis rotamers have lower triplet energies compared to the s-trans rotamers. Therefore, the rotational isomers around the single bonds are assumed to play an important role in the quantum chain process. The observation that the quantum chain process proceeds in several directions indicates that the energy minimum exists not only at $^3$tt*, but also at $^3$ct* and $^3$cc*.

cis-23 $\rightleftharpoons$ trans-23

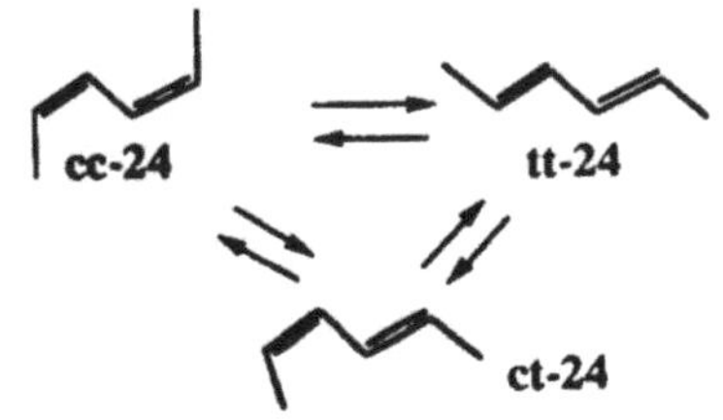

cc-24 $\rightleftharpoons$ tt-24

ct-24

The quantum chain process has also been observed in benzophenone-sensitized isomerization of 2,6-dimethyl-2,4,6-octatriene (25) [120]. The effect of azulene on the photostationary state isomer composition suggests that both $^3tt^*$ and $^3tc^*$ are the stable conformers in the triplet state and are equilibrated. However, equilibration of all excited intermediates is not complete within the lifetime of the excited triplet state ($\approx 50$ ns). The triplet lifetime of 1,3,5-hexa-triene (26) is reported as $\approx 100$ ns (Table 5) [121].

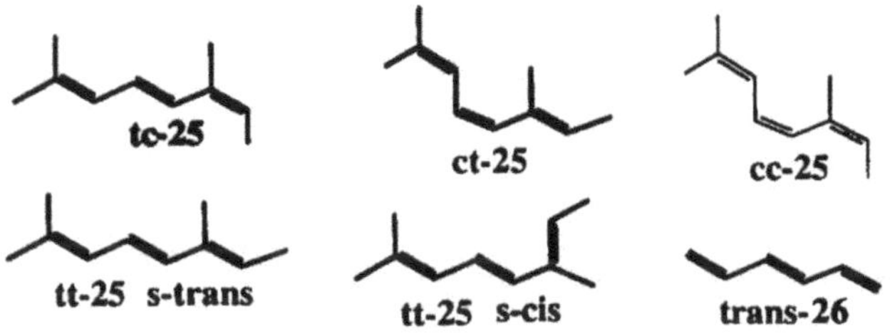

Substitution with phenyl group changes the potential energy surfaces so much that one-way isomerization takes place. On benzophenone sensitization 1,4-diphenyl-1,3-butadiene (27), 1,6-diphenyl-1,3,5-hexatriene (28), and 1,8-diphenyl-1,3,5,7-octatetraene (29) exhibited T-T absorption spectra with $\lambda_{max}$ at 380–440 nm and $\tau_T$ of 10–100 μs [122] (Table 5). The relatively long triplet lifetimes indicate that the planar trans triplet exists as the most stable conformer in the excited triplet state. Furthermore, a 10-μs lifetime of 27 indicates two-way isomerization with the quantum chain process as observed for 8b and 100-μs lifetimes for 27 and 28 indicate the potential energy surface similar to the one-way isomerizing olefin such as 2 (cf. Table 2).

**Table 5**  T-T Absorption Maxima and the Triplet Lifetimes of Polyenes

|    | $\lambda_{max}$(T-T), nm | $\tau_T$, μs | Refs. |
|----|--------------------------|--------------|-------|
| 26 |                          | $\approx 0.1$ | 121   |
| 27 | 392                      | 10           | 122   |
| 28 | 400,425                  | 100          | 122   |
| 29 | 419,444                  | 100          | 122   |
| 30 | 450,610                  | 8            | 124   |
| 31 | 380                      | 13           | 127   |
| 32 | 526                      | 6.3          | 132   |

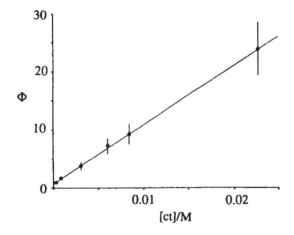

Actually, the quantum yield of ct → tt isomerization of **27** on 9-fluorenone sensitization rises well above 1 (Fig. 14) and the photostationary state isomer ratio ([tt]/[cc])$_{pss}$ increases with increasing the total concentration of **27** [123].

Substitution of a 2-anthryl group on the terminal carbon of dienes results in one-way cis-trans isomerization in **30**. On benzil sensitization the quantum yield of ct → tt isomerization is determined to be 24 at $5 \times 10^{-3}$ M of ct-**30** [124].

## B.  Retinal and β-Carotene

Retinal (**31**) undergoes isomerization to give 7-cis (7C), 9-cis (9C), 11-cis (11C), 13-cis (13C), and all-trans (T) isomers depending on the starting isomers

**Figure 14**  Initial quantum yield of ct → tt photoisomerization of **27** by sensitization with 9-fluorenone as a function of [ct]. (Reprinted from Ref. 123; copyright 1998 American Chemical Society.)

[125,126]. The photophysics and photochemistry of several retinal isomers have been studied with respect to the potential energy surfaces of isomerization by transient spectroscopy as well as product analyses [127–129].

The isomerization of **31** proceeds by a quantum chain process for several directions [125,126,130]. For example, all-trans retinal undergoes isomerization giving the 9-cis and 13-cis isomers in the excited triplet state; the quantum yield of isomerization of the all-trans isomer increased from 0.13 at $1.1 \times 10^{-5}$ M to 0.84 at $1.14 \times 10^{-3}$ M [126]. In addition, with an increase in the total concentration, the composition at the photostationary state of the all-trans and 11-cis isomers decreased, but that of 9-cis increased. Since the s-cis conformers are likely to have lower triplet energies than s-trans conformers, the reaction scheme involving the 12-(S)-cis conformer was proposed in the quantum chain process in the triplet state.

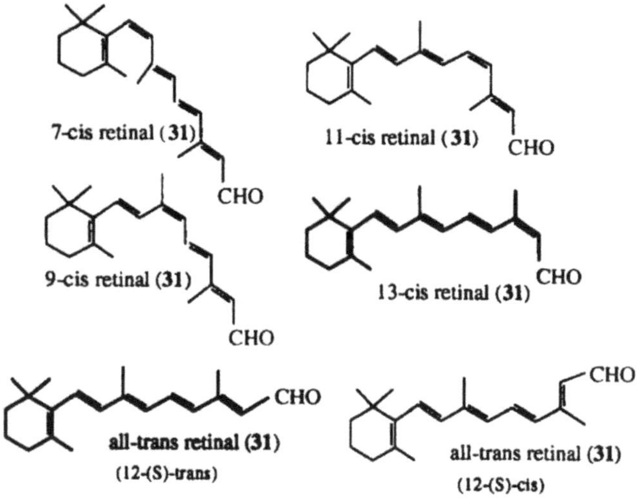

7-cis retinal (**31**)

11-cis retinal (**31**) CHO

9-cis retinal (**31**)

CHO

13-cis retinal (**31**) CHO

CHO

all-trans retinal (**31**)
(12-(S)-trans)

CHO

all-trans retinal (**31**)
(12-(S)-cis)

Anthracene-sensitized isomerization of β-carotene (**32**) gave a photostationary mixture of all-trans (85%), 7-cis (0%), 9-cis (10%), 13-cis (4%), and 15-cis (1%) [131]. The equilibrium among the four isomers except 7-cis in the triplet state is possible and the quantum chain process can be expected for several directions. However, a detailed analysis indicates that true equilibrium may not be reached among the isomers.

all-trans β-carotene (**32**)

**Table 6** cis-trans Isomerization of ArCH=CHR in the Triplet State

| Example | 4 | 1 | 8b | 2 | 9 | 11 |
|---|---|---|---|---|---|---|
| Mode Diabatic or adiabatic | Two-way Diabatic | Two-way Diabatic | Two-way Diabatic and adiabatic | One-way Adiabatic | One-way Adiabatic | Two-way Adiabatic |
| Triplet lifetime | $\approx 20$ ns | 60 ns | 27 $\mu$s | $\approx 100$ $\mu$s | $\leq 1$ ns | 50 $\mu$s |
| Potential minimum | $^3p^*$ | $^3p^*$ and $^3t^*$ | $^3p^*$ and $^3t^*$ | $^3c^*$ and $^3t^*$ | $^3c^*$ and $^3t^*$ | $^3c^*$ and $^3t^*$ |
| $K_{tp}$ | $<10^2$ | $\approx 10^1$ | $\approx 10^{-3}$ | $\leq 10^{-5}$ | $\leq 10^{-5}$ | $\leq 10^{-11}$ |
| $K_{tc}^{a}$ | $\leq 10^{-3}$ | $\leq 10^{-3}$ | $\leq 10^{-3}$ | $\leq 10^{-3}$ | $\leq 10^{-3}$ | $\approx 10^{-1}$ |
| Deactivation | $^3p^*$ | $^3p^*$ | $^3p^*$ and $^3t^*$ | $^3t^*$ | $^3c^*$ and $^3t^*$ | $^3c^*$ and $^3t^*$ |

$^a$ $K_{tc} = [^3c^*]/[^3t^*]$.

The triplet lifetimes of **31** and **32** are also reported as 13 and 6.3 µs, respectively (Table 5) [127,132].

## XIV.  CONCLUSION

The cis-trans isomerization of various olefins in the triplet state was reviewed. The mode of cis-trans isomerization is correlated to the triplet energy of the aryl substituent (ArH), since only the planar triplet energy decreases with decreasing $E_T$ of ArH. In addition, the energy minima as well as the stable conformations in the triplet state were discussed based on the triplet lifetime, the efficiency of isomerization, and the quenching experiments. Several examples of quantum chain process in olefins and polyenes were also described.

Table 6 summarizes the features of one-way and two-way isomerization of arylethenes depending on the aryl substituents. Thus, the typical features of one-way and two-way isomerizations shown in Table 1 are modified, but one-way isomerization does not necessarily accompany the quantam chain process and two-way isomerization does not necessarily involve the diabatic process. For some compounds two-way isomerization can take place as an adiabatic process by mutual conversion of $^3c^*$ and $^3t^*$ followed by deactivation to the ground state. The short triplet lifetime due to the heavy atom effect may cause a very inefficient one-way isomerization. Furthermore, the quantum chain processes can be observed in many compounds when proper experimental conditions can be established.

The isomerization mechanism can be estimated from the structure of the compounds on the basis of the above-mentioned effect of the substituent on the mode of isomerization. Furthermore, the triplet lifetime and the efficiency of isomerization could be estimated for some olefinic compounds.

## ACKNOWLEDGMENT

Preparation of this chapter was supported by Research Foundation for Opto-Science and Technology and Grant-in-Aid for Scientific Research (B) from the Ministry of Education, Science, Sports and Culture, Japan No. 10440166.

## REFERENCES

1.  Arai, T; Karatsu, T.; Sakuragi, H.; Tokumaru, K. *Tetrahedron Lett.*, **1983**, *24*, 2873–2876.
2.  Arai, T.; Tokumaru, K. *Chem. Rev.*, **1993**, *93*, 23–39.
3.  Tokumaru, K.; Arai, T. *J. Photochem. Photobiol. A: Chem.*, **1992**, *65*, 1–13.
4.  Tokumaru, K.; Arai, T. *Bull. Chem. Soc. Jpn.*, **1995**, *68*, 1065–1087.
5.  Arai, T.; Tokumaru, K. *Adv. Photochem.*, **1995**, *20*, 1–57.

6.  Görner, H; Kuhn, H.J. *Adv. Photochem.*, **1995**, *19*, 1–117.
7.  Hamaguchi, H. *J. Mol. Struct.* **1985**, *126*, 125–132.
8.  Sandros, K.; Becker, H.-D. *J. Photochem.* **1987**, *39*, 301–315.
9.  Becker, H.-D. *Chem. Rev.* **1993**, *93*, 145–172.
10. Becker, H.-D. *Adv. Photochem.* **1989**, *15*, 139–227.
11. Görner, H. *J. Photochem. Photobiol. A: Chem.*, **1988**, *43*, 263–289.
12. Sandros, K.; Becker, H.-D. *J. Photochem. Photobiol. A: Chem.*, **1988**, *43*, 291–292.
13. Saltiel, J.; D'Agostino, J.; Megarity, E.D.; Metts, L.; Neuberger, K.R.; Wrighton, M.; Zafiriou, O.C. *Organic Photochemistry*; Chapman, O.L., Ed.; Marcel Dekker: New York, 1973; Vol. 3, Pp. 1–113.
14. Saltiel, J., Charlton, M.L. *Rearangement in Ground and Excited States*; de Mayo, P., Ed.; Academic Press: New York, 1980; Vol. 3, Pp. 25–89.
15. Saltiel, J.; Sun, Y.-P. *Photochromism. Molecules and Systems*; Dürr, H.; Bouas-Laurent, H., Eds.; Elsevier: Amsterdam, 1990; Pp. 64–164.
16. Saltiel, J.; Waller, A.S.; Sun, Y.-P.; Sears, D.F., Jr. *J. Am. Chem. Soc.*, **1990**, *112*, 4580–4581.
17. Saltiel, J.; Waller, A.S.; Sears, D.F., Jr. *J. Photochem. Photobiol. A: Chem.*, **1992**, *65*, 29–40.
18. Saltiel, J.; Waller, A.S.; Sears, D.F., Jr. *J. Am. Chem. Soc.*, **1993**, *115*, 2453–2465.
19. Olson, A.R. *Trans. Faraday Soc.* **1931**, *27*, 69–76. Olson, A.R.; Hudson, F.L. *J. Am. Chem. Soc.* **1933**, *55*, 1410–1424. Olson, A.R.; Maroney, W. *J. Am. Chem. Soc.* **1934**, *56*, 1320–1322.
20. Lewis, G.N.; Magel, T.T.; Lipkin, D. *J. Am. Chem. Soc.* **1940**, *62*, 2973–2980.
21. Förster, Th. *Z. Elektrochem.* **1952**, *56*, 716–722.
22. Stegemeyer, H. *J. Phys. Chem.* **1962**, *66*, 2255–2260.
23. Gegiou, D.; Muszkat, K.A.; Fischer, E. *J. Am. Chem. Soc.* **1968**, *90*, 12–18. Gegiou, D.; Muszkat, K.A.; Fischer, E. *J. Am. Chem. Soc.* **1968**, *90*, 3907–3918.
24. Saltiel, J.; Chang, D.W.L.; Megarity, E.D.; Rousseau, A.D.; Shannon, P.T.; Thomas, B.; Uriarte, A.K. *Pure Appl. Chem.* **1975**, *41*, 559–579.
25. Waldeck, D.H. *Chem. Rev.* **1991**, *91*, 415–436.
26. Saltiel, J.; Megarity, E.D. Kneipp, K.G. *J. Am. Chem. Soc.* **1966**, *88*, 2336–2338.
27. Saltiel, J.; Megarity, E.D. *J. Am. Chem. Soc.* **1969**, *91*, 1265–1267.
28. Saltiel, J.; Megarity, E.D. *J. Am. Chem. Soc.* **1972**, *94*, 2742–2749.
29. Marinari, A.; Saltiel, J. *Mol. Photochem.* **1976**, *7*, 225–249.
30. Saltiel, J.; Marinari, A.; Chang, D.W.L.; Mitchener, J.C.; Megarity, E.D.; *J. Am. Chem. Soc.* **1979**, *101*, 2982–2996.
31. Dyke, R.H.; McClure, D.S. *J. Chem. Phys.* **1962**, *36*, 2326–2345.
32. Malkin, S.; Fischer, E. *J. Phys. Chem.* **1962**, *66*, 2482–2486.
33. Hammond, G.S.; Saltiel, J.; Lamola, A.A.; Turro, N.J.; Bradshaw, J.S.; Cowan, D.O.; Counsell, R.C.; Vogt, V.; Dalton, C. *J. Am. Chem. Soc.*, **1964**, *86*, 3197–3217.
34. Lamola, A.A.; Hammond, G.S. *J. Chem. Phys.* **1965**, *43*, 2129–2135.
35. Herkstroeter, W.G.; Hammond, G.S. *J. Am. Chem. Soc.* **1966**, *88*, 4769–4777.
36. Hammond, G.S.; DeMeyer, D.E.; Williams, J.L.R. *J. Am. Chem. Soc.* **1969**, *91*, 5180–5181.

37.  Saltiel, J.; Rousseau, A.D.; Thomas, B. *J. Am. Chem. Soc.* **1983**, *105*, 7631–7637.
38.  Saltiel, J. Marchand, G.R.; Kirkor-Kaminska, E.; Smothers, W.K.; Mueller, W.B.; Charlton, J.L. *J. Am. Chem. Soc.* **1984**, *106*, 3144–3151.
39.  Saltiel, J.; Ganapathy, S.; Werking, C.; *J. Phys. Chem.* **1987**, *91*, 2755–2758.
40.  Ni, T.; Caldwell, R.A.; Melton, L.A. *J. Am. Chem. Soc.* **1989**, *111*, 457–464.
41.  Görner, H.; Schulte-Fröhlinde, D. *J. Phys. Chem.* **1981**, *85*, 1835–1841.
42.  Karatsu, T. Arai, T.; Sakuragi, H.; Tokumaru, K. *Chem. Phys. Lett.* **1985**, *115*, 9–15.
43.  Hamaguchi. H.; Tasumi, M.; Karatsu, T.; Arai, T.; Tokumaru, K.*J. Am. Chem. Soc.* **1986**, *108*, 1698–1699.
44.  Arai, T.; Karatsu, T.; Tsuchiya, M.; Sakuragi, H.; Tokumaru, K. *Chem Phys. Lett.* **1988**, *149*, 161–166.
45.  Karatsu, T.; Tsuchiya, M.; Arai, T.; Sakuragi, H.; Tokumaru, K. *Chem. Phys. Lett.* **1990**, *169*, 36–42.
46.  T. Karatsu, Tsuchiya, M.; Arai, T.; Sakuragi, H.; Tokumaru, K. *Bull. Chem. Soc. Jpn.* **1994**, *67*, 3030–3039.
47.  Momicchioli, F. Baraldi, I.; Fischer, E. *J. Photochem. Photobiol. A. Chem.* **1989**, *48*, 95–107.
48.  Orlandi, G.; Negri, F.; Mazzucato, U.; Bartocci, G. *J. Photochem. Photobiol. A. Chem.* **1990**, *55*, 37–42.
49.  Bartocci, G.; Masetti, F.; Mazzucato, U.; Spalletti, A.; Orlandi, G.; Poggi, G. *J. Chem. Soc. Faraday Trans. 2* **1988**, *84*, 385–399.
50.  Görner, H.; Elisei, F.; Alosisi, G.G. *J. Chem. Soc., Faraday Trans.* **1992**, *88*, 29–34.
51.  Ramamurthy, V.; Liu, R.S.H. *J. Am. Chem. Soc.* **1976**, *98*, 2935–2942.
52.  Caldwell, R.A.; Sovocool, G.W.; Peresie, R.J. *J. Am. Chem. Soc.* **1973**, *95*, 1496–1502.
53.  Caldwell, R.A.; Cao, C.V. *J. Am. Chem. Soc.* **1982**, *104*, 6174–6180.
54.  Bonneau, R. *J. Photochem.* **1979**, *10*, 439–449.
55.  Bonneau, R. *J. Am. Chem. Soc.* **1982**, *104*, 2921–2923.
56.  Rockley, M.G.; Salisbury, K. *J. Chem. Soc. Perkin II*, **1973**, 1582–1585. Crosby, P.M.; Dyke, J.M.; Metcalfe, J.; Rest, A.J.; Salisbury, K.; Sodeau, J.R. *J. Chem. Soc. Perkin II*, *1977*, 182–185. Ghiggino, K.P.; Hara, K.; Mant, G.R.; Phillips, D.; Salisbury, K. Steer, R.P.; Swords, M.D. *J. Chem. Soc. Perkin II*, **1978**, 88–91.
57.  Arai, T.; Sakuragi, H.; Tokumaru, K. *Chem. Lett.* **1980**, 261–264. Arai, T.; Sakuragi, H.; Tokumaru, K. *Bull. Chem. Soc. Jpn.* **1982**, *55*, 2204–2207.
58.  Arai, T.; Sakuragi, H.; Tokumaru, K. *Chem. Lett.* **1980**, 1335–1338.
59.  Arai, T.; Karatsu, T.; Sakuragi, H.; Tokumaru, K. *Chem. Lett.* **1981**, 1377–1380.
60.  Arai, T.; Sakuragi, H.; Tokumaru, K.; Sakaguchi, Y.; Nakamura, J.; Hayashi, H. *Chem. Phys. Lett.* **1983**, *98*, 40–44.
61.  Saltiel, J.; Eaker, D.W. *Chem. Phys. Lett.* **1980**, *75*, 209–213.
62.  Görner, H.; Eaker, D.W.; Saltiel, J. *J. Am. Chem. Soc.* **1981**, *103*, 7164–7169.
63.  Karatsu, T.; Hiresaki, T.; Arai, T.; Sakuragi, H.; Tokumaru, K.; Wirz, J. *Bull. Chem. Soc. Jpn.* **1991**, *64*, 3355–3362.
64.  Arai, T.; Kuriyama, Y. Karatsu, T.; Sakuragi, H.; Tokumaru, K.; Oishi, S. *J. Photochem.* **1987**, *36*, 125–130.

65. Furuuchi, H.; Arai, T.; Kuriyama, Y.; Sakuragi, H.; Tokumaru, K. *Chem. Phys. Lett.* **1989**, *162*, 211–216. Furuuchi, Y.; Kuriyama, Y.; Arai, T.; Sakuragi, H.; Tokumaru, K. *Bull. Chem. Soc. Jpn.* **1991**, *64*, 1601–1606.

66. Arai, T.; Okamato, H.; Sakuragi, H.; Tokumaru, K. *Chem. Phys. Lett.* **1989**, *157*, 46–50. Okamoto, H.: Arai, T.; Sakuragi, H.; Tokumaru, K. *Bull. Chem. Soc. Jpn.* **1990**, *63*, 2881–2890. Okamoto, H.: Arai, T.; Sakuragi, H.; Tokumaru, K.; Kawanishi, Y. *Bull. Chem. Soc. Jpn.* **1991**, *64*, 216–220.

67. Wismontski-Knittel, T.; Das, P.K. *J. Phys. Chem.* **1984**, *88*, 1168–1173.

68. Kikuchi, O.; Segawa, K.; Takahashi, O.; Arai, T.; Tokumaru, K. *Bull. Chem. Soc. Jpn.* **1992**, *65*, 1463–1465.

69. Mazzucato, U. *Gazz. Chim. Ital.* **1987**, *117*, 661–665.

70. Richards, J.H.; Pisker-Trifunac, N. *J. Paint Technol.* **1969**, *41*, 363–371.

71. Arai, T.; Ogawa, Y.; Sakuragi, H.; Tokumaru, K. *Chem. Phys. Lett.* **1992**, *196*, 145–149.

72. Castel;, N; Fischer, E.; Strauch, M.; Niemeyer, M.; Lüttke, W. *J. Photochem. Photobiol. A: Chem.* **1991**, *57*, 301–315.

73. Arai, T.; Takahashi, O.; Asano, T.; Tokumaru, K.; *Chem. Lett.*, **1994**, 205–208.

74. Arai, T.; Takahashi, O. *J. Chem. Soc., Chem. Commun.* **1995**, 1837–1838.

75. Görner, H. *J. Phys. Chem.* **1989**, *93*, 1826–1832.

76. Saltiel, J.; Atwater, B.W. *Adv. Photochem.* **1988**, *14*, 1–90.

77. Valentine, D. Jr.; Hammond, G.S. *J. Am. Chem. Soc.* **1972**, *94*, 3449–3454.

78. Saltiel, J. *J. Am. Chem. Soc.* **1968**, *90*, 6394–6400.

79. Saltiel, J.; D'Agostino, J.T.; Herkstroeter, W.G.; Saint-Ruf, G.; Buu-Hoi, N.P. *J. Am. Chem. Soc.* **1973**, *95*, 2543–2549.

80. Lazare, S.; Lapouyade, R.; Bonneau, R. *J. Am. Chem. Soc.* **1985**, *107*, 6604–6609.

81. Malkin, S.; Fischer, E. *J. Phys. Chem.* **1964**, *68*, 1153–1163.

82. Schulte-Fröhlinde, D.; Görner, H. *Pure Appl. Chem.* **1979**, *51*, 279–297.

83. Görner, H.; Schulte-Fröhlinde, D. *Ber. Bunsenges. Phys. Chem.* **1984**, *88*, 1208–1216. Bent, D.V.; Schulte-Fröhlinde, D. *J. Phys. Chem.* **1974**, *78*, 446–454.

84. Smit, K.J. *J. Phys. Chem.* **1992**, *96*, 6555–6558.

85. Görner, H.; Schulte-Fröhlinde, D. *J. Photochem.* **1978**, *8*, 91–102.

86. De Haas, M.P.; Warman, J.M. *Chem. Phys.* **1982**, *73*, 35–53.

87. Görner, H. *J. Chem. Soc. Faraday Trans.* **1993**, *89*, 4027–4033.

88. Elisei, F.; Mazzucato, U.; Görner, H. *J. Chem. Soc. Faraday Trans. 1*, **1989**, *85*, 1469–1483.

89. Bortolus, P.; Galiazzo, G. *J. Photochem.* **1974**, *2*, 361–370.

90. Görner, H.; Elisei, F.; Mazzucato, U.; Galiazzo, G. *J. Photochem. Photobiol. A: Chem.* **1988**, *43*, 139–154.

91. Whitten, D.G.; McCall, M.T. *J. Am. Chem. Soc.* **1969**, *91*, 5097–5103.

92. Whitten. D.G.; Lee, Y.L. *J. Am. Chem. Soc.* **1972**, *94*, 9142–9148.

93. Herkstroeter, W.G.; McClure, D.S. *J. Am. Chem. Soc.* **1968**, *90*, 4522–4527.

94. Whitten, D.G.; Wildes, P.D.; DeRosier, C.A. *J. Am. Chem. Soc.* **1972**, *94*, 7811–7823.

95. Mercer-Smith, J.A.; Whitten, D.G. *J. Am. Chem. Soc.* **1978**, *100*, 2620–2625.

96. Sundahl, M.; Wennerström, O.; Sandros, K.; Arai, T.; Tokumaru, K. *J. Phys. Chem.* **1990**, *94*, 6731–6734.

97.  Sundahl, M.; Wennerström, O. *J. Photochem. Photobiol. A: Chem.* **1996**, *98*, 117–120.

98.  Brink, M.; Jonson, H.; Sundahl, M. *J. Photochem. Photobiol. A: Chem.* **1998**, *112*, 149–153.

99.  Ebbesen, T.W.; Tokumaru, K. *Appl. Opt.* **1986**, *25*, 4618–4621.

100. Krongauz, V.; Castel, N.; Fischer, E. *J. Photochem.* **1987**, *39*, 285–300.

101. Laarhoven, W.H.; Cuppen, Th. J.H.M.; Castel, N.; Fischer, E. *J. Photochem. Photobiol. A: Chem.* **1989**, *49*, 137–141.

102. Jaworska-Augustyniak, A.; Karolczak, J.; Maciejewski, A.; Wojtczak, J. *Chem. Phys. lett.* **1987**, *137*, 134–138. Maciejewski, A.; Jaworska-Augustyniak, A.; Szeluga, Z.; Wojtczak, J.; Karolczak, J. *Chem. Phys. lett.* **1988**, *153*, 227–232.

103. Mazzucato, U.; Spalletti, A.; Bartocci, G. Coord. *Chem. Rev.* **1993**, *125*, 251–260.

104. Saltiel, J.; Zhang, Y.; Sears, D.F., Jr. *J. Am. Chem. Soc.* **1997**, *119*, 11202–11210.

105. Jacobs, H.J.C.; Havinga, E. *Adv. Photochem.* **1979**, *11*, 305–373.

106. Mazzucato, U.; Momicchioli, F. *Chem. Rev.* **1991**, *91*, 1679–1719.

107. Brearley, A.M.; Stanjord, A.J.G.; Flom, S.R.; Barbara, P.F. *Chem. Phys. Lett.* **1987** *113*, 43–48. Flom, S.R.; Nagarajan, V.; Barbara, P.F. *J. Phys. Chem.* **1986**, *9'*, 2085–2092. Brearly, A.M.; Flom, S.R.; Nagarajan, V.; Barbara, P.F. *J. Phys. Chem.* **1986**, *90*, 2092–2099. Barbara, P.F.; Jarzeba, W. *Acc. Chem. Res.* **1988**, *21*, 195–199.

108. Arai, T. Karatsu, T, Sakuragi, H.; Tokumaru, K.; Tamai, N.; Yamazaki, I. *Chem. Phys. Lett.* **1989**, *158*, 429–434. Arai, T.; Karatsu, T.; Sakuragi, H.; Tokumaru, K.; Tamai, N.; Yamazaki, I. *J. Photochem. Photobiol. A.: Chem.* **1992**, *65*, 41–51.

109. Karatsu, T.; Yoshikawa, N.; Kitamaura, A.; Tokumaru, K. *Chem. Lett.* **1994**, 381–384.

110. Arai, T.; Furuya, Y.; Tokumaru, K. *J. Phys. Chem.* **1994**, *98*, 9945–9949.

111. Kobayashi, T.; Yokota, K.; Nagakura, S. *Bull. Chem. Soc. Jpn.* **1975**, *48*, 412–415.

112. Caldwell, R.A.; Riley, S.J.; Gorman, A.A.; McNeeney, S.P.; Unett, D.J. *J. Am. Chem. Soc.* **1992**, *114*, 4424–4426.

113. Gorman, A.A.; Lambert, C.; Prescott, A. *Photochem. Photobiol.* **1990**, *51*, 29–35. Gorman, A.A.; Hamblett, I.; Reddoes, R.L.; Hamblett, I.; McNeeney, S.P.; Prescott, A.L.; Unett, D.J. *J. Chem. Soc. Chem. Commun.* **1991**, 963–964.

114. Arai, T.; Okamoto, H.; Tokumaru, K.; Ni, T.; Caldwell, R.A.; Ueno, K. *J. Photochem. Photobiol. A: Chem.* **1993**, *75*, 85–90.

115. Karatsu, T.; Kitamura, A.; Zeng, H.; Arai, T.; Sakuragi, H.; Tokumaru, K. *Bull Chem. Soc. Jpn.* **1995**, *68*, 920–928.

116. Karatsu, T.; Kitamura, A.; Zeng, H.; Arai, T.; Sakuragi, H.; Tokumaru, K. *Chem. Lett.* **1992**, 2193–2196.

117. Karatsu, T.; Misawa, H.; Nojiri, M.; Nakahigashi, N.; Watanabe, S.; Kitamura, A.; Arai, T.; Sakuragi, H.; Tokumaru, K. *J. Phys. Chem.* **1994**, *98*, 508–512.

118. Hyndman, H.L.; Monroe, B.M.; Hammond, G.S. *J. Am. Chem. Soc.* **1969**, *91*, 2852–2859. Hurley, R.; Testa, A.C. *J. Am. Chem. Soc.* **1970**, *92*, 211–212. Caldwell, R.A. *J. Am. Chem. Soc.* **1970**, *92*, 3229–3230.

119. Saltiel, J.; Townsend, D.E.; Sykes, A. *J. Am. Chem. Soc.* **1973**, *95*, 5968–5973.

120. Butt, Y.C.C.; Singh, A.K.; Baretz, B.H.; Liu, R.S.H. *J. Phys. Chem.* **1981**, *85*, 2091–2097.
121. Langkilde, F.W.; Jensen, N.-H., Wilbrandt, R. *J. Phys. Chem.* **1987**, *91*, 1040–1047. Langkilde, F.W.; Wilbrandt, R.; Moller, S.; Brouwer, A.M.; Negri, F.; Orlandi, G. *J. Phys. Chem.* **1991**, *95*, 6884–6894. Negri, F.; Orlandi, G.; Brouwer, A.M.; Langkilde, F.W.; Moller, S.; Wilbrandt, R. *J. Phys. Chem.* **1991**, *95*, 6895–6904.
122. Görner, H. *J. Photochem.* **1982**, *19*, 343–356.
123. Yee, W.A.; Hug, S.J.; Kliger, D.S. *J. Am. Chem. Soc.* **1988**, *110*, 2164–2169.
124. Gong, Y.; Arai, T.; Tokumaru, K. *Chem. Lett.* **1993**, 753.
125. Jensen, N.-H.; Wilbrandt, R.; Bensasson, R.V. *J. Am. Chem. Soc.* **1989**, *111*, 7877–7888.
126. Ganapathy, S.; Liu, R.S.H. *J. Am. Chem. Soc.* **1992**, *114*, 3459–3464.
127. Becker, R.S. *Photochem. Photobiol.* **1988**, *48*, 369–399.
128. Tahara, T.; Toleutaev, B.N.; Hamaguchi, H. *J. Chem. Phys.* **1994**, *100*, 786–796.
129. Freis, A.; Wegewijs, B.; Gärtner, Braslavsky, S.E. *J. Phys. Chem. B* **1997**, *101*, 7620–7627.
130. Mukai, Y.; Hashimoto, H. Koyama, Y. *J. Phys. Chem.* **1990**, *94*, 4042–4051.
131. Hashimoto, H.; Koyama, Y. *J. Phys. Chem.* **1988**, *92*, 2101–2108.
132. Dallinger, R.F.; Farquharson, S.; Woodruff, W.H.; Rodgers, M.A.J. *J. Am. Chem. Soc.* **1981**, *103*, 7433–7440. Mathis, P.; Kleo, J. *Photochem. Photobiol.* **1973**, *18*, 343–346. Truscott, T.G.; Land, E.J.; Sykes, A. *Photochem. Photobiol.* **1973**, *17*, 43–51.

# 4

# Photochemical cis-trans Isomerization from the Singlet Excited State

**V. Jayathirtha Rao**
Indian Institute of Chemical Technology, Hyderabad, India

## I.  INTRODUCTION

Photochemical cis-trans isomerization is a major area of interest in modern photo-chemical research and is also studied as part of organic photochemistry. Photo-chemical cis-trans isomerization has a major role in many photobiological phe-nomena, such as vision (rhodopsin) [1], ATP synthesis (bacteriorhodopsin) [2], phototaxis (*Chlamydomonas*) [3], and other allied processes. It has practical ap-plication in industry [4–6], i.e., vitamin A and D processes. Furthermore, it is a likely candidate for many optoelectrical and optomechanical switching and stor-age devices [7]. In this chapter, mainly various aspects of cis-trans isomerization originating from the singlet excited state will be discussed.

## II.  PHOTOCHEMICAL TRANS-CIS ISOMERIZATION INVOLVING SINGLET EXCITED STATE

Photoisomerization of the ethylenic double bond via a singlet excited state may be formulated in a simplest way as described below. The primary event is the light absorption and population of the singlet excited state ($t^{1*}$) of trans isomer.

The rotation or twisting (90°) of the double bond leads to an excited perpendicular ($p^1*$) state. This excited perpendicular ($p^1*$) singlet state gets deactivated to the ground state (internal conversion). The deactivation of $p^1*$ involving completion of 180° rotation will lead to cis isomer, whereas rotation involving going back to 0° will lead to trans isomer.

Olson was the first to postulate that optical excitation of the ethylenic double bond involves rotation around a double bond in its excited state and that this rotation leads to an observable photoisomerization process [8–10]. Olson dealt with this aspect in terms of potential energy curves and mentioned the possibility of adiabatic photoisomerization process. Later, Lewis and co-workers [11] studied the photoisomerization process of *trans*-stilbene with great interest but could not detect the *cis*-stilbene fluorescence. More recently, more detailed fluorescence studies carried out by Saltiel and co-workers [12–15] revealed that *cis*-stilbene fluoresces very weakly ($\Phi_{flu} \approx 0.0001$) and shows an inefficient adiabatic isomerization process. The singlet mechanism currently proposed by Saltiel [16] is supported by quenching studies [17–20]. The extensive studies carried out on stilbene and its analogs have already been reviewed [21–23]. Here the nature of the singlet excited state involved in the trans-cis isomerization process is dealt with.

## III. THEORETICAL STUDIES ON THE NATURE OF THE SINGLET EXCITED STATE OF OLEFIN

Mulliken [24] had pointed out that an optically excited ethylene may have a highly polarized or zwitterionic character [25]. Then the remarkable polar behavior [26] of excited state cyclohexene and cycloheptene was discussed. Dauben and others [27–29] put forward a proposal that a zwitterionic excited state is responsible for the observed stereospecific photocyclization. Indeed, these observations triggered many schools to conduct theoretical studies on the nature of the optically excited olefin. Soon after these proposals were made, Wulfman and Kumei [30] independently came up with a note that optically excited ethylene achieves a highly polarized nature upon twisting in its singlet excited state. Later, Salem carried out [31] a direct ab initio calculation (using an all-electron, self-consistent field, restriction open shell method [32]) of the diradical and zwitterionic states of s-*cis*, s-*trans*-hexatriene twisted to 90° around its central double bond. Furthermore, they found that charge separation peaks sharply at 90° with a $\pm 2°$ window and this phenomenon was termed "sudden polarization." When the same calculations [33] were carried out on s-*cis*,s-*cis* and s-*trans*,s-*trans*-hexatriene, it was found that there is no charge separation indicating the role of conformation in forming zwitterionic excited state and determining the photochemical reactivity. The sudden polarization phenomenon was extended to explain the visual transduction process [34]. The 11-*cis*-retinal bound to protein opsin (supramolecular assembly rhodopsin) in the form of a protonated Schiff base undergoes twisting upon light absorption resulting in charge separation, which ultimately triggers the transduction process [35–37].

## IV.  VARIOUS OLEFINS STUDIED FOR THEIR EXCITED STATE SUDDEN POLARIZATION PHENOMENON

Sudden polarization phenomenon was investigated in ethylene using nonempirical molecular electronic structural theory [38]. The studies predicted that $D_{2d}$ symmetry of the ground state ethylene undergoes twisting in the excited state to attain a pyramidalized shape.

Theoretical studies on the ground and excited states of styrene were conducted using PPP-CI (Pariser-Parr-Pople–configuration interaction) with geometry optimization and CNDO/S-CI (complete neglect of differential overlap/spectroscopic–configuration interaction) techniques [39]. Excitation energies, natural orbitals, and oscillator strengths for the ground and excited states were calculated and further energy-minimized equilibrium geometries of lowest singlet electronic excited states were examined [39]. Another study on the styrene excited states [40] also revealed that zwitterionic states are relatively stabilized compared to the diradicaloid states.

Energy surfaces of the excited states of butadiene were determined employing ab initio large-scale CI treatments [41]. At the 90° twist, the excited singlet state of butadiene acquires the polar zwitterionic structure. It was mentioned that the polarity of the zwitterionic excited state developed gradually over a large twist angle. Furthermore, it was predicted that vibronic coupling may play a role in the formation of zwitterionic excited singlet states [41].

The polarized nature of the singlet excited state of allene was studied by performing ab initio FORS (full optimized reaction space) MCSCF (multiconfigurational self-consistent field) and CI calculation [42]. The theoretical studies on the allene system revealed that singlet excited state is highly polarized in nature and also the polarization is developed gradually through bending and twisting motions. The theoretical studies were supported by the phenylallene photochemistry, where methanol addition takes place to phenylallene upon excitation from its singlet excited state [43].

The theoretical studies carried out to understand the photochemical behavior of β-t-butylstyrene, using open shell SCF methods involving semiempirical MINDO/3 approximation [44], indicate the highly polarized or zwitterionic nature of the singlet excited state. The model is also successful in explaining the β-t-butylstyrene photochemistry [45].

A model exact solution of PPP model was utilized to study the nature of the singlet excited state of push-pull polyenes [46]. Indeed, the studies revealed that the push-pull polyenes acquire a highly polarized or zwitterionic nature in their singlet excited state. A large change in the dipole moment was also noted in their states at rotational angle of 90°.

Theoretical investigations carried out, right from the beginning of the 1930s to the present, on several olefinic systems clearly indicate that an optically excited double bond acquires a highly polarized or zwitterionic nature in its singlet excited state and it is coupled with carbon-carbon (C=C) double-bond twisting as well as rotations. The methanol (solvent) addition to the singlet excited state of stilbenes [47], allenes [43], styrenes [42], cyclohexene [48], cycloheptenes and cyclooctenes [49] is believed to be via a highly polarized or zwitterionic excited singlet state. The same highly polarized zwitterionic excited state has been implicated as an intermediate in many organic photochemical reactions, including trans-cis photoisomerization. The following information on photochemical trans-cis isomerization of a variety of substrates and in a variety of environments deals with the singlet excited state of possible highly polarized or zwitterionic character.

## V.  HIGHLY POLARIZED OR ZWITTERIONIC EXCITED STATE OF TETRAPHENYLETHYLENE, STUDIED BY SPECTROSCOPIC TECHNIQUES

Transient absorption studies carried out using picosecond laser pulse on tetraphenylethylene (TPE) revealed that it has two absorption bands [50], one at 430 nm and another at 630 nm. It was assigned that the 630-nm band is a vertically excited singlet state and the 430-nm band is a transition from the nonfluorescent twisted singlet excited state. These assignments on the TPE excited state were soon found to be consistent with time-resolved emission experiments carried out on TPE [51]. Several substituted derivatives of TPE showed light-induced trans-cis isomerization [52] accounting for over 95% of excitation deactivation. Other excited state processes like intersystem crossing, photocyclization to phenanthrene derivative, and fluorescence were found to be small [53–58].

Lifetime measurements were carried out on TPE in various solvents utilizing picosecond absorption spectroscopy [59,60]. Two absorption bands, 430 nm and 630 nm, were found for TPE excitation. The absorption band at 630 nm was found to be very weak and the decay was very rapid, within the experimental conditions. The absorption band around 430 nm was taken up for detail studies and its decay was studied in various solvents [61]. The solvent polarity is found to have a profound role in the decay of the 430 nm absorption band of excited TPE [61] and furthermore a linear relationship between log $k$ (decay rate constant) and solvent polarity parameter $E_T$ was established. Based on these results, the authors mentioned that the 430-nm absorption band of singlet excited TPE has the "twisted zwitterionic" structure (as depicted in 1) and is nonfluorescent [61].

1

In another investigation, picosecond absorption studies were conducted on TPE singlet excited state in supercritical fluids. Studies were also carried out on ethyl-($N$,$N$-dimethylamino)benzoate TICT (twisted intramolecular charge transfer) state for comparison with the TPE studies in supercritical fluids [62]. The decay rates of excited singlet of TPE were found to be correlatable with the solvent-induced changes. The results of this study indicated that the singlet excited state of TPE is twisted and is associated with a polar or zwitterionic character.

A picosecond optical calorimetric technique was employed to understand the singlet excited state of TPE. Picosecond optical calorimetry provides the energetics involved in the solvent polarity–dependent nature of the decay of a polar or zwitterionic excited state. It was found that the dipole moment of the singlet excited TPE is around 6.3D and energy of excited TPE decreases with increase in solvent polarity. These results on the TPE singlet excited state dipole moment, decay rate, and decrease in the energy as the solvent polarity increases are cited as evidence in favor of the polar or zwitterionic character associated with the twisted TPE ($p^{1*}$) excited state [63,64].

The same technique, picosecond optical calorimetry, was employed to study the kinetics of $p^{1*}$ (twisted zwitterionic singlet excited state) decay in various TPE derivatives [64]. The results obtained indicate that two excited states are involved: 1.) the vertically excited fluorescent state and 2.) the twisted zwitterionic excited state ($p^{1*}$). Furthermore, they indicated that these two states are in equilibrium in nonpolar solvents.

Time-resolved microwave conductivity technique has been employed to study the photoexcited state behavior of TPE and tetramethoxy TPE. The change in dipole moment upon excitation was found to be in the order of 7.5D for these compounds [65]. This has been cited as direct evidence for the twisted dipolar or zwitterionic nature of the singlet excited state (phantom state).

Piotrowak and co-workers [65b] reported zwitterionic nature of the singlet excited state of biphenanthrylidene (bi-4H-cyclopenta[def]phenanthren-4-ylidine; abbreviated as BPH) is an analog of TPE. The illustration is given in Scheme 1. This is the first direct observation reported on a singlet diradical to zwitterion transition. The assignment of zwitterionic state stems from the dc and microwave time–resolved conductivity experiments.

Scheme: 1

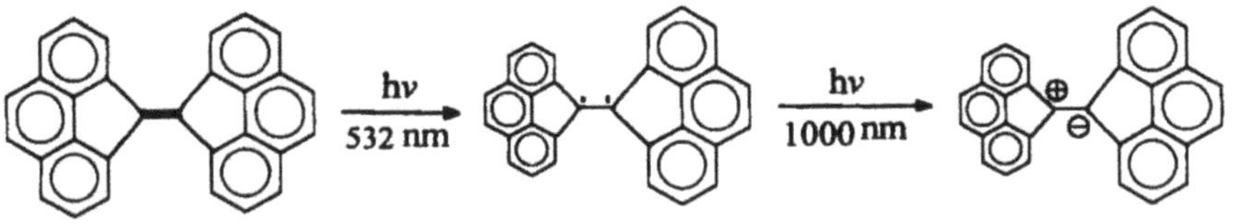

## VI.   NATURE OF SINGLET EXCITED STATE OF OLEFINS STUDIED BY MEASUREMENT OF DIPOLE MOMENTS

Mathies and Stryer have determined the ground and excited state dipole moments of all-*trans*-retinal (**3**), 11-*cis*-retinal (**4**), *n*-butylamine Schiff bases of retinals (**5, 6**) [66] by adopting the intense electric field–induced perturbation of UV-visible absorption spectrum [67,68]. They observed that there is a big change in dipole moment upon excitation ($\Delta\mu$) to singlet excited state (Exhibit 1). They proposed that a negative charge moves toward the aldehyde or Schiff base linkage part of the molecule upon excitation. They further proposed that the change in dipole moment upon excitation ($\Delta\mu$) has a prominent role to play in the trans-cis isomerization as well as chromophore–protein interactions leading to visual transduction process.

Similarly, the retinol acetate was taken up for dipole moment and fluorescence studies [69]. The change in dipole moment upon excitation was found to be 2.7D for retinol acetate. It was interpreted that the change in dipole moment upon excitation makes the retinol acetate molecule to interact more efficiently

**Exhibit 1**

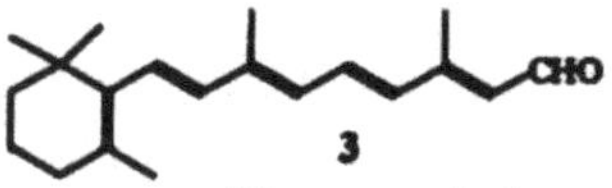

**3**

All trans retinal

$\mu g = 3.5\,D$

$\Delta\mu = 15.6\,D$

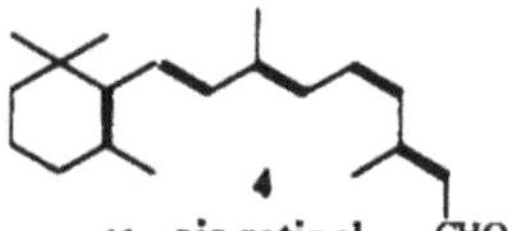

**4**

11–cis retinal   CHO

$\mu g = 10.3\,D$

$\Delta\mu = 12.7\,D$

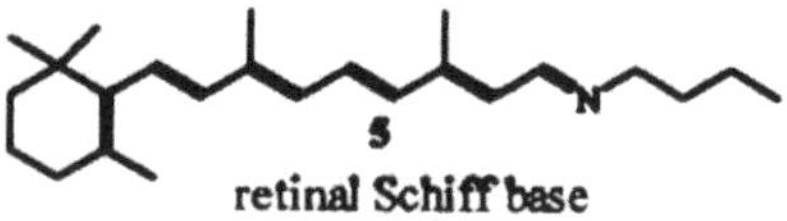

**5**

retinal Schiff base

$\mu g = 1.3\,D$

$\Delta\mu = 9.9\,D$

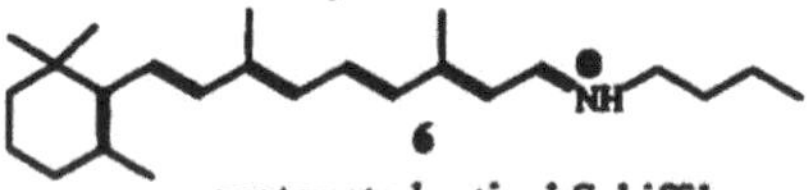

**6**

protonated retinal Schiff base

$\mu g = 6.2\,D$

$\Delta\mu = 12.0\,D$

$\mu g =$ ground state dipole-moment

$\Delta\mu =$ change in dipole moment upon excitation

with the polar medium and hence the decrease in the fluorescence and also decreases the barrier for the trans-cis isomerization.

Change in the dipole moment and polarizability upon electronic excitation
was determined for various linear polyenes by adopting electric field–induced
changes in the optical absorption spectrum [70]. Polyenes studied (Exhibit 2)
were diphenylbutadiene (DPB; **7**), diphenylhexatriene (DPH; **8**), diphenyloctate-
traene (DPO; **9**), diphenyldecapentaene (DPD; **10**), and all-*trans*-retinal (**3**). Exhibit 2 describes the experimental values determined. It is proposed that (based
on the results obtained) the excited state dipole moments determined on these
polyene systems have a role to play in the mechanism of trans-cis photoisomerization.

**Exhibit 2**

7

DPB

$\Delta\mu = 1.4$ D

$\Delta\bar{\alpha} = 53.2$

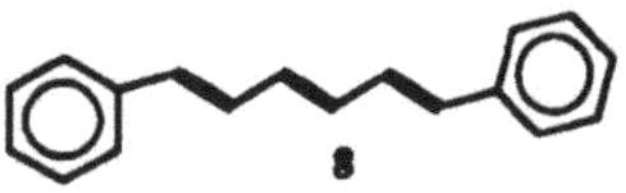

8

DPH

$\Delta\mu = 1.8$ D

$\Delta\bar{\alpha} = 91$

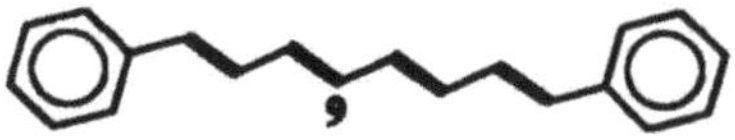

9

DPO

$\Delta\mu = 2.9$ D

$\Delta\bar{\alpha} = 116$

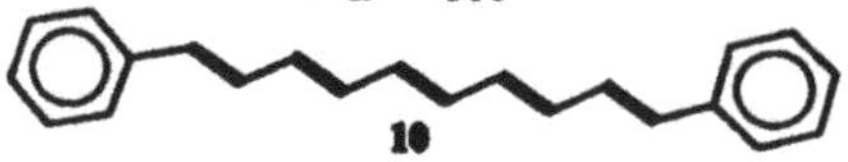

10

DPD

$\Delta\mu = 3.5$ D

$\Delta\bar{\alpha} = 140$

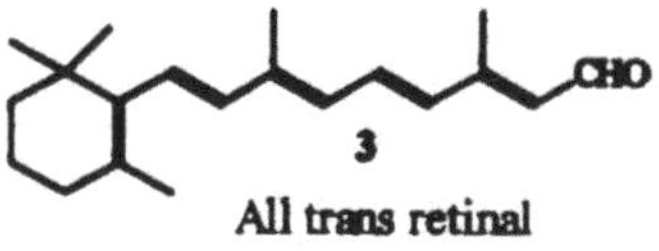

3

All trans retinal

$\Delta\mu = 13.2$ D

$\Delta\bar{\alpha} = 180$

$\Delta\mu$ = change in dipole-moment upon excitation

$\Delta\bar{\alpha}$ = change in polarizability upon excitation

**Exhibit 3**

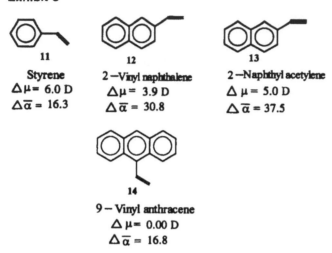

|  |  |  |
|---|---|---|
| **11** | **12** | **13** |
| Styrene | 2 – Vinyl naphthalene | 2 – Naphthyl acetylene |
| $\Delta\mu = 6.0$ D | $\Delta\mu = 3.9$ D | $\Delta\mu = 5.0$ D |
| $\Delta\bar{\alpha} = 16.3$ | $\Delta\bar{\alpha} = 30.8$ | $\Delta\bar{\alpha} = 37.5$ |

**14**

9 – Vinyl anthracene

$\Delta\mu = 0.00$ D

$\Delta\bar{\alpha} = 16.8$

   A different set of chromophores [71], styrene (**11**), 2-vinylnaphthalene (**12**), naphthylacetylene (**13**), and 9-vinylanthracene (**14**) were taken up to determine the excited state dipole moments and polarizabilities using the method of electric field–induced changes in the absorption spectrum and the values determined are given in Exhibit 3. The results obtained on these determinations suggested that the singlet excited state acquires a higher degree of charge transfer nature and this may be responsible for the fast protonation reactions in the excited state. Excited state dipole moments were determined for various molecules falling into the category of twisted intramolecular charge transfer excited state and are discussed at length in several reviews [72–75].

## VII.  OLEFINS UNDERGOING TRANS-CIS PHOTOISOMERIZATION VIA A HIGHLY POLARIZED SINGLET EXCITED STATE

Polyenes (**15–30**) shown in Exhibit 4 exhibits an interesting aspect of photoisomerization. They undergo solvent polarity dependent trans-cis isomerization [76–87]. In a nonpolar solvent like hexane, photoisomerization is found to be restricted to trisubstituted double bond (formation of 13-cis, 9-cis) isomers in vitamin A series, whereas in a polar solvent like MeOH or EtOH or acetonitrile, both trisubstituted and disubstituted double bonds (formation of 7-cis, 11-cis, 9-cis, and 13-cis) undergo isomerization. It is clear from these studies that solvent polarity plays a role in interacting with the singlet excited state of these polyenes

Exhibit 4

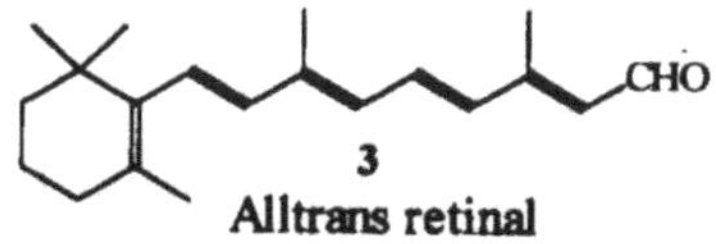

**3**
Alltrans retinal

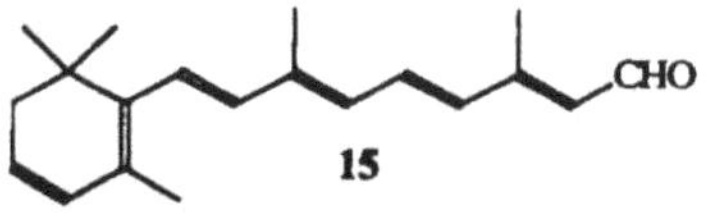

**15**
3,4 —Dehydro all trans retinal

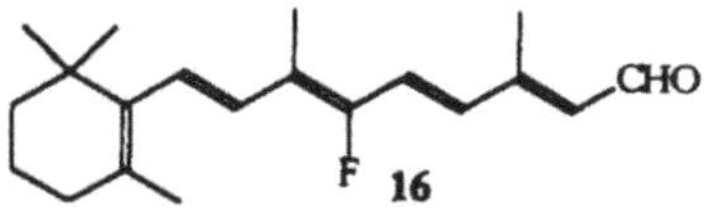

**16**
10 — Fluoro retinal

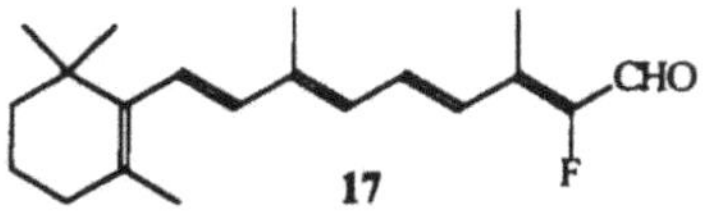

**17**
14 — Fluoro retinal

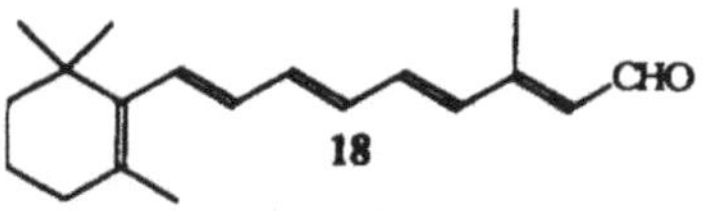

**18**
9—Demethyl retinal

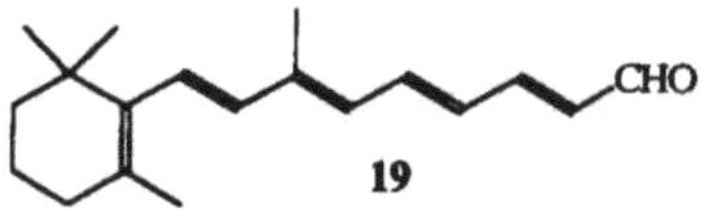

**19**
13 —Demethyl retinal

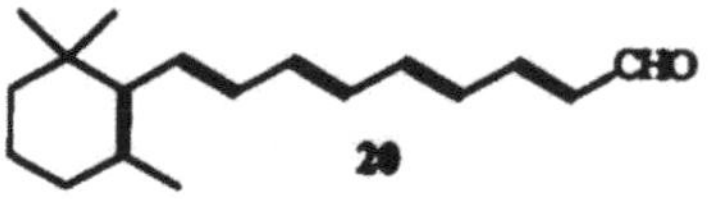

**20**
9,13 — Didemethyl retinal

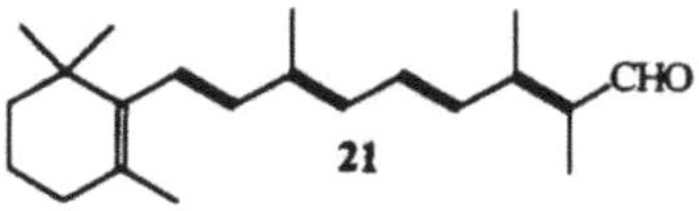

**21**
14—Methyl retinal

**22 - 27**

**28**
Methylretinoate

R = phenyl;o-tolyl;piperonyl;2-chloro-
fluorophenyl;3,4,5-trimethoxyphenyl

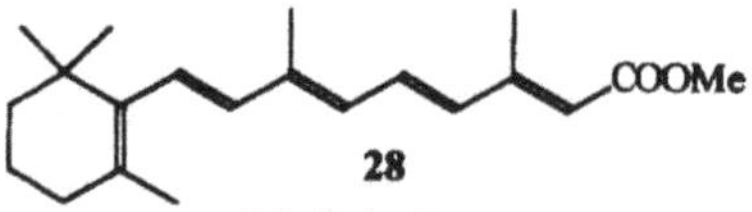

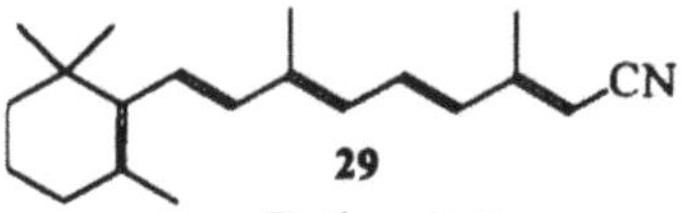

**29**
Retinonitrile

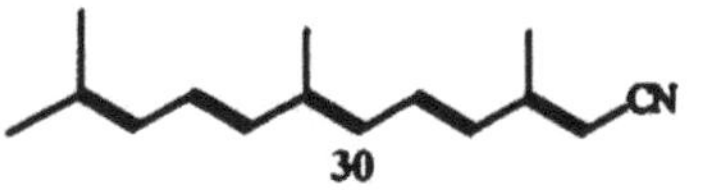

**30**
Pentenenitrile

and this interaction has an effect on the twisting process of the double bonds. Liu and co-workers [88] suggested a rationale that the singlet excited state of these polyenes acquires a highly polarized or zwitterionic character, and furthermore its interaction with medium will influence the isomerization process (Scheme 2).

Regioselective photoisomerization is identified in several of the substrates

Scheme: 2

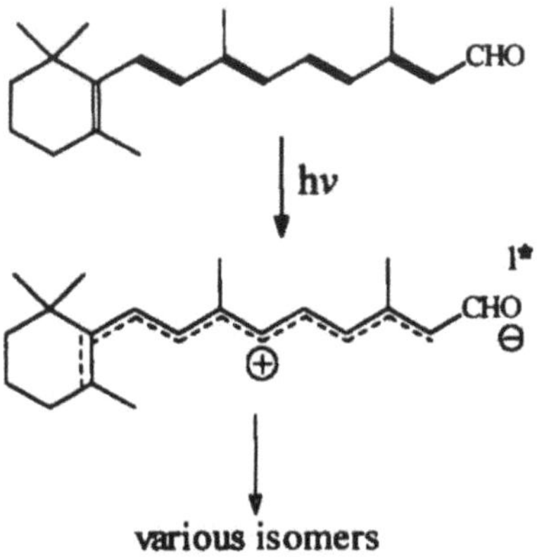

various isomers

studied (as shown in Exhibit 5) [89–93]. The photoisomerization observed in all
of these polyene substrates (Exhibit 5) was explained by involving a zwitterionic
singlet excited state. Furthermore, the regioselectivity was explained based on
the relative ease of formation and relative stabilization of the zwitterionic excited
state intermediate, shown as **46–50** in Exhibit 5.

## VIII.  PHOTOISOMERIZATION IN ARYL-SUBSTITUTED ETHYLENE DERIVATIVES

Novel "one-way" cis → trans isomerization is discussed by Arai and Tokumaru
[94] in anthryl and other aryl-substituted ethylenes and will not be discussed
here. The trans → cis isomerization generally does not occur from the triplet
excited state [94] in anthryl ethylene derivatives and it may be possible from the
singlet excited state. A variety of anthryl-substituted ethylene derivatives have
been prepared (Exhibits 6 and 7) to study the photoisomerization process. 4-
Substituted styrylanthracene (Exhibit 6) was found to undergo trans-to-cis and
cis-to-trans photoisomerization [95], whereas the parent styrylanthracene
[94,96,97] undergoes only cis-to-trans isomerization. Evidently, it is the 4-substi-
tution on the phenyl ring of styrylanthracene that has the effect in bringing about
the trans-to-cis and cis-to-trans photoisomerization. All of these trans-to-cis isomerizations were interpreted as involving singlet excited state with highly polarized
or charge transfer character. Quite a few compounds (Exhibit 6; shown in Table
1) are found to exhibit a solvent polarity effect on fluorescence [95a,98,99,100]
and also on isomerization. In all of these styrylanthracene derivatives, it is proposed that the singlet excited state acquires highly polarized or charge transfer

**Exhibit 5**

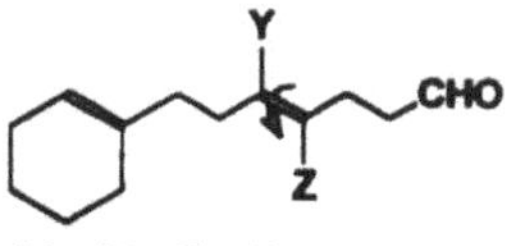

31 = Y = Z = H
32 = Y = H; Z = Me
33 = Y = Me; Z = H

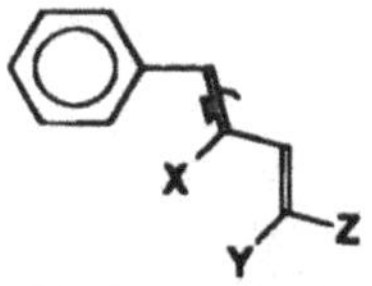

34 = X = Y = H; Z = CHO
35 = X = H; Y = F; Z = COOR
36 = X = F; Y = H; Z = COOR
37 = X = F; Y = F; Z = COOR

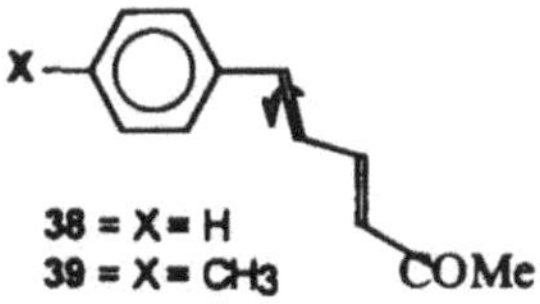

38 = X = H
39 = X = CH3
40 = X = OMe
41 = X = NMe₂

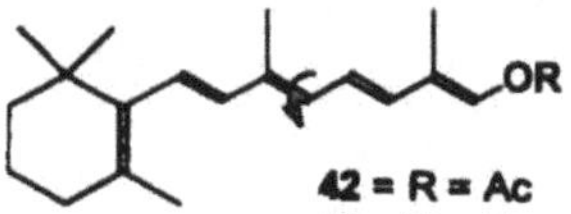

42 = R = Ac
43 = R = H

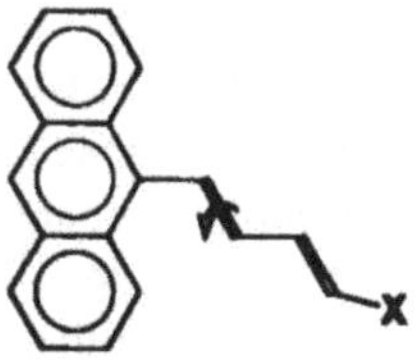

44 = X = COOMe
45 = X = CN

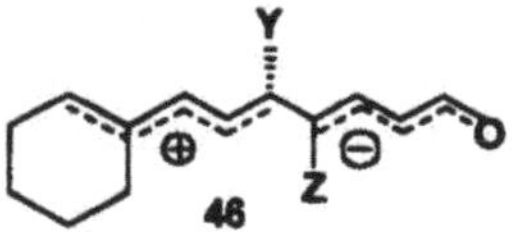

46

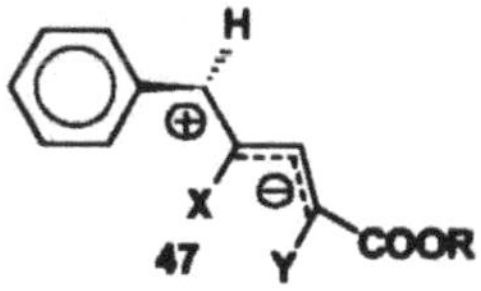

47

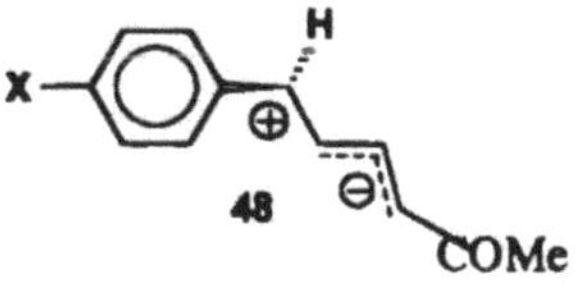

48

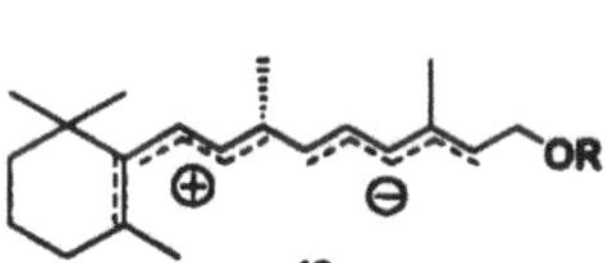

49

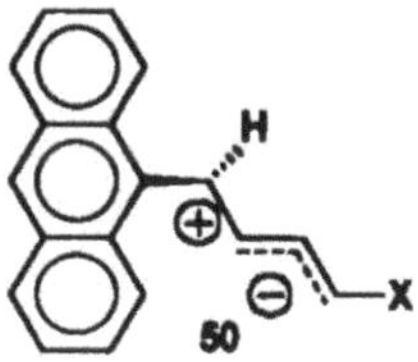

50

(The arrow represents the site of isomerization; Zwitterionic singlet excited
state proposed for the above molecules undergoing *cis-trans* isomerization)

**Exhibit 6**

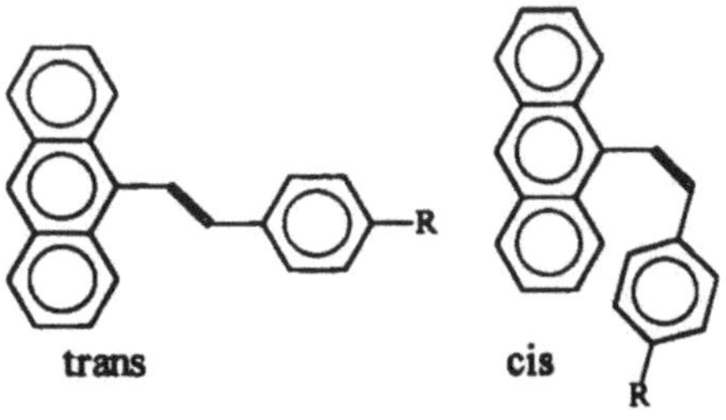

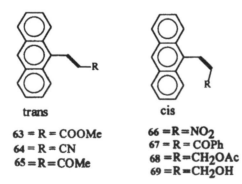

trans                                      cis

51 = R = NMe₂                    57 = R = CN
52 = R = OMe                     58 = R = Br
53 = R = NH₂                     59 = R = Benzoyl
54 = R = Me                      60 = R = 4- sulfonylphenyl
55 = R = Ph                      61 = R = NO₂
56 = R = Cl                      62 = R = Formyl

character and this rationale is supported by the substituent effect on the fluorescence and isomerization data (Table 1) generated. Many anthryl derivatives having electron acceptor or donor groups are found to form intramolecular excited state complexes (intramolecular exciplexes) leading to highly polar or charge transfer excited state [101,103]. Indeed, Gorner and co-workers [104] proposed a highly polar excited state involving intramolecular electron transfer in a stilbene derivative and discussed its possible role in isomerization process. The trans-to-cis photoisomerization is more interesting in the case of 9-anthrylethylene derivatives (Exhibit 7). In these derivatives, the electron withdrawing end group (Exhibit 7; R = COOMe; CN; COCH₃; NO₂) determines the trans-to-cis isomeriza-

**Exhibit 7**

trans                                      cis

63 = R = COOMe                   66 = R = NO₂
64 = R = CN                      67 = R = COPh
65 = R = COMe                    68 = R = CH₂OAc
                                 69 = R = CH₂OH

**Table 1**   Fluorescence and Isomerization Data on Anthrylstyrene
Derivatives in Nonpolar and Polar Solvents [95a,98–100]

| Compound | Solvent | $\Phi_{fluo}$ | $\Phi_{iso}$ |
|---|---|---|---|
| 51 | Cyclohexane | 0.28 | 0.24 |
| 51 | Acetonitrile | 0.007 | 0.22 |
| 52 | Cyclohexane | 0.49 | 0.02 |
| 52 | Acetonitrile | 0.004 | 0.33 |
| 53 | Methylcyclohexane | 0.23 | <0.01 |
| 53 | Acetonitrile | 0.002 | 0.18 |
| 54 | Cyclohexane | 0.46 | <0.01 |
| 54 | Acetonitrile | 0.38 | 0.13 |
| 55 | Cyclohexane | 0.54 | <0.01 |
| 55 | Acetonitrile | 0.35 | 0.41 |
| 56 | Cyclohexane | 0.47 | — |
| 56 | Acetonitrile | 0.13 | — |
| 57 | Cyclohexane | 0.48 | 0.2 |
| 57 | Acetonitrile | 0.024 | 0.24 |
| 58 | Methylcyclohexane | 0.27 | — |
| 58 | Acetonitrile | 0.24 | — |
| 59 | Cyclohexane | 0.47 | 0.2 |
| 59 | Acetonitrile | 0.1 | 0.24 |
| 60 | Cyclohexane | 0.51 | 0.06 |
| 60 | Acetonitrile | 0.026 | 0.37 |
| 61 | Cyclohexane | 0.27 | 0.13 |
| 61 | Acetonitrile | (<0.001) | 0.02 |
| 62 | Cyclohexane | 0.57 | 0.02 |
| 62 | Acetonitrile | 0.045 | 0.38 |

tion [105,106]. The wavelength-dependent very high cis (>94%) isomer formation (Exhibit 8) in these compounds (**63** to **67**) was explained based on the preferential light absorption and excitation of trans isomer. Triplet sensitization in these compounds leads only cis-to-trans isomerization but not trans-to-cis, thereby indicating the role of singlet excited state in bringing trans-to-cis isomerization. The fluorescence and the photoisomerization of these compounds carrying electron-withdrawing groups (**63–67**) are found to be sensitive to solvent polarity. The fluorescence solvatochromism, decrease in the $\Phi_{fluo}$ and increase in the $\Phi_{iso}$ are cited in favor of the involvement of a polar singlet excited state [106b] in these compounds. Compounds **68** and **69** do not have electron-withdrawing group and do not respond to solvent polarity, do not exhibit fluorescence solvatochromism, and do not undergo trans-to-cis isomerization. Compounds **68** and **69** exhibit very high fluorescence ($\Phi_{fluo} = 0.9$), indicating that fluorescence is the

Exhibit 8

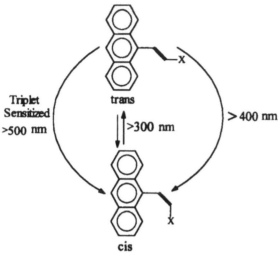

Triplet
Sensitized
>500 nm

trans

>300 nm

> 400 nm

cis

X=COOMe; CN; COMe; NO$_2$

main excited state deactivation pathway. But these compounds **68** and **69** undergo cis-to-trans isomerization upon direct excitation or by triplet sensitization. The room temperature fluorescence [106b,107] of cis and trans isomers (Fig. 1) of **63** and **64** are found to be almost identical and lead to the proposal that [106b] upon excitation trans or cis isomer will have a common excited state.

Mataga and co-workers [108] studied the effect of the intramolecular charge transfer interaction on trans-to-cis photoisomerization of 4-substituted β-(1-pyrenyl)-styrenes (Exhibit 9). In these compounds 4-substituted phenyl group acts as donor and pyrenyl group acts as acceptor. The quantum yield of isomerization, fluorescence quantum yields, lifetimes, and transient absorption measurements were carried out in various solvents on these substrates (Exhibit 9). All of these arguments were cited to show the involvement of singlet excited state in these substrates (**70, 71,** and **72**) leading to formation of cis isomer. The increase in the trans-to-cis $\Phi_{iso}$ combined with the increase in the electron donating capability of 4-substituent with the increase in solvent polarity, the decrease in the fluorescence as the increase in solvent polarity, and the picosecond transient absorption studies (absorption at ~700 nm) indicated the strong charge transfer nature associated with the singlet excited state of these pyrenyl derivatives. All of these results were explained by formulation of an excited state structure as shown **73**. Thus the postulated **73**, intramolecular charge transfer singlet excited

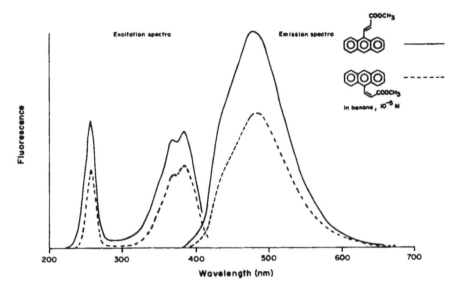

**Figure 1**  Fluorescence and fluorescence excitation spectra of trans and cis isomers of **63**. (From Ref. 106G.)

state can interact efficiently with the polar environment, thereby decreasing (lowering) the barrier for the twisting process leading to perpendicular configuration, which causes the increase in the isomerization yield and decrease in the fluorescence.

Effects of substituents and solvents on the cis-to-trans isomerization of styrylpyrene were also studied [109]. The styrylpyrene derivatives that were prepared

**Exhibit 9**

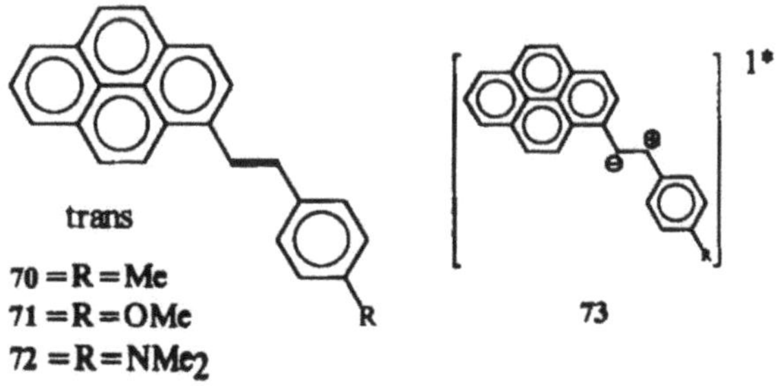

trans

70 = R = Me
71 = R = OMe
72 = R = NMe₂

73

**Exhibit 10**

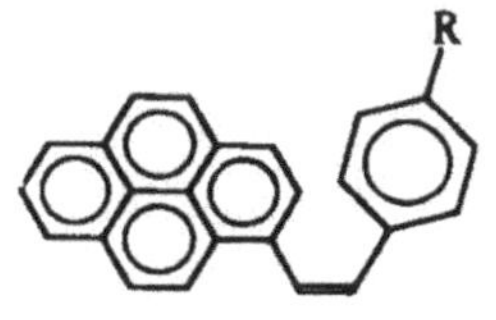

74 = R = H
75 = R = CN
76 = R = OMe

cis

are given in Exhibit 10. Fluorescence spectroscopy (steady-state and time-re-solved) was utilized to study the role of polar solvent and polar substituent on cis-to-trans isomerization involving singlet excited state. The parent styrylpyrene undergoes cis-to-trans isomerization adiabatically from the singlet excited state [109,110]. Introduction (Scheme 3) of a polar substituent (75 and 76) and use of polar solvent resulted in the cis-to-trans isomerization in the d.. )atic route. The rationale suggested is that the relative stabilization of polar p$^{1}$* (perpendicu-lar excited state) resulting from the cis$^{1}$* using polar environment and polar sub-stituent results in an energy minimum for p$^{1}$*. This p$^{1}$* quickly deactivates to ground state trans isomer as shown in Scheme 3 and as a consequence there is a decrease in fluorescence.

Pyrenylethylene derivatives were prepared (Exhibit 11) to study their pho-toisomerization process [111]. The "R" groups in these are carbomethoxy (77),

**Scheme: 3**

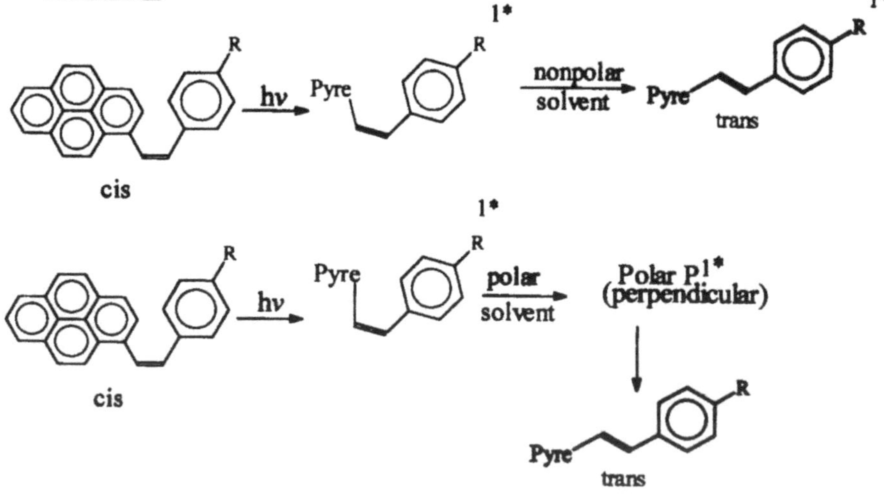

**Exhibit 11**

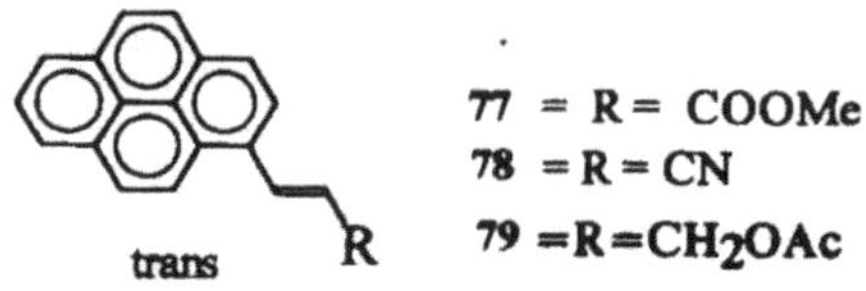

77 = R = COOMe

78 = R = CN

79 = R = CH$_2$OAc

trans    R

nitrile (**78**), and acetoxymethyl (**79**). The two groups, carbomethoxy (**77**) and nitrile (**78**), are electron withdrawing in nature. The trans-to-cis photoisomerization is found to be possible from the singlet excited state in **77** and **78**, whereas **79** did not undergo photoisomerization upon direct excitation or by triplet sensitization. The $\Phi_{iso}$ determined for **77** and **78** was found to be higher in polar solvents. The steady-state fluorescence recorded for **77** is shown in Fig. 2. The solvent polarity has profound influence on the fluorescence of **77**; in hexane it is structured fluorescence whereas in acetonitrile it is broad, structureless fluorescence. The results are interpreted by involving a polar p$^1$* state in these isomerizations.

Another interesting report in the styrylpyrene series concerns 4-nitrostyrylpyrene [112]. The 4-nitrostyrylpyrene undergoes trans-to-cis isomerization in

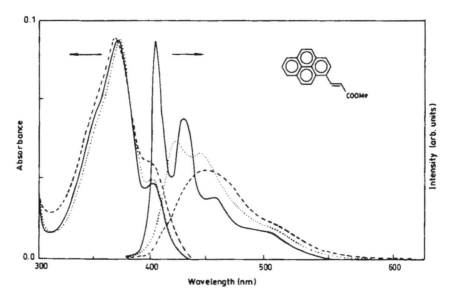

**Figure 2** Absorption and fluorescence spectra of **77** in various solvents: in hexane (—); 1,4-dioxane (······); and acetonitrile (------). (From Ref. 111.)

Scheme: 4

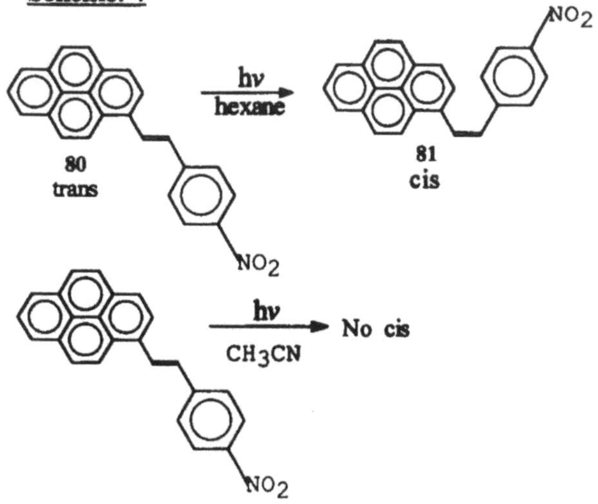

nonpolar solvents like hexane upon direct excitation, whereas in polar acetonitrile solvent it does not (Scheme 4). The 4-nitrostyrylpyrene exhibits fluorescence solvatochromism and based on these observations it was suggested that isomerization in **80** is via a charge transfer singlet excited state.

## IX.  ONE-WAY TRANS-CIS ISOMERIZATION

There are very few citations for the photochemical trans → cis one-way isomerization process. In this aspect, some findings are discussed below to foster understanding about the role of various factors governing this trans-to-cis one-way isomerization. The effect of wavelength, conformation, configuration, hydrogen bonding, and other factors was taken into account to highlight the one-way trans → cis isomerization. Most heterocyclic analogs fall into this category. Lewis and co-workers [113] studied the photoisomerization process in furyl- and imidazole-substituted ethylenes. Methylfurylacrylate (**82**), methylurocanate (**83**), and methylcinnamate (**84**) were prepared to study photoisomerization in the presence and absence of Lewis acid (Table 2; Scheme 5). Perusal of Table 2 indicates that direct excitation of compounds **82**, **83**, and **84** leads to trans-to-cis isomerization with almost a 1:1 photostationary state composition. The same reaction was conducted in the presence of Lewis acids (Lewis acid–olefin complex), which shows a high preference (Table 2) for the formation of cis isomer. The selective trans-to-cis isomerization observed in these substrates is explained by considering the following factors: 1.) Lewis acid complex formation with trans isomer is more

**Table 2** Photostationary State and Quantum Yields of Isomerization [113]

| Compound | Lewis acid | $\lambda_{ex}$, nm | % cis | $\Phi_{trans\text{-}cis}$ |
|---|---|---|---|---|
| **84**-trans | None | 313 | 46 | 0.3 |
| **84**-trans | BF$_3$/etherate | 313 | 88 | 0.7 |
| **84**-cis | None | 313 | — | 0.29 |
| **84**-cis | BF$_3$/etherate | 313 | — | 0.28 |
| **82**-trans | None | 313 | 52 | 0.5 |
| **82**-trans | BF$_3$/etherate | 365 | 94 | 0.69 |
| **82**-trans | EtAlCl$_2$ | 365 | 92 | — |
| **82**-cis | None | 313 | — | 0.45 |
| **82**-cis | BF$_3$/etherate | 365 | — | 0.33 |
| **83**-trans | None | 254 | 85 | — |
| **83**-trans | BF$_3$/etherate | 254 | 95 | — |

Scheme: 5

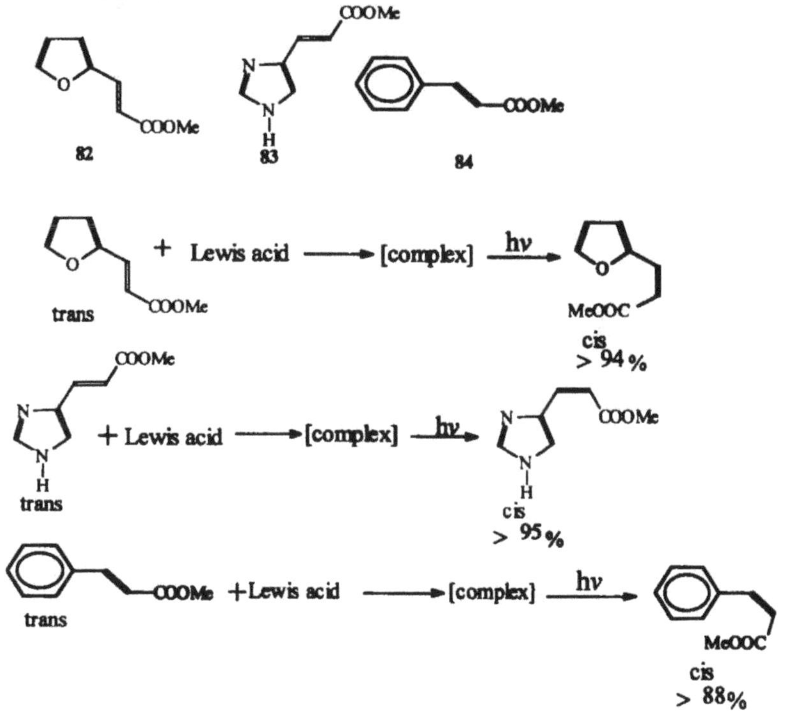

82      83      84

trans + Lewis acid ⟶ [complex] $\xrightarrow{h\nu}$ cis > 94%

trans + Lewis acid ⟶ [complex] $\xrightarrow{h\nu}$ cis > 95%

trans + Lewis acid ⟶ [complex] $\xrightarrow{h\nu}$ cis > 88%

facile than with cis isomer; 2.) trans isomer absorbs more light than cis isomer; 3.) there is higher $\Phi_{iso}$ for the trans isomer complex; 4.) and in the case of imidazol derivative, the cis isomer forms a better H-bonding.

Pyridyl pyrrolinones are reported to be resistant to thermal and photochemical isomerization in polar solvents like alcohol and acids. It was presumed that hydrogen bonding plays an important role in these one-way trans-to-cis isomerizations [114]. Lewis and co-workers [115] prepared several pyridylpropenamide derivatives (Exhibit 12) to understand the role of hydrogen bonding and conformation on the trans-to-cis one-way photoisomerization. Photoisomerization results on all compounds are listed in Table 3. Compounds **85–88** are capable of forming intramolecular hydrogen bonds whereas **89** and **90** are not. Persual of Table 3 indicates that cis isomer content is high for the intramolecular hydrogen-bonding compounds and not for others, thereby highlighting the importance of intramolecular hydrogen bonding. By changing solvent to dimethylsulfoxide, it was found that cis-to-trans isomerization (Table 3) was observable. The rationale suggested is that dimethylsulfoxide disrupts intramolecular hydrogen bonding, leading the molecule to attain a different conformation. From there the molecule undergoes cis-to-trans isomerization. The rationale suggested is supported by the NOE studies carried out. Other heterocyclic compounds capable of forming intramolecular hydrogen bonding in cis isomer were investigated by Arai and Tokumaru [116]. They prepared the compounds shown in Exhibit 13 for studying the trans-to-cis one-way isomerization involving hydrogen bonding as a governing factor. Indeed, both compounds exhibited highly selective trans-to-cis isomerization upon direct excitation. Triplet sensitization studies [116a] indicated that a

**Exhibit 12**

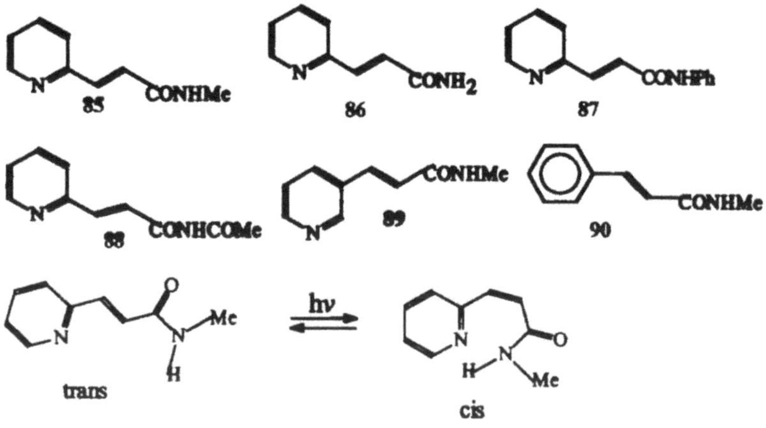

**Table 3** Photostationary State Composition and Quantum Yield of Isomerization Data on Compounds **85–90** [115]

| Compound | Solvent | % cis | $\Phi_{trans\text{-}cis}$ | $\Phi_{cis\text{-}trans}$ |
|---|---|---|---|---|
| 85 | $CH_2Cl_2$ | 99 | 0.31 | <0.001 |
| 85 | $CH_3CN$ | 98 | 0.22 | <0.001 |
| 85 | $H_2O$ | 99 | — | <0.001 |
| 85 | Tetrahydrofuran | — | 0.05 | 0.07 |
| 85 | EtOH | — | 0.12 | 0.04 |
| 85 | MeOH | — | 0.14 | 0.011 |
| 85 | $(CH_3)_2SO$ | 77 | — | 0.06 |
| 86 | $CH_2Cl_2$ | 93 | 0.18 | 0.01 |
| 86 | MeOH | — | 0.11 | — |
| 86 | $(CH_3)_2SO$ | — | 0.09 | — |
| 88 | $CH_2Cl_2$ | 93 | 0.59 | <0.01 |
| 88 | MeOH | — | 0.31 | — |
| 88 | $CH_2Cl_2$ | 65 | 0.01 | 0.001 |
| 89 | $CH_2Cl_2$ | 65 | 0.16 | 0.28 |
| 89 | $H_2O$ | 62 | — | — |
| 89 | $(CH_3)_2SO$ | 77 | 0.02 | 0.08 |
| 90 | $CH_2Cl_2$ | 40 | 0.034 | 0.27 |
| 90 | MeOH | — | 0.14 | — |
| 90 | $(CH_3)_2SO$ | — | — | 0.019 |

**Exhibit 13**

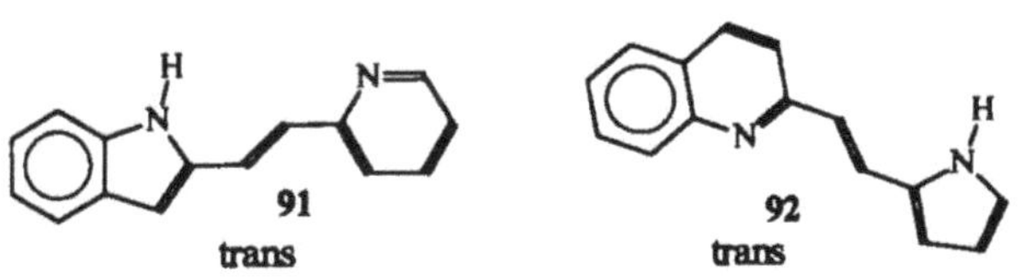

91 trans

92 trans

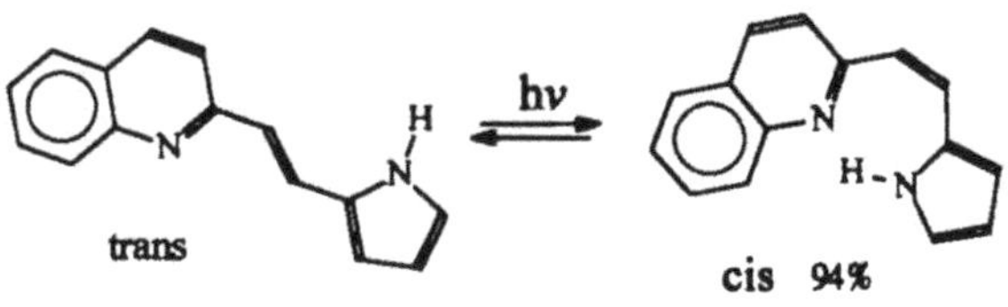

trans    $\xrightarrow{h\nu}$    cis  94%

singlet excited state is involved in these isomerizations. Interestingly, the fluorescence of compound **92**-cis isomer is red-shifted relative to the trans isomer, indicating that the molecule experiences drastic changes in its excited singlet state. The very high formation of cis isomer in these compounds is due to the formation of intramolecular hydrogen bonding in the cis isomer and further stabilizing the cis isomer relative to trans isomer. Furthermore, they suggested that the intramolecular hydrogen bond in cis isomer can cause deactivation from its singlet excited state by hydrogen atom transfer mechanism [116b].

Lewis and co-workers [117] reinvestigated one-way trans-to-cis photo-isomerization in α-pyridyl pyrrolinones (Exhibit 13). A very efficient trans-to-cis isomerization is observed. It was found that both trans and cis isomers exhibit fluorescence and the reactive state assigned is the $\pi\pi^*$ singlet excited state. The low reactivity ($\Phi_{iso}$) of cis isomer was attributed earlier due to the presence of an intramolecular hydrogen bonding. The cis isomer was found to display a red-shifted fluorescence and this observation was interpreted as meaning that the

**Exhibit 14**

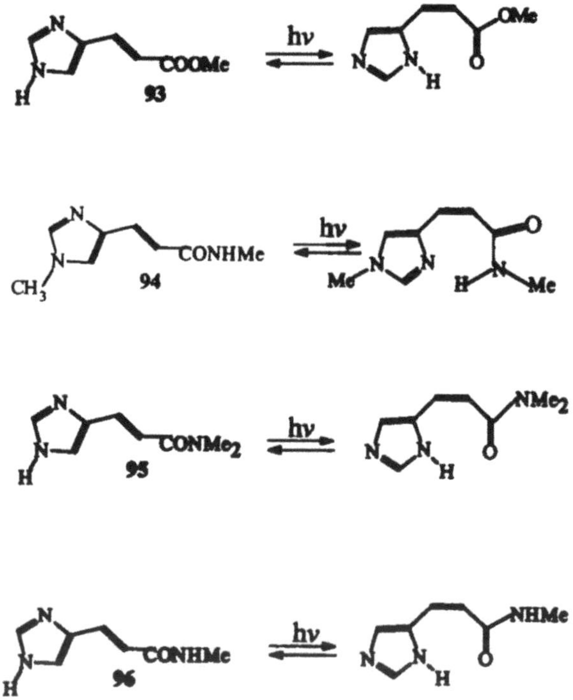

fluorescing species is a tautomer formed adiabatically. Furthermore, the rationale was supported by the transient absorption studies. Based on these observations they concluded that the low quantum yield of isomerization found in cis isomer is due to the competing adiabatic hydrogen transfer reaction [118].

Uraconate derivatives [119] were prepared to study the photoisomerization process with an emphasis on the role of hydrogen bonding. Exhibit 14 gives the uraconate derivatives taken up for study. The results of photoisomerization of these substrates are arranged in Table 4. All of the substrates are capable of forming intramolecular hydrogen bonding (Exhibit 14). The trans-to-cis $\Phi_{iso}$ and cis isomer content in photostationary state composition for all of the substrates (Table 4) indicates that cis-to-trans isomerization is also an efficient process. These results reveal that intramolecular hydrogen bonding present in cis isomer is not the only influencing factor for the high cis isomer formation observed in other cases discussed here.

A series of indolylpropeonate derivatives [120] (Exhibit 15) were taken up to study trans-to-cis isomerization and to understand the role of conformation and intramolecular hydrogen bonding capabilities toward the selective formation of cis isomer. The $\Phi_{iso}$ and photostationary state composition data generated are arranged in Table 5. All of these isomerizations are reported to be originating from the singlet excited state. The decrease in the cis isomer content and decrease in the $\Phi_{cis-trans}$ isomerization in the solvents (Table 5), like dimethylsulfoxide and EtOH, is attributed to their intramolecular hydrogen bonding disruptive capabilities leading to different conformers capable of undergoing cis-to-trans isomerization. The possible conformers participating in these isomerizations were supported by NOE and fluorescence. The structureless emission observed at 77 K in the case of 100-cis isomer was attributed to an excited state tautomer formed via intramolecular hydrogen transfer as shown in Exhibit 15.

Anthrylethylene derivatives were prepared (Exhibit 7) to study trans-cis isomerization [105,106]. Compounds 63–66 underwent very efficient trans-to-cis isomerization from the singlet excited state. The triplet sensitization is effective in bringing only cis-to-trans isomerization (Exhibit 8). It is interesting to mention

**Table 4** Photoisomerization Data on Uraconate Derivatives [119]

| Compound | Solvent | $\lambda_{irr}$, nm | % cis | $\Phi_{trans-cis}$ |
|---|---|---|---|---|
| 93 | CH$_3$CN | 254 | 85 | 0.94 |
|  | CH$_3$CN | 313 | 25 | 0.67 |
| 94 | CH$_2$Cl$_2$ | 313 | 76 | 0.48 |
|  | CH$_3$CN | 313 | 52 | — |
| 96 | CH$_3$CN | 254 | 89 | — |

**Exhibit 15**

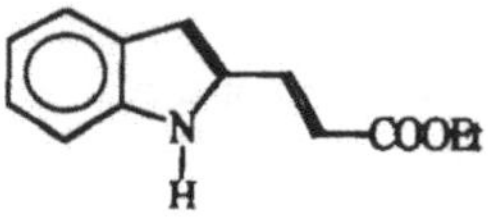


97

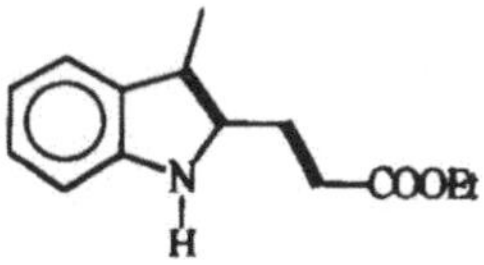


98

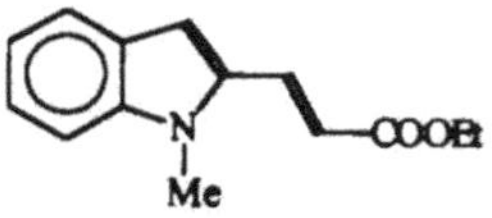


99

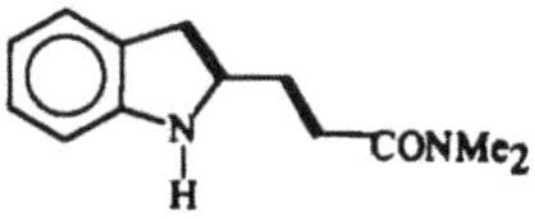


100

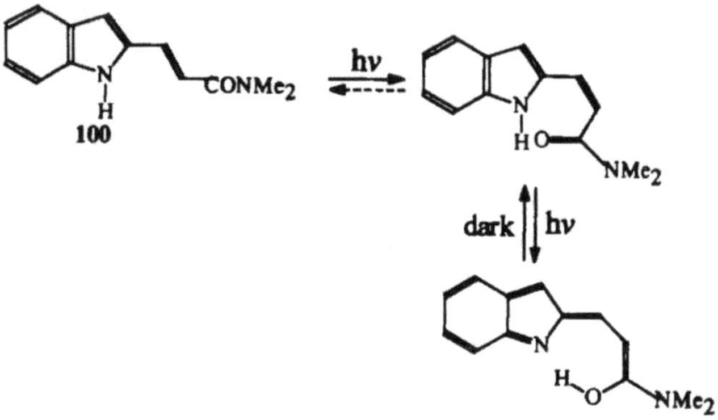


**Table 5** Photoisomerization Data of Indoloylpropeonates [120]

| Compound | Solvent | % cis | $\Phi_{trans\text{-}cis}$ | $\Phi_{cis\text{-}trans}$ |
|---|---|---|---|---|
| **97** | $C_6H_6$ | 95 | 0.65 | 0.047 |
| | $CH_3CN$ | 72 | 0.61 | 0.10 |
| | DMSO | 51 | 0.37 | 0.19 |
| | EtOH | 67 | 0.22 | 0.075 |
| **98** | $C_6H_6$ | 98 | 0.47 | 0.02 |
| | EtOH | 68 | 0.06 | 0.005 |
| **99** | $C_6H_6$ | 22 | 0.034 | 0.34 |
| | EtOH | 20 | 0.012 | 0.12 |
| **100** | $C_6H_6$ | 96 | 0.77 | 0.0049 |
| | $CH_3CN$ | 96 | 0.60 | 0.056 |
| | DMSO | 82 | 0.25 | — |
| | EtOH | 85 | 0.30 | — |

Scheme: 6

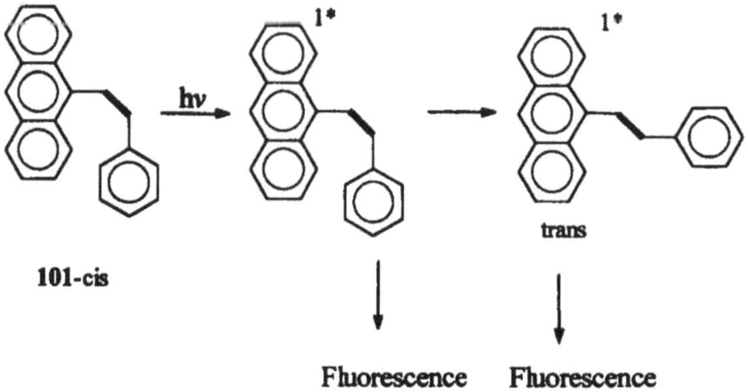

101-cis

trans

Fluorescence     Fluorescence

that these anthryl derivatives can't form hydrogen bonding (intra- or intermolecular) and do not have conformational equilibrium, but yet they display very high cis isomer (94%) formation.

## X. ADIABATIC ISOMERIZATION IN THE SINGLET EXCITED STATE

Olson [8–10] was first to postulate that photochemical trans-cis isomerization in olefins may be an adiabatic reaction. Hammond and co-workers [121] and others [122–124] reported the quantum chain cis-to-trans isomerization process from the triplet excited state of olefin. They postulated the energy transfer process from the trans isomer adiabatically formed to cis isomer in the ground state. Later Arai and Tokumaru [94] with their extensive investigations showed that cis-to-trans adiabatic photoisomerization resulted in quantum chain process from the triplet excited state in a variety of olefins. Nevertheless, the adiabatic photoisomerization in olefins originating from the singlet excited state is less studied and the same is highlighted.

Sandros and Becker [107] studied the cis-to-trans photoisomerization process in 9-styrylanthracenes from the singlet excited state. They found that decay of fluorescence emission from the cis isomer is biexponential and they explained the biexponential decay as involving cis and trans singlet excited states (Scheme 6). Furthermore, they noticed that the trans isomer formed adiabatically can undergo intersystem crossing to the triplet state and can initiate the quantum chain process. The adiabatic cis-to-trans isomerization from the singlet excited state of styrylpyrene derivatives [109,110] (Chart 10; Scheme 3) is already discussed.

Sandros and co-workers [126] prepared substituted p-styrylstilbene and studied the cis-to-trans photoisomerization process. They found that excitation of *cis, cis-p*-styrylstilbene gave both isomers cis,trans and trans,trans. The isomerization process was found to be from the singlet excited state and also adiabatic in nature (Scheme 7). Support for the proposed singlet excited state adiabatic photoisomerization process stems from the quantum yield of isomerization and time-resolved fluorescence measurements. Furthermore, they have performed quantum mechanical calculations to generate surface profiles to support the proposed adiabatic photoisomerization mechanism. This reaction is an interesting example showing twofold adiabatic photoisomerization from the singlet excited state.

Sandros and Sundahl [127] have synthesized all six isomers of 4,4-bis (3,5-di-*tert*-butylstyryl) stilbene (Exhibit 16) to study photoisomerization. Isomerization is preferred from cis-to-trans conversion and single isomerizations are more dominant. The increase in the quantum yield of isomerization resulting from an increase in the solvent polarity is interpreted to mean that a twisted singlet excited $(p^1*)$ state is polar in nature and its interaction with polar medium lowers its energy, thereby facilitating the isomerization process. They have found that **103** with central double bond in cis configuration displayed adiabatic cis-to-trans isomerization.

Detailed and careful fluorescence analysis carried out by Saltiel and co-workers [12–15] on *cis*-stilbene demonstrated that cis-to-trans isomerization is indeed adiabatic in nature. *cis*-Stilbene fluorescence is found to be composed of emission from *cis*- and *trans*-stilbene-excited singlet states.

Scheme: 7

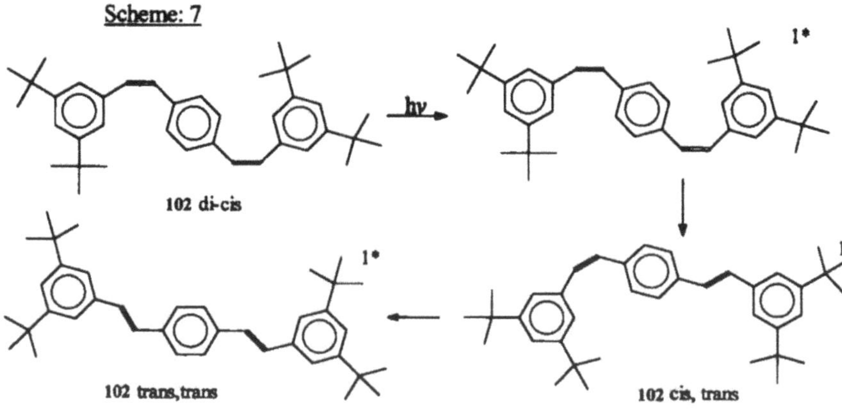

102 di-cis

102 trans,trans

102 cis, trans

Exhibit 16

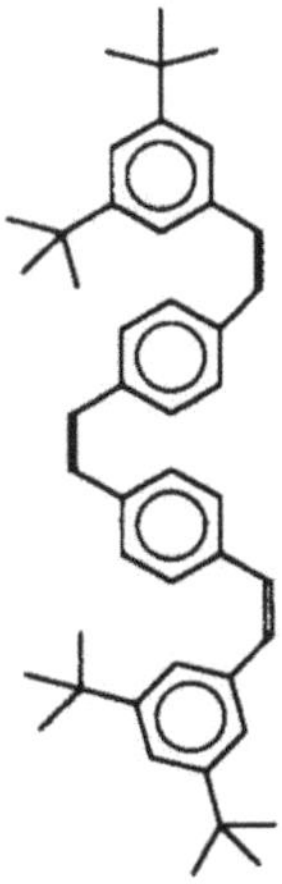

103 — tri cis 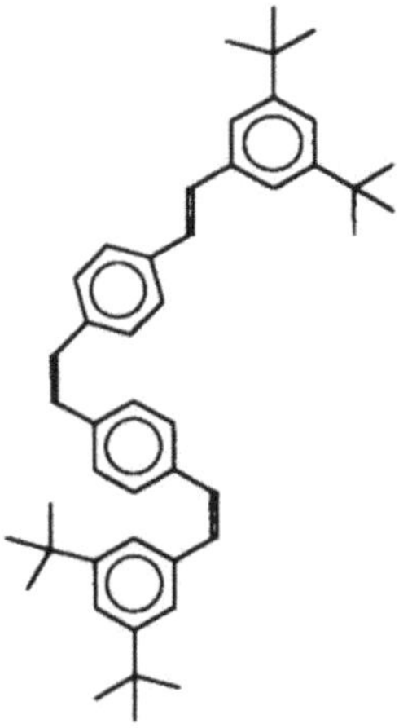          103 — cis,cis,trans 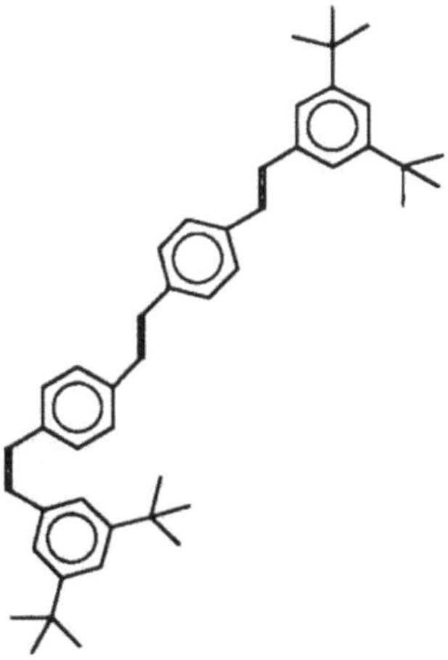          103 — cis,trans,trans

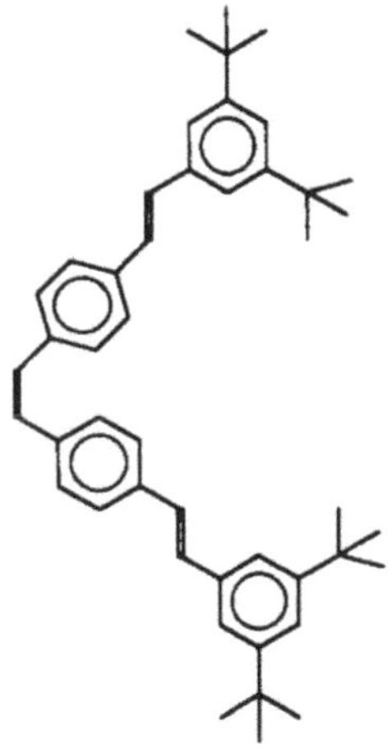

103 — trans,cis,trans 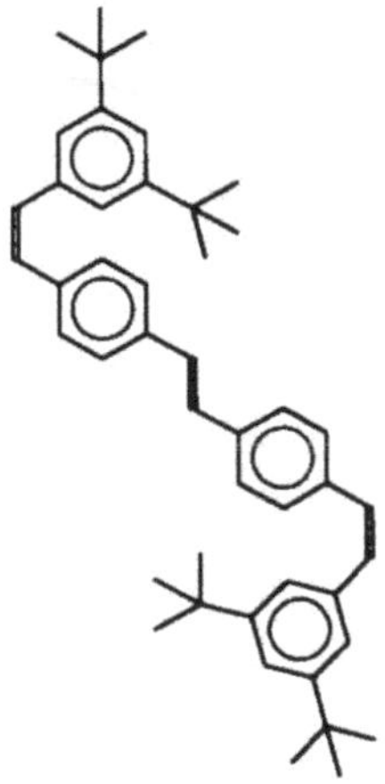          103 — cis,trans,cis 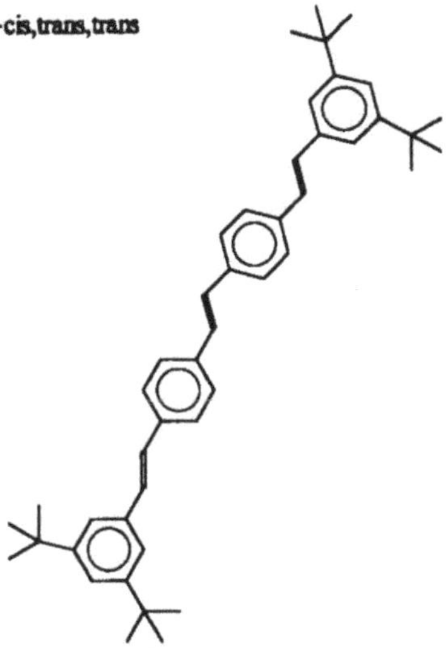          103 — all trans

2-Naphthyl- and 2-anthrylstyrenes were prepared by Saltiel and co-workers [128] to study the isomerization in terms of adiabaticity and the role of conformation. The investigations carried out on 2-naphthyl and 2-anthrylstyrenes (Scheme 8) revealed that cis-to-trans isomerization is adiabatic in nature and furthermore it is found to be conformer-specific (Scheme 8). The interesting trend observed

Scheme: 8

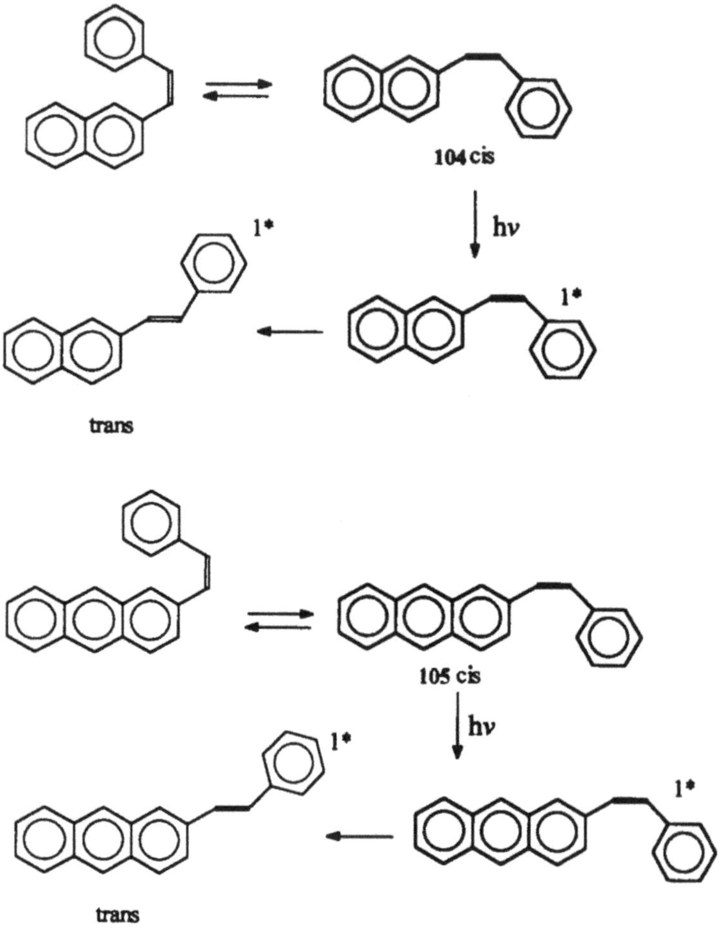

in these systems studied by Saltiel and co-workers [12–15,128] is that cis-to-trans adiabatic process in the singlet excited state becomes more efficient with anthryl as the largest aryl group.

Anthrylethylene derivatives displayed trans-to-cis isomerization from the singlet excited state as an adiabatic process (Scheme 9) [106,129]. The cis isomer formed adiabatically is shown to involve in energy transfer process to the ground state trans molecule leading to quantum chain isomerization process originating from the singlet excited state. Fluorescence, fluorescence lifetime [130], and

Scheme: 9

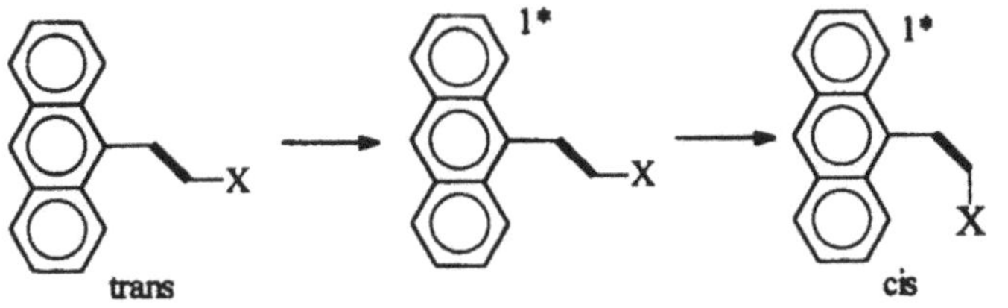

trans                                    cis

X = COOMe
X = CN

quantum yield of isomerization data are cited in favor of the mechanism proposed. This is the first report showing singlet excited trans isomer forming singlet excited cis isomer and further excited cis isomer involves in energy transfer process.

## XI. TRANS-CIS ISOMERIZATION IN NATURAL AND ARTIFICIAL SUPRAMOLECULAR ASSEMBLIES

Photochemical cis-trans (geometrical) isomerization of retinal and its analogs has been studied in great detail in solution phase [76–87,131–133]. For retinyl compounds photochemical cis-trans isomerization is found to be solvent polarity–dependent: higher selectivity is observed in nonpolar solvent like hexane and cyclohexane, leading to selective conversion of trisubstituted trans double bonds to cis double bonds, whereas in polar solvents, such as acetonitrile and methanol, all of the double bonds undergo trans-to-cis isomerization exhibiting no selectivity. The situation is totally different when the same retinal molecule is transfered and bound to protein (supramolecule) residue. A remarkable regiospecificity is observed in these supramolecular assemblies. Nature has engineered these supramolecular assemblies to perform biological functions. The highest degree of selectivity achieved in these natural supramolecular assemblies involving photochemical cis-trans isomerization is highlighted below with a few examples. The origin of control or cause in these highly selective photochemical cis-trans isomerizations in supramolecular assemblies is a debatable subject.

## XII. RHODOPSIN

Rhodopsin is present in the eye and primarily responsible for vision. Rhodopsin is a polytopic integral membrane protein, spanning the membrane seven times,

Scheme 10:

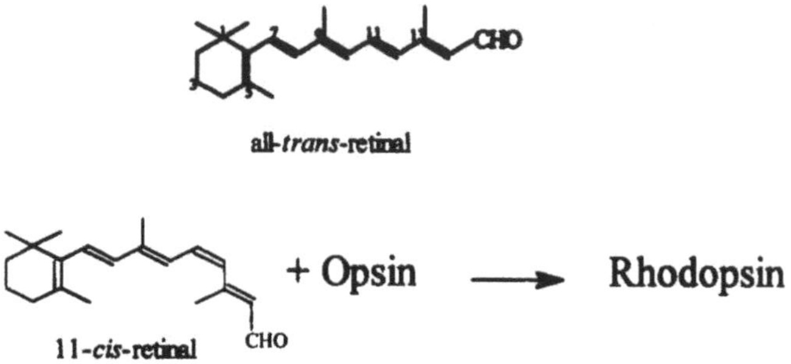

all-*trans*-retinal

11-*cis*-retinal     CHO

+ Opsin ⟶ Rhodopsin

containing 348 amino acids with an 11-*cis*-retinal as a chromophore linked to the supramolecule (opsin protein) at Lys-296 via a protonated Schiff base [134–136]. The chromophore retinal is embedded in the trans membrane protein segments and it is held inside the supramolecule with hydrophobic interactions and also via a protonated Schiff base linkage [137–140]. The red shift observed in the chromophore absorption maxima of the supramolecule (rhodopsin) is rationalized based on the influence of the polarity of the medium and also the proximity of the charged species originating from the supramolecule (protein) and assembled around the chromophore [141–144]. 11-*cis*-Retinal is found to be the perfect fit in forming rhodopsin (Scheme 10) and is the natural selection [145]. Other cis isomers and di-cis isomers are also found to form pigments (supramolecular assemblies) [137,138]. However, all-*trans*-retinal does not form pigment (supramolecular assembly). The photochemical reaction involved is the conversion of 11-*cis*-retinal, bound to supramolecule, to all-trans isomer (Scheme 11). The analog pigments formed by the binding of cis (other than 11-cis) and di-cis isomers of retinal to opsin also undergo photochemical cis-to-trans isomerization. The quan-

Scheme: 11

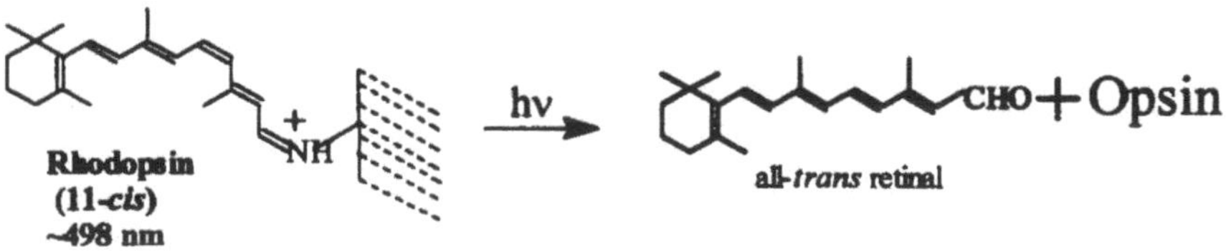

Rhodopsin
(11-*cis*)
~498 nm

hv

all-*trans* retinal + Opsin

tum yield of photoisomerization is higher (0.67) [146] for the 11-*cis*-retinal bound to opsin compared to solution phase quantum yield (0.2) of isomerization [147]. In almost all visual pigments (rhodopsins) it is found to be 11-cis to all-trans [148] isomerization except in cephalopod retinoid pigment. Rhodopsin and other analog rhodopsins exhibit one photon–one bond isomerization through a possible singlet excited state [149,150]. The dynamics of cis-trans photoisomerization of rhodopsin using time-resolved techniques reveals that excited state dynamics appear about 20 fs after photon absorption [151].

## XIII.  BACTERIORHODOPSIN

Bacteriorhodopsin is the pigment present in the membranes of the *Halobacterium halobium* (halophilic bacteria), a light-driven proton pump that synthesizes ATP [2]. Bacteriorhodopsin is a polytopic integral membrane protein, spanning the membrane seven times, containing 248 amino acids with all-*trans*-retinal as chromophore linked to Lys-216 in the form of a protonated Schiff base [152]. The structure of the bacteriorhodopsin is assigned based on the diffraction data [152–154]. The large red shift observed in the bound chromophore absorption band of the supramolecule is attributed to the influence of the charged groups originating from the supramolecule (protein) and pointed toward the chromophore around certain specific sites [155–157]. The photochemical reaction involved is the all-trans to 13-cis isomerization [158] (Scheme 12).

Scheme: 12

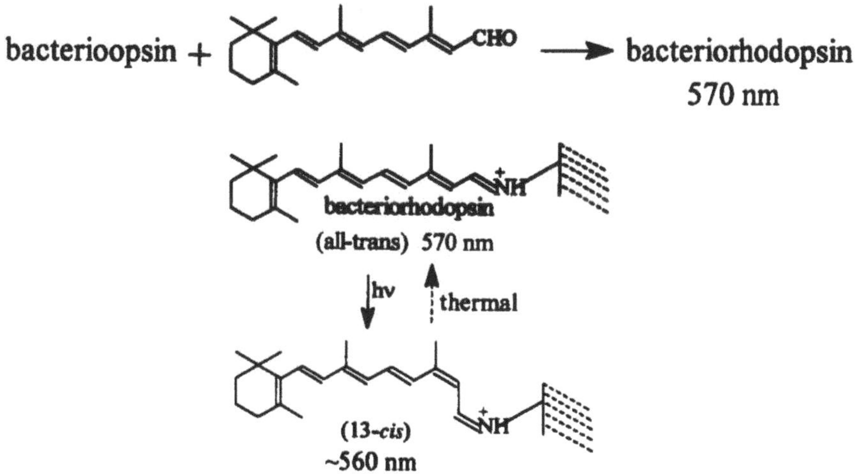

bacterioopsin +  →  bacteriorhodopsin
                                         570 nm

bacteriorhodopsin
(all-trans) 570 nm

hv      thermal

(13-*cis*)
~560 nm

Scheme: 13

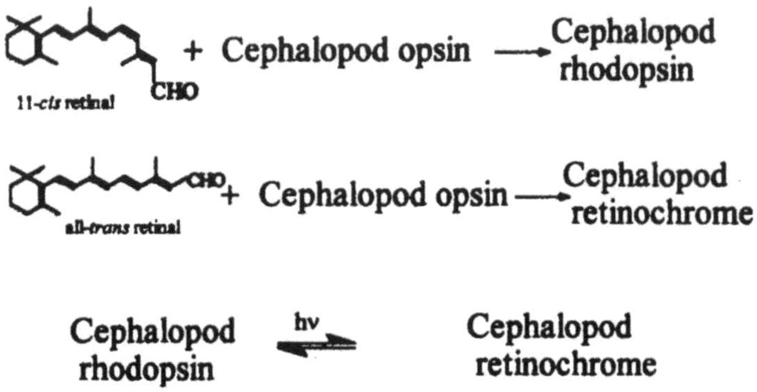

11-*cis* retinal + Cephalopod opsin ⟶ Cephalopod rhodopsin

all-*trans* retinal + Cephalopod opsin ⟶ Cephalopod retinochrome

Cephalopod rhodopsin ⇌ (hv) Cephalopod retinochrome

## XIV.  THE TWO PIGMENTS OF RETINOCHROME

Pigments isolated from cephalopod retina are found to be very interesting [159,160]. They contain two pigments, rhodopsin and retinochrome. The two pigments have the same supramolecule (protein), have common chromophore retinal, but they differ in their bound chromophore stereochemistry (Scheme 13). Cephalopod rhodopsin has 11-*cis*-retinal as chromophore where as retinochrome has all-*trans*-retinal as chromophore [161,162], and when exposed to light they get interconverted (Scheme 13). [163]. More interestingly, the cephalopod retinochrome has been used to achieve one-way 11-cis isomerization induced by light. Retinal all-trans, 13-cis, and 9-cis isomers were mixed with retinochrome and irradiated at ~390 nm to get specifically 11-*cis*-retinal (Scheme 14) [164,165].

Scheme: 14

Retinal + Retinochrome $\xrightarrow[\sim390 \text{ nm}]{hv}$ 11-*cis* retinal
(all-*trans*
13-*cis*
& 9-*cis*)

Scheme: 15

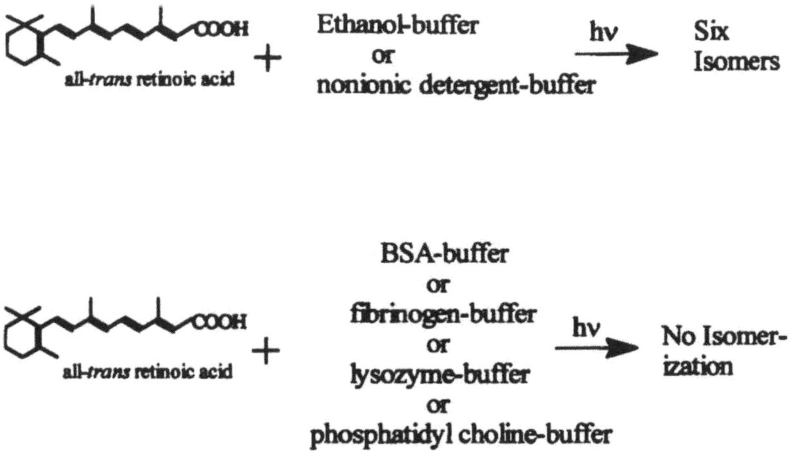

## XV. PHOTOISOMERIZATION OF RETINOIC ACID IN PHYSIOLOGIC-LIKE SOLUTIONS

Photochemical cis-trans isomerization of retinoic acid was conducted in ethanol-buffer, nonionic detergent-buffer, BSA-buffer, fibrinogen-buffer, lysozyme-buffer, and phosphatidylcholine-buffer media [166]. In solution phase [167], ethanol-buffer, and nonionic detergent–buffer media, the photoisomerization process of retinoic acid was found to be efficient and gave six isomeric mixture (Scheme 15) composition as monitored by high-performance liquid chromatography (HPLC). However, the photoisomerization of retinoic acid is completely inhibited in BSA-buffer, fibrinogen-buffer, lysozyme-buffer, and phosphatidylcholine-buffer media (Scheme 15). The observation was termed as photoprotection of retinoic acid by the supramolecular assembly. The photoisomerization process observed is rationalized as being due to the amphiphilic nature of medium and its interaction with the bound chromophore.

## XVI. LACTOGLOBULIN MEDIATED PHOTOISOMERIZATION OF RETINAL AND ITS RELATED COMPOUNDS

Retinal and other related compounds were complexed with the β-lactoglobulin (supramolecule; protein) and the complexes [168–170] were characterized. Thus prepared β-lactoglobulin/retinal (other compounds) complexes were photolyzed to study cis-trans isomerization process and to understand the role of supramolecule (protein). The photostationary state composition (Scheme 16) derived from the complex revealed the selectivity toward forming 11-cis isomer [168]. The suggested rationale for the observed selectivity in the photoisomerization reaction is the result of β-lactoglobulin environment exerting as polar medium, arranging the specific hydrophobic interactions and also due to the possible protonation capability.

The photochemical cis-trans isomerization of retinal chromophore observed in rhodopsin (11-cis to all-trans), retinochrome (all-trans to 11-cis), and bacterio-rhodopsin (all-trans to 13-cis) are regiospecific in nature. The supramolecule protein structure, alignment of chromophore within the supramolecule, hydrophobic interactions, polarity of the protein, and location of charged amino acids around the bound chromophore may be controlling the regiospecificity observed in these systems. The model reactions carried out in BSA and β-lactoglobulin environments have pointed out the same rationale. The geometrical isomerization involving 180° rotation around the double bond of the chromophore in a supramolecular assembly is more interesting. There are a few models [36,37,171–175] put forward on the rotation of the double bond of the chromophore in a supramolecular assembly, e.g., bicycle pedal model [171–173] involving stepwise or concerted rotation of alternate bonds, and the concerted-twist (CT-n) model [174] involving concerted rotation of adjacent bonds. Nevertheless, understanding the mechanism of photochemical cis-trans isomerization in these supramolecular assemblies is a challenging exercise.

## XVII. PHOTOISOMERIZATION IN ORGANIZED ASSEMBLIES

Whitten carried out cis-trans photoisomerization of thioindigo dyes in organized monolayer assemblies [176]. He pointed out that the membranelike monolayer

Scheme: 16

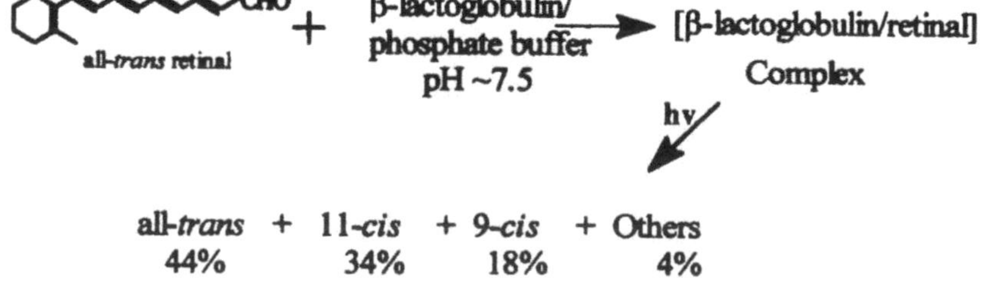

all-*trans* retinal

+ β-lactoglobulin/
phosphate buffer
pH ~7.5

→ [β-lactoglobulin/retinal]
Complex

hν ↙

all-*trans*  +  11-*cis*  +  9-*cis*  +  Others
44%            34%           18%          4%

environment has a profound influence on photochemical cis-trans isomerization [177]. The selective cis-to-trans photoisomerization of thioindigo compounds in the assemblies prepared was rationalized by considering two factors: 1) the influence of constrained environment and 2.) the differing sizes of cis and trans isomers. Further supramolecular aggregates of azobenzene phospholipids [178] and stilbene fatty acids [179] are prepared to understand the photoisomerization process.

Jiang and Akida [180] prepared various dendrimers containing azobenzene chromophore. They found that infrared radiation can be utilized to bring out cis-trans isomerization in azobenzene, as a core molecule in the highly branched dendrimer. They proposed that azobenzene as a core molecule in the dendrimer entity is insulated against collisional energy scattering and infrared absorption can excite photoisomerization by multiphoton energy transfer. Furthermore, they mentioned that these dendrimer assemblies can be used for harvesting low-energy photons. These cis-trans isomerizations in supramolecular assemblies have the scope for utilizing the information for optical data processing purposes.

## XVIII. CONCLUSIONS

Photochemical trans-cis isomerization from the singlet excited state is discussed in this chapter. The possible highly polarized or zwitterionic nature of singlet excited state undergoing trans-cis isomerization is highlighted. Theoretical studies on ethylene and other olefins, as well as spectroscopic investigations on tetraphenylethylene and excited state dipole moments measured on various olefins bring light to the highly polarized or zwitterionic nature of the singlet excited state. The solvent polarity effect on a variety of olefins undergoing trans-cis photoisomerization indicates that a polar singlet excited state interacts with polar medium efficiently. The effect of solvent polarity on fluorescence and fluorescence lifetimes is also a test for the involvement of polar excited state. The photoisomerization process in natural and artificial supramolecular assemblies is also discussed. The role of the polar nature of the singlet excited state in the photoisomerization process in solution and in supramolecular assemblies will be interesting to study in future. Studies of one-way isomerization and adiabatic isomerization originating from the singlet excited state have great academic scope. It will be interesting to follow developments on the blending of this trans-cis photoisomerization, involving polar excited state, with optical data processing materials.

## ACKNOWLEDGMENTS

I thank all of my students and colleagues for their help and contributions. I sincerely thank Director, Dr. K. V. Raghavan, and Deputy Director, Dr. J. Madhusudana Rao, for their encouraging support. I also thank Mr. K. Mani Bushan, Dr. V. Raj Gopal, Dr. T. Soujanya, Dr. E. T. Ayodele, Dr. B. Chinna Raju, and Dr. V. V. Narayana Reddy for their help in preparing this chapter. I am very grateful to DST, New Delhi for financial support.

## REFERENCES

1. Wald, G. Science, **1968**, *162*, 230.
2. a) Oesterhelt, D.; Stoeckenius, W. *Nature* (London) *New Biol*, **1971**, *233*, 149; b) Oesterhelt, D.; Stoeckenius, W. *Proc. Natl. Acad. Sci. U.S.A.*, **1973**, *70*, 2835.
3. Foster, K.W.; Saranak, J.; Patel, N.; Zarrilli, G.; Okabe, M.; Kline, T.; Nakanishi, K. *Nature* (London), **1984**, *311*, 756.
4. Braun, A.M.; Maurette, M.T.; Oliveros, E. *Photochemical Technology*, Wiley, **1991**, Chapter 12, p 500.
5. Kirk-Othmer Encyclopedia of Chemical Technology, 4th Ed., Wiley, **1996**, vol. 18, p 799.
6. Ullmans Encyclopedia of Industrial Chemistry, 5th Ed., VCH, **1991**, vol. A 19, p 573.
7. Photochromism, Molecules and Systems, Durr, H. and Bouas-Laurant, H. (Eds.), Elsevier, Amsterdam, **1990**.
8. Olson, A.R. *Trans. Faraday Soc.* **1931**, *27*, 69.
9. Olson, A.R.; Hudson, F.L. *J. Am. Chem. Soc.*, **1933**, *55*, 1410.
10. Olson, A.R.; Maroney, W. *J. Am. Chem. Soc.*, **1934**, *56*, 1320.
11. Lewis, G.N.; Magel, T.T.; Lipkin, D. *J. Am. Chem. Soc.*, **1940**, *62*, 2973.
12. Saltiel, J.; Waller, A.S.; Sears, D.F. *J. Am. Chem. Soc.*, **1993**, *115*, 2453.
13. Saltiel, J.; Waller, A.S.; Sears, D.F. *J. Photochem. Photobiol. A: Chem.*, **1992**, *65*, 29.
14. Saltiel, J.; Waller, A.S.; Sears, D.F.; Garrett, C.Z. *J. Phys. Chem.*, **1993**, *97*, 2516.
15. Saltiel, J.; Waller, A.S.; Sun, Y.-P.; Sears, D.F. *J. Am. Chem. Soc.*, **1990**, *112*, 4580.
16. Saltiel, J. *J. Am. Chem. Soc.*, **1967**, *89*, 1036.
17. Saltiel, J.; Megarity, E.D. *J. Am. Chem. Soc.*, **1972**, *94*, 2742.
18. Saltiel, J.; Agostino, D.; Megarity, E.D.; Metts, L.; Newberger, K.R.; Wrighton, M.; Zafirinou, O.C. *Org. Photochem.*, **1973**, *3*, 1.
19. Saltiel, J.; Charlton, J.L. in *Rearrangements in Ground and Excited States*, vol. 3, Paul deMayo, Ed., Academic Press, New York, **1980**, p 25.
20. Saltiel, J.; Sun, Y.-P. *Photochromism, Molecules and Systems*, Durr, H. and Bouas-Laurent, H., Eds., Elsevier, Amsterdam, **1990**, p 64.
21. Goerner, H.; Kuhn, H.J. *Adv. Photochem.*, **1995**, *19*, 1.
22. Allen, M.T.; Whitten, D.G. *Chem. Rev.*, **1989**, *89*, 1691.
23. Waldeck, D.H. *Chem. Rev.*, **1991**, *91*, 415.
24. Mulliken, R.S. *Phys. Rev.*, **1932**, *41*, 751.
25. Jaffe, H.H.; Orchin, M. *Theory and Applications of Ultraviolet Spectroscopy*, Academic Press, New York, **1962**, pp 96–98.
26. Marshall, J. *Science*, **1970**, *170*, 137.

27. Dauben, W.G.; Ritscher, J.S. *J. Am. Chem. Soc.*, 1970, *92*, 2925.
28. Dauben, W.G.; Kellog, M.S.; Seeman, J.I.; Wietmeyer, N.D.; Wendschuh, P.H. *Pure Appl. Chem.*, **1973**, *33*, 197.
29. Padwa, A.; Brodsky, L.; Clough, S. *J. Am. Chem. Soc.*, **1972**, *94*, 6767.
30. Wulfman, C.E.; Kumei, S. *Science*, **1971**, *172*, 1061.
31. a) Salem, L. *Acc. Chem. Res.*, **1979**, *12*, 87; b) Salem, L.; Rowland, C. *Angew. Chem. Intl. Ed. Eng.*, **1972**, *11*, 92; c) Bonacic-Koutecky, V. *J. Am. Chem. Soc.*, **1978**, *100*, 396.
32. a) Salem .; Leforestier, C.; Segal, G.; Wetmore, R. *J. Am. Chem. Soc.*, **1975**, *97*, 479.
33. Bonacic-Koutecky, V.; Bruckman, P.; Hiberty, P.; Koutecky, J.; Leforestier, C.; Salem, L. *Angew. Chem. Int. Ed. Eng.*, **1975**, *14*, 575.
34. Salem, L.; Bruckman, P. *Nature (London)*, **1975**, *258*, 526.
35. *Chemical Engineering News*, **1997**, *75(37)*, p 30.
36. Lewis, A.; *Proc. Natl. Acad. Sci. U.S.A.*, **1978**, *75*, 549.
37. Honig, B.; Ebrey, T.; Callender, R.R.; Dinur, U.; Ottolenghi, M.; *Proc. Natl. Acad. Sci. U.S.A.*, **1979**, *76*, 2503.
38. Brooks, B.R.; Scafer, F. *J. Am. Chem. Soc.*, **1979**, *101*, 307.
39. Hemley, R.J.; Dinur, U.; Vaida, V.; Karplus, M. *J. Am. Chem. Soc.*, **1985**, *107*, 836.
40. Orlandi, G.; Palmieri, P.; Poggi, G. *J. Chem. Soc. Faraday Trans. 2*, **1981**, *77*, 71.
41. Bonacic-Koutecky, V.; Persico, M.; Dohnert, D.; Sevin, A. *J. Am. Chem. Soc.*, **1982**, *104*, 6900.
42. Lam, B.; Johnson, R.P. *J. Am. Chem. Soc.*, **1983**, *105*, 7479.
43. Klet, M.W.; Johnson, R.P. *Tetrahedron Lett.*, **1983**, *24*, 3107.
44. Kikuchi, O.; Yoshida, H. *Bull. Chem. Soc. Jpn.*, **1985**, *58*, 131.
45. a) Hixon, S. *J. Am. Chem. Soc.*, **1976**, *98*, 1271; b) Hixon, S. *J. Am. Chem. Soc.*, **1975**, *97*, 1981.
46. Albert, I.D.L.; Ramasesha, S. *J. Phys. Chem.*, **1990**, *94*, 6540.
47. Woning, J ; Oudenmampsen, A.; Laorhoven, W.H. *J. Chem. Soc. Perkin Trans. II*, **1989**, 2147.
48. Marshall, J.A. *Acc. Chem. Res.*, **1969**, *2*, 33.
49. a) Nelson, S.F.; Hintz, P.J. *J. Am. Chem. Soc.*, **1969**, *91*, 6190; b) Kato, H.; Kawanishi, M. *Tetrahedron Lett.*, **1970**, *11*, c) 865; c) Miyamoto, N.; Kawanishi, M.; Nazaki, H. *Tetrahedron Lett.*, **1971**, *12*, 2565; Hixon, S.S. *J. Am. Chem. Soc.*, **1972**, *94*, 2505.
50. Greene, B.I. *Chem. Phys. Lett.*, **1981**, *59*, 3061.
51. Barbara, P.F.; Rand, S.D.; Rentzipis, P.M. *J. Am. Chem. Soc.*, **1981**, *103*, 2156.
52. Leigh, W.J.; Arnold, D.R. *Can. J. Chem.*, **1981**, *59*, 3061.
53. Goerner, H. *J. Phys. Chem.*, **1982**, *86*, 2028.
54. Stegmeyer, H. *Ber. Bunsenges. Phys. Chem.*, **1972**, *72*, 335.
55. Sharafy, S.; Muszkat, K.A. *J. Am. Chem. Soc.*, **1971**, *93*, 4119.
56. Klingenberg, H.H.; Rapp, W.Z. *Physik. Chem. NF*, **1973**, *84*, 92.
57. Klingenberg, H.H.; Lippert, E.; Rapp, W.Z. *Chem. Phys. Lett.*, **1973**, *18*, 417.
58. Olson, R.J.; Buckler, R.E. *J. Photochem.*, **1979**, *10*, 215.
59. Hilinski, E.F.; Rentzpis, P.M. *Anal. Chem.*, **1983**, *55*, 1121.

60. Schmidt, J.A.; Hilinski, E.F.; Bouchard, D.A.; Hill, C.A. *Chem. Phys. Lett.*, **1987**, *138*, 346.

61. Schilling, C.L.; Hilinski, E.F. *J. Am. Chem. Soc.*, **1988**, *110*, 2296.

62. Sun, Y.-P.; Fox, M.A. *J. Am. Chem. Soc.*, **1993**, *115*, 747.

63. Morais, J.; Jangseok, M.; Zimmt, M.B. *J. Phys. Chem.*, **1991**, *95*, 3885.

64. a) Zimmt, M.B. *Chem. Phys. Lett.*, **1989**, *160*, 564; b) Ma, J.; Zimmt, M.B. *J. Am. Chem. Soc.*, **1992**, *114*, 9723.

65. Schuddeboom, W.; Jonker, S.A.; Warman, J.M.; Deltas, M.P.; Vermulen, M.J.W.; Jager, W.F.; deLange, B.; Feringa, B.L.; Fessenden, R.W. *J. Am. Chem. Soc.*, **1993**, *115*, 3286; b) Piotrawak, P.; Strati, G.; Smirnov, N.S.; Warman, J.M.; Schuddeboom, W. *J. Am. Chem. Soc.*, **1996**, *118*, 8981.

66. Mathies, R.; Stryer, L. *Proc. Natl. Acad. Sci. U.S.A.*, **1976**, *73*, 2169.

67. Liptay, W. in *Excited States*, Lim, E.C., Ed., Academic Press, New York, **1974**, p 129.

68. Liptay, W.; Czekalla, Z. *Naturforsch.*, **1960**, *A15*, 1072.

69. Bonderav, S.L.; Belkov, M.V.; Pavlenko, V.B. (U.S.S.R.) *Zh. Prikl. Spektrosk.*, **1985**, *42*, 213 (Chem. Abstr., **1985**, *102*, 204125d).

70. Ponder, M.; Mathies, R. *J. Phys. Chem.*, **1983**, *87*, 5090.

71. Sinha, H.K.; Thopson, P.C.P.; Yates, K. *Can. J. Chem.*, **1990**, *68*, 1507.

72. Rettig, W.; *Topics in Current Chemistry*, Springer-Verlag, New York, **1994**, vol. 169, p 253.

73. Barbara, P.F. *Adv. Photochem.*, **1990**, *15*, 1.

74. Bhattacharya, K.; Choudhary, M. *Chem. Rev.*, **1993**, *93*, 507.

75. Lippert, E.; Rettig, W.; Bonacik-Koutecky, V.; Heisel, F.; Meihe, J.A. *Adv. Chem. Phys.*, **1987**, *68*, 1.

76. Kropf, A.; Hubbard, R. *Photochem. Photobiol.*, **1970**, *12*, 249.

77. Denny, M.; Liu, R.S.H. *J. Am. Chem. Soc.*, **1977**, *99*, 4865.

78. Liu, R.S.H.; Asato, A.E.; Denny, M. *J. Am. Chem. Soc.*, **1977**, *99*, 4865.

79. Tsukida, K.; Masahara, R.; Ito, M. *J. Chromatogr.*, **1977**, *134*, 331.

80. Liu, R.S.H.; Denny, M.; Grodowski, M.; Asato, A.E. *Nou. J. Chem.*, **1979**, *3*, 503.

81. Broek, A.D.; Muradin-Szweykowska, M.; Courtin, J.M.L.; Lugtenberg, *J. Recl. Trsv. Chim. Pays-Bas*, **1983**, *102*, 46.

82. Waddel, W.H.; Hopkins, D.L.; Vemura, M.; West, J.L. *J. Am. Chem. Soc.*, **1978**, *100*, 1970.

83. Waddel, W.H.; West, J.L. *J. Phys. Chem.*, **1980**, *84*, 134.

84. Gartner, W.; Hopf, H.; Hull, W.E.; Oesterhelt, D.; Scheutzow, D.; Towner, P. *Tetrahedron Lett.*, **1980**, *21*, 347.

85. Matsumoto, H.; Asato, A.E.; Denny, M.; Baretz, B.; Yen, Y.-P.; Tong, D.; Liu, R.S.H. *Biochemistry*, **1980**, *19*, 4589.

86. Halley, B.A.; Nelson, E.C. *Int. J. Vit. Nutr. Res.*, **1979**, *49*, 347.

87. Englert, G.; Weber, S.; Klaus, M. *Helv. Chim. Acta.*, **1978**, *61*, 2679.

88. Jayathirtha Rao, V.; Fenstemacher, J.; Liu, R.S.H. *Tetrahedron Lett.*, **1984**, *25*, 1115.

89. Arjunamn, P.; Liu, R.S.H. *Tetrahedron Lett.*, **1988**, *29*, 853.

90. Muthuramu, K.; Liu, R.S.H. *J. Am. Chem. Soc.*, **1987**, *109*, 6510.

91. Jayathirtha Rao, V.; Bhalerao, U.T. *Tetrahedron Lett.*, **1990**, *31*, 3441.

92. Jayathirtha Rao, V. *J. Photochem. Photobiol. A: Chem.*, **1994**, *83*, 211.
93. Raj Gopal, V.; Jayathirtha Rao, V. *Proceedings of Trombay Symposium on Radiation and Photochemistry*, **1996**, *1*, 248.
94. a) Tokumaru, K.; Arai, T. *Bull. Chem. Soc. Jpn.*, **1995**, *68*, 1065; b) Arai, T.; Tokumaru, K. *Adv. Photochem.*, **1995**, *20*, 1; c) Arai, T.; Tokumaru, K. *Chem. Rev.*, **1993**, *93*, 23; d) Arai, T.; Karatsu, H.; Misawa, H.; Kuriyama, Y.; Okamoto, H.; Hiresaki, T.; Furuchi, H.; Sakuragi, H.; Tokumaru, K. *Pure Appl. Chem.*, **1988**, *60*, 989.
95. a) Becker, H.-D. *Adv. Photochem.*, **1990**, *15*, 139; b) Becker, H.-D. *Chem. Rev.*, **1993**, *93*, 145.
96. Becker, H.-D.; Anderson, K.; Sandros, K. *J. Org. Chem.*, **1985**, *50*, 3913.
97. a) Bartocci, G.; Mazzucato, U.; Spalletti, A.; Orlandi, G.; Poggi, G. *J. Chem. Soc. Faraday Trans.*, **1992**, *88*, 2155; b) Mazzu-cato, U.; Aloisi, G.G.; Elisei, F. *Proc. Ind. Acad. Sci. Chem. Sci.*, **1993**, *105*, 475.
98. Sun, L.; Gorner, H. *Chem. Phys. Lett.*, **1993**, *208*, 43.
99. Sun, L.; Gorner, H. *J. Phys. Chem.*, **1993**, *97*, 11186.
100. Aloisi, G.G.; Elisei, F.; Latterni, L.; Passerimi, M.; Galiazzo, G. *J. Chem. Soc. Faraday Trans. 2*, **1996**, 92.
101. a) Wang, Y.; Crawford, M.C.; Eisenthal, K.B. *J. Am. Chem. Soc.*, **1982**, *104*, 5874; b) Crawford, M.B.; Wang, Y.; Eisenthal, K.B. *Chem. Phys. Lett.*, **1981**, *79*, 529.
102. a) Okada, T.; Mataga, N.; Baumann, W.; Siemiarczuk, A. *J. Phys. Chem.*, **1987**, *971*, 4490; b) Okada, T.; Fujita, T.; Kubota, M.; Masaki, S.; Mataga, N.; Ide, R.; Sakata, Y. *Chem. Phys. Lett.*, **1972**, *14*, 563.
103. a) Jones II, G.; Farhat, S.M. *Adv. Electron Transfer Chem.*, **1993**, *3*, 1; b) Zhang, S.; Lang, M.J.; Goodman, S.; Durnell, C.; Fildar, V.; Flemming, G.R.; Yang, N.C. *J. Am. Chem. Soc.*, **1996**, *118*, 9042.
104. Gruen, H.; Gorner, H. *J. Phys. Chem.*, **1989**, *93*, 7144.
105. a) Becker, H.-D.; Anderson, K. *J. Org. Chem.*, **1983**, *48*, 4542; b) Becker, H.-D.; Anderson, K.; Sandros, K. *J. Org. Chem.*, **1985**, *50*, 3913; c) Becker, H.-D.; Sorenson, H.; Sandros, K. *J. Org. Chem.*, **1986**, *51*, 3223.
106. a) Raj Gopal, V.; Jayathirtha Rao, V. *Proc. Trombay Symposium on Radiation and Photochemistry*, **1994**, *1*, 386; b) Raj Gopal, V.; Mahipal Reddy, A.; Jayathirtha Rao, V. *J. Org. Chem.*, **1995**, *60*, 7966; c) Mahipal Reddy, A.; Raj Gopal, V.; Jayathirtha Rao, V. *Rad. Phys. Chem.*, **1997**, *49*, 119.
107. Sandros, K.; Becker, H.-D. *J. Photochem.*, **1987**, *39*, 301.
108. Maeda, Y.; Okada, T.; Mataga, N. *J. Phys. Chem.*, **1984**, *88*, 2114.
109. Kikuchi, Y.; Okamoto, H.; Arai, T.; Tokumaru, K. *Chem. Phys. Lett.*, **1994**, *229*, 564.
110. Spalletti, A.; Bartocci, G.; Mazzucato, U. *Chem. Phys. Lett.*, **1991**, *186*, 297.
111. Raj Gopal, V.; Jayathirtha Rao, V.; Saroja, G.; Samanta, A. *Chem. Phys. Lett.*, **1997**, *270*, 593.
112. Kikuchi, Y.; Okamoto, H.; Arai, T.; Tokumaru, K. *Chem. Lett.*, **1993**, 1811.
113. Lewis, F.D.; Howard, D.K.; Oxman, J.D.; Upthagrove, A.L.; Quillen, S.L. *J. Am. Chem. Soc.*, **1986**, *108*, 5964.
114. a) Lightner, D.A.; Park, Y.-T. *J. Hetrocycl. Chem.*, **1977**, *14*, 415; b) Falk, H. Neufingerl, J.A. *Monatsh. Chem.*, **1979**, *110*, 1243.

115. Lewis, F.D.; Yoon, B.A. *J. Org. Chem.*, **1994**, *59*, 2537.
116. a) Arai, T.; Iwasaki, T.; Tokumaru, K. *Chem. Lett.*, **1993**, 691; b) Arai, T.; Moriyama, M.; Tokumaru, K. *J. Am. Chem. Soc.*, **1994**, *116*, 3171.
117. Lewis, F.D.; Yoon, B.A. *J. Photochem. Photobiol. A; Chem.*, **1995**, *87*, 193.
118. Lewis, F.D.; Yoon, B.A.; Arai, T.; Iwasaki, T.; Tokumaru, K. *J. Am. Chem. Soc.*, **1994**, *116*, 3171.
119. Lewis, F.D.; Yoon, B.A. *Res. Chem. Intermed.*, **1995**, *21*, 749.
120. Lewis, F.D.; Yang, J.-S. *J. Phys. Chem.*, **1996**, *100*, 14560.
121. Hyndman, H.L.; Monroe, B.; Hammond, G.S. *J. Am. Chem. Soc.*, **1969**, *91*, 2852.
122. Hurley, R.; Testa, A.C. *J. Am. Chem. Soc.*, **1970**, *92*, 211.
123. Saltiel, J.; Townsend, D.E.; Sykes, A. *J. Am. Chem. Soc.*, **1973**, *95*, 5968.
124. Butt, Y.C.C.; Singh, A.K.; Baretz, B.H.; Liu, R.S.H. *J. Phys. Chem.*, **1981**, *85*, 2091.
125. a) Goerner, H. *J. Photochem. Photobiol. A: Chem.*, **1988**, *43*, 263; b) Sandros, K.; Becker, H.-D. *J. Photochem. Photobiol. A: Chem.*, **1988**, *43*, 291; c) Larhoven, W.H.; Cuppen, Th. J.H.M.; Castel, N.; Fischer, E. *J. Photochem. Photobiol. A: Chem.*, **1989**, *49*, 137.
126. Sandros, K.; Sundhal, M.; Wennerstrom, O. *J. Am. Chem. Soc*, **1990**, *112*, 3082.
127. Sandros, K.; Sundhal, M.; Wennerstrom, O. *J. Phys. Chem.*, **1993**, *97*, 5291.
128. a) Saltiel, J.; Tarkalanov, N.; Sears, D.F. Jr. *J. Am. Chem. Soc.*, **1995**, *117*, 5586; b) Saltiel, J.; Zhang, Y.; Sears, D.F. Jr. *J. Am. Chem. Soc.*, **1996**, *118*, 2811; c) Saltiel, J.; Zhang, Y.; Sears, D.F. Jr. *J. Am. Chem. Soc.*, **1997**, *119*, 11202.
129. Raj Gopal, V. Ph.D. Thesis dissertation, Osmania University, Hyderabad 500 007, India, **1997**.
130. Raj Gopal, V.; Mani Bushan, K.; Jayathirtha Rao, V. Unpublished results.
131. Fredman, K.; Becker, R.S. *J. Am. Chem. Soc.*, **1986**, *108*, 1245.
132. Child, R.F.; Shaw, G.S. *J. Am. Chem. Soc.*, **1988**, *110*, 3013.
133. Jenson, N.; Wilbrandt, R.; Bensasson, R. *J. Am. Chem. Soc.*, **1989**, *111*, 7877.
134. Carless, J.M.; McCaslin, D.R.; Scott, B.L. *Proc. Natl. Acad. Sci. U.S.A.*, **1982**, *79*, 1116.
135. Hargrave, P.A.; McDowell, J.H.; Feldman, R.J.; Atkinson, P.H.; Mohana, R.; Agros, P. *Vision Res.*, **1984**, *24*, 1487.
136. Findlay, J.B.C.; Pappin, D.J.C. *Biochemical J.*, **1986**, *238*, 625.
137. Balogh Nair, V.; Nakanishi, K. in *Chemistry and Biology of Synthetic Retinoids*, Dawson, M.; Okumura, W., eds., CRC Press, Boca Raton, FL, **1990**, p 147.
138. Liu, R.S.H.; Asato, A.E. in *Chemistry and Biology of Synthetic Retinoids*, Dawson, M.; Okumura, W., eds., C.R.C. Press, Boca Raton, FL, **1990**, p 51.
139. Matsumoto, H.; Yoshizawa, T. *Nature* (London), **1975**, *258*, 523.
140. Kropf, A. *Nature* (London), **1976**, *264*, 92.
141. Honig, B.; Dinur, U.; Nakanishi, K.; Balogh Nair, V.; Gawinowicz, M.A.; Arnaboldi, M.; Motto, M.G. *J. Am. Chem. Soc.*, **1979**, *101*, 7084.
142. Blatz, P.E.; Mohler, J.H.; Navangul, H.V. *Biochemistry*, **1972**, *11*, 848.
143. Suzuki, H.; Komatsu, T.; Kitazima, M. *J. Phys. Soc. Jpn.*, **1974**, *37*, 177.
144. Irving, C.S.; Byers, G.W.; Leermakers, P.A. *Biochemistry*, **1970**, *9*, 858.
145. Wald, G. *Annu. Rev. Biochem.*, **1953**, *22*, 497.
146. Becker, R.S.; Fredman, K. *J. Am. Chem. Soc.*, **1985**, *107*, 1477.

147.  Dartnall, H.J.A. *Vision Res.*, **1968**, *8*, 339.
148.  Shichi, H. *Biochemistry of Vision*, Academic Press, New York, **1983**.
149.  Crouch, R.; Purvin, V.; Nakanishi, K.; Ebrey, T. *Proc. Natl. Acad. Sci., U.S.A.*, **1972**, *72*, 1538.
150.  Denny, M.; Liu, R.S.H. *Biochemistry*, **1988**, *27*, 6495.
151.  a) Kakitani, K.; Akiyama, R.; Hatano, Y.; Schichida, Y.; Verdegen, P.; Lugtenberg, J. *J. Phys. Chem.*, **1998**, *102*, 1334; b) Verven, T.; Bernardi, F.; Garvelli, M.; Olivucci, M.; Robb, M.A.; Schlegel, H.B. *J. Am. Chem. Soc.*, **1997**, *119*, 12687.
152.  Methods in Enzymology, Biomembranes Part I, Visual Pigments and Purple Membranes, II, vol. 88, Packer, L., Ed., Academic Press, **1982**.
153.  Henderson, R.; Unwin, P.N.T. *Nature* (London), **1975**, *257*, 28.
154.  Henderson, R.; Baldwin, J.M.; Ceska, T.A.; Zemlin, F.; Beckman, E.; Downing, K.H. *J. Mol. Biol.*, **1990**, *213*, 899.
155.  Nakanishi, K.; Balogh Nair, V.; Arnaboldi, M.; Tsujimoto, K.; Honig, B. *J. Am. Chem. Soc.*, **1980**, *102*, 7945.
156.  Jayathirtha Rao, V.; Derguini, F.; Nakanishi, K.; Taguichi, T.; Hosada, A.; Hanzawa, Y.; Kobayashi, Y.; Pande, C.M.; Callender, R.R. *J. Am. Chem. Soc.*, **1986**, *108*, 6077.
157.  Khorana, H.G. *Proc. Natl. Acad. Sci. U.S.A.*, **1993**, *90*, 1166.
158.  a) Gai, F.; Hassan, K.C.; McDonald, J.C.; Anfinrud, P.A. *Science*, **1998**, *279*, 1886; b) Elsayed, M.A.; Logunov, S. *Pure Appl. Chem.*, **1997**, *69*, 749.
159.  Hara, T.; Hara, R. *Nature* (London), **1965**, *206*, 1331.
160.  Hara, T.; Hara, R. *J. Gen. Physiol.*, **1976**, *67*, 791.
161.  Hara, T.; Hara, R. *Nature* (London), **1967**, *214*, 573.
162.  Hara, T.; Hara, R. *Nature* (London), **1968**, *219*, 450.
163.  a) Seki, T.; Hara, T.; Hara, R. *Photochem. Photobiol.*, **1980**, *32*, 469; b) Tsujimoto, K.; Shirasaka, Y.; Mizukami, T.; Ohashi, M. *Chem. Lett.*, **1997**, 813.
164.  Hara, T.; Hara, R. *J. Gen. Physiol.*, **1980**, *75*, 1.
165.  Hara, T.; Hara, R. *Nature* (London), **1973**, *242*, 39.
166.  Curley, Jr. R.W.; Fowble, J.W. *Photochem. Photobiol.*, **1988**, *47*, 831.
167.  McKenzie, R.M.; Hellwege, D.M.; McGregor, M.L.; Rockley, N.L.; Riquetti, P.J.; Nelson, E.C. *J. Chromatogr.*, **1978**, *155*, 379.
168.  Li, X.Y.; Asato, A.E.; Liu, R.S.H. *Tetrahedron Lett.*, **1990**, *31*, 4841.
169.  Fugaste, R.D,; Song, P. *Biochim. Biophys. Acta*, **1980**, *625*, 28.
170.  Horwitz, J; Heller, J. *J. Biol. Chem.*, **1974**, *249*, 4712.
171.  Warshel, A. *Nature*, **1976**, *260*, 679.
172.  Warshel, A.; Baroby, N. *J. Am. Chem. Soc.*, **1982**, *104*, 1469.
173.  Warshel, A. *Proc. Natl. Acad. Sci. U.S.A.*, **1978**, *75*, 2558.
174.  Liu, R.S.H.; Asato, A.E. *Proc. Natl. Acad. Sci. U.S.A.*, **1985**, *82*, 259.
175.  Birge, R.R.; Hubbard, L.M. *J. Am. Chem. Soc*, **1980**, *102*, 2195.
176.  Whitten, D.G. *J. Am. Chem. Soc.*, **1974**, *96*, 594.
177.  Whitten, D.G. *Angew. Chem. Int. Ed. Eng.*, **1979**, *18*, 440.
178.  Song, X.; Perlstein, J.; Whitten, D.G. *J. Am. Chem. Soc.*, **1997**, *119*, 9144.
179.  Song, S.; Geiger, C.; Farahat, M.; Perlstein, J.; Whitten, D.G. *J. Am. Chem. Soc.*, **1997**, *119*, 12481.
180.  Jiang, D.-L.; Aida, T. *Nature*, **1997**, *388*, 454.

# 5

# Photochemical Cleavage Reactions of Benzyl–Heteroatom Sigma Bonds

**Steven A. Fleming**

Brigham Young University, Provo, Utah

**James A. Pincock**

Dalhousie University, Halifax, Nova Scotia, Canada

## I. INTRODUCTION

Nucleophilic substitution reactions by solvolysis at a carbon atom with a leaving group, Eq. (1), are well enough understood that they are often used in introductory organic chemistry textbooks as an instructional foundation for mechanistic concepts. Information on how variables such as the structure, stereochemistry, the leaving group (LG), and the nucleophilicity of the solvent (SOH) control the reactivity is so extensive that prediction of results for new cases can be made with considerable confidence.

$$SOH \; + \; -\overset{|}{\underset{|}{C}}-LG \;\longrightarrow\; -\overset{|}{\underset{|}{C}}-OS \; + \; ^+H \; + \; LG^- \qquad (1)$$

Studies on the analogous photochemical reactions have been limited almost exclusively to cases where the carbon is benzylic ($PhCH_2$—LG) or, more generally, arylalkyl (Ar—$CR_2$—LG). The aryl group provides the necessary chromo-

phore for photochemical excitation that results in some cases in rate enhance-
ments of many orders of magnitude relative to the corresponding ground state
process. An important difference in the photochemical reactions is that almost
invariably competition occurs between two pathways proceeding through ion
pairs (photosolvolysis), Eq. (2a), and radical pairs, Eq. (2b). For simplicity in
Eq. (2) the leaving group is shown as negative (i.e., neutral when attached to the
carbon) but examples of neutral leaving groups (i.e., positive when attached) have
also been extensively studied. The yield of each pathway is typically determined
easily because the products obtained are distinctive; solvent-trapped products for
path 2a and radical coupling, atom transfer, or disproportionation products for
path 2b. In addition to the reaction variables listed above for the ground state
reactions, the photochemical reactions add another important feature, namely, the
multiplicity (singlet versus triplet) of the excited state.

$$Ar\!-\!\overset{|}{\underset{|}{C}}\!-\!LG \xrightarrow[\text{SOH}]{h\nu} \left[ Ar\!-\!\overset{|}{\underset{|}{C}}\!-\!LG \right]^{*} \begin{cases} Ar\!-\!\overset{|}{\underset{|}{C}}^{\oplus} \; {}_{,}LG^{\ominus} \longrightarrow \begin{array}{c}\text{Ion-derived} \\ \text{products}\end{array} & (2a) \\[2em] Ar\!-\!\overset{|}{\underset{|}{C}}{}^{\cdot} \; {}^{\cdot}LG \longrightarrow \begin{array}{c}\text{Radical-derived} \\ \text{products}\end{array} & (2b) \end{cases}$$

The last extensive review article on this class of photochemical reactions
was written by Cristol and Bindel and appeared in 1983 in Vol. 6 of the predeces-
sor of this series [1]. The general mechanistic conclusions reached at that time
have not changed significantly over the intervening years and a modern summary
of them appears in Scheme 1 [2,3]. The most obvious feature of this scheme is
its complexity even without the inclusion of the several types of ion pairs (contact,
solvent-separated, and fully solvated). For example, the yield of ion-derived products is determined by the relative magnitude of six rate constants, three for forma-
tion ($k_{het}^{S1}$, $k_{hom}$, and $k_{etri}$) and three for the destruction ($k_{etir}$, $k_{icom}$, and $k_{SOH}$) of the
ion pair. In most cases, the factors that control these processes are not yet well
enough understood to allow a reliable prediction of product distribution.

Not surprisingly, and for a variety of reasons, considerable progress has
been made since 1983. These reasons include an interest in the fundamental problem of the competition between homolytic ($k_{hom}^{S1}$) and heterolytic ($k_{het}^{S1}$) cleavage
in excited singlet states [4–8]; attempts to observe and determine the reactivity
of transient arylalkyl radicals and cations by laser flash photolysis [9,10]; the
design of photolabile protecting groups [11] including "photocages" [12,13] and
photoaffinity labels [14] of biologically relevant molecules; and photoacid and
photobase generation [15]. The aim of this chapter is to review these more recent
results; the literature is covered up to May 1998.

Section II includes a general discussion of the principles of how reaction
conditions and structural variables in the substrate affect these processes, Sec. III

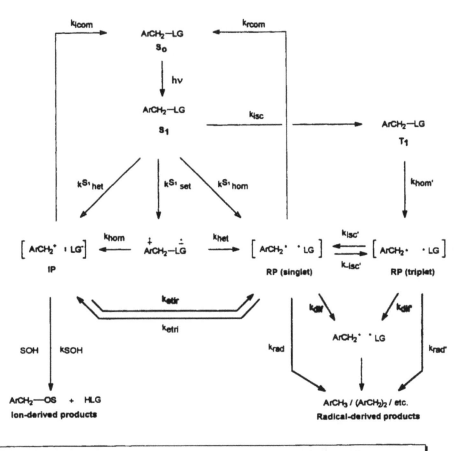

**Scheme 1** General mechanisms for the photochemistry of arylmethyl compounds.

gives a brief qualitative description of theoretical approaches, and Sec. IV gives a systematic review of recent results by leaving group. This latter material will be organized by ascending order of the atomic number of the leaving group atom which is attached to the arylmethyl carbon. The contents will be limited to reactions from excited states and will therefore exclude cleavages at arylalkyl carbons induced by photochemical electron transfer generation of radical cations [16,17] or anions [17,18]. For each leaving group discussed, the coverage will not be comprehensive but rather a few key recent references will be summarized. The interested reader can use these references as a starting point for more complete bibliographies.

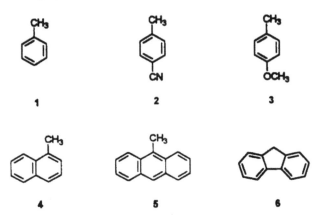

## II.  GENERAL PRINCIPLES

### A.  The Chromophore

The excited state properties, in polar solvent where available, for a few of the common chromophores, 1 to 6, that have been found to induce photochemical sigma bond cleavage reactions at benzylic carbons are given in Table 1 [19,20]. These are all $\pi,\pi^*$ states and they serve as good models for the photoreactive substrates because usually the leaving group in ArCH$_2$-LG does not significantly perturb the chromophore.

The singlet ($E_{S1}$) and triplet ($E_{T1}$) excited state energies for 1, 2, and 3 exhibit only a small effect when the substituent on the aromatic ring is changed. In contrast, the substituents significantly increase the molar absorptivity ($\varepsilon_{max}$) and shorten the singlet lifetime ($\tau_S$). Extending the conjugation of the ring using fused aromatics, 4 and 5, lowers both the singlet and triplet excited state energies considerably. As shown by the $\lambda_{0,0}$ and the approximate $\lambda_{max}$ values, the absorption band moves to a longer wavelength; this effect can be used to separate the reactive chromophore from other potential UV-absorbing functional groups elsewhere in

**Table 1** Excited State Properties of Selected Aryl Chromophores

| Aryl | $E_{S1}$[a] (kcal/ mol) | $E_{T1}$[a] (kcal/ mol) | $\Phi_{ISC}$[a] (kcal/ mol) | $\Phi_F$[a] (kcal/ mol) | $\tau_s$[a] (ns) | $\lambda_{0,0}$[b] (nm) | $\varepsilon_{max}(\lambda_{max})$[c] M$^{-1}$ cm$^{-1}$ (nm) |
|---|---|---|---|---|---|---|---|
| 1: toluene | 106 | 83 | 0.53 | 0.14 | 35 | 270 | 200(260) |
| 2: 4-methylbenzonitrile | 101 | 76 | — | 0.16 | 9.1 | 280 | 1000(271) |
| 3: 4-methylanisole[d] | 103 | 81 | 0.64 | 0.24 | 7.5 | 278 | 1480(269) |
| 4: 1-methylnaphthalene | 90 | 61 | 0.58 | 0.19 | 97 | 317 | 5000(280) |
| 5: 9-methylanthracene | 73 | 41 | 0.67 | 0.33 | 5.8 | 392 | 8000(360) |
| 6: fluorene | 45 | 68 | 0.32 | 0.68 | 10 | 301 | 11,000(260) |

[a] Ref. 19.
[b] Calculated from $E_{S1}$.
[c] Ref. 20.
[d] Some data for anisole rather than 4-methylanisole.

the molecule. Depending on the strength of the bond being broken, the shifting of the absorption band to longer wavelengths and consequently to a lower excitation energy eventually will not provide the necessary driving force for bond cleavage. For instance, benzylic ammonium salts are reactive from both $S_1$ and $T_1$ [21], but 1-naphthylmethylammonium salts only react from the higher energy $S_1$ state [22].

## B. The Solvent

Because the substituted benzene chromophores absorb in the 205- to 280-nm range and have low $\varepsilon$ values, the solvents used must be transparent down to at least 250 nm. This requirement is unnecessary for the more conjugated naphthalene and anthracene chromophores. The usual polar, nucleophilic solvents that have been used to observe ions and ion-derived products are various alcohols, water, or water mixed with a cosolvent such as dioxane for solubility reasons. Recently, and particularly for the observation of the intermediate carbocations by laser flash photolysis (LFP) methods, the strongly ionizing (high $Y_{OTS}$ values) but weakly nucleophilic (low $N$ values) alcohols, 2,2,2-trifluoroethanol (TFE) and 1,1,1,3,3,-hexafluoro-2-propanol (HFIP), have been more commonly used. A limited list of polar solvents and their properties is given in Table 2 [23,24].

Arylmethyl cleavage reactions have also been studied in less polar solvents like saturated hydrocarbons (hexane, cyclohexane), ethers (diethyl ether, tetrahydrofuran, dioxane), and methylene chloride. In these solvents, ion pair chemistry is less likely and valuable information can be obtained about the photogenerated radical pairs.

**Table 2**   Properties of Selected Polar Solvents for Arylmethyl Photochemistry

| Solvent | $Y_{OT_s}$[a] | $N_{OT_s}$[a] | $k_{BP}$[b] $(M^{-1} s^{-1})$ |
|---|---|---|---|
| Water | 4.1 | −0.44 | |
| 1,1,1,3,3,3-Hexafluoro-2-propanol (HFIP) | 3.82 | — | |
| HFIP (97%)/water | 3.61 | −4.27 | |
| TFE (97%)/water | 1.83 | −2.79 | |
| 2,2,2-Trifluoroethanol (TFE) | 1.77 | −3.07 | |
| AN (50%)/water | 1.2 | — | |
| Methanol | −0.92 | −0.04 | $3 \times 10^5$ |
| Dioxane (80%)/water | −1.30 | −0.29 | |
| Ethanol | −1.96 | 0.06 | $8 \times 10^5$ |
| 2-Propanol | −2.83 | 0.12 | $2 \times 10^6$ |
| Acetonitrile (AN) | −3.21 | — | $1.3 \times 10^2$ |
| t-Butanol | −3.74 | — | |

[a] Ref. 23.
[b] Bimolecular rate constant for hydrogen abstraction by benzophenone, Ref. 24.

## C.   Multiplicity of the Reactive Excited State

The intersystem crossing efficiency ($\Phi_{ISC}$) for all of the chromophores in Table 1 is quite high and therefore triplet reactivity must be considered even in direct irradiations. As outlined in Scheme 1, the triplet excited state can only react by homolytic cleavage to give the triplet radical pair ($k_{hom'}$). Ion-derived products are therefore unlikely if intersystem crossing back to the singlet is slow, which is likely unless there are heavy atoms present to induce spin-orbit coupling. Direct irradiations using quenching with dienes as a test for triplet reactivity is complicated by the fact that most dienes quench both excited single [25,26] and triplet states [24] of aromatics with rates that are close to the diffusion limit. Because all of these aromatics show efficient fluorescence ($\Phi_F$), the quenching of their singlet excited state by a diene can be determined by standard Stern-Volmer fluorescence methods. For substituted benzenes, however, competitive absorption by the diene may create problems in these experiments because of the low $\epsilon$ values for the aromatics and the contrasting high values for the dienes. If the triplet state can be independently observed by laser flash photolysis (LFP) and its lifetime in solution determined, and if energy transfer to the diene is exergonic, then the extent of triplet quenching with respect to diene concentration can be estimated by assuming that the quenching rate is at the diffusion limit. Even if the triplet lifetime is not known, the reasonable assumption can be made that it will be longer lived than the excited singlet state. Differences between the quenched and unquenched photolyses can then be used to assess triplet reactivity.

Triplet-sensitized reactions have been examined for many cases, although this may be problematic for the substituted benzene derivatives because there are few sensitizers with triplet energies higher than 80 kcal/mol. Often acetone ($E_T$ = 79 kcal/mol) [27] has been used, but because of its low ε value, it must be used in high concentrations or even as the solvent to ensure that it is absorbing all of the light. Even with acetone, energy transfer to aromatic chromophores will usually not be exergonic and will therefore be slow. Complications due to chemical sensitization or radical ion formation must be considered as has been reported for 1-naphthylmethyl halides [28].

Determining the multiplicity of the reactive excited state is easier for the conjugated aromatics, which absorb at longer wavelengths and with higher ε values. Quenching experiments with a diene are more straightforward in these cases because competitive absorption by the diene is not normally a problem. Common triplet sensitizers such as benzophenone ($E_T$ = 69 kcal/mol) [29] or acetophenone ($E_T$ = 74 kcal/mol) [30] can now be used because energy transfer will be exergonic. However, because these sensitizers form n,π* triplet states with unit efficiency, they cannot be used in the presence of good hydrogen atom donors, such as methanol, ethanol, or 2-propanol, which undergo hydrogen atom abstraction (see $k_{BP}$, Table 2) [24]. To our knowledge, the reactivities of the fluorinated alcohols TFE and HFIP toward the triplet state of benzophenone have not been determined. Xanthone ($E_T$ = 74 kcal/mol), with a π,π* triplet state, has been used as a triplet sensitizer for naphthalene derivatives [31] in methanol but reacts rapidly ($2.2 \times 10^5$ M$^{-1}$ s$^{-1}$) with the more reactive hydrogen atom donor, 2-propanol [32].

## D.  Energy of the Radical Pair and the Ion Pair

Thermochemical analyses similar to that outlined here have appeared previously [33,34]. Bond dissociation energies (BDEs) [35] can be used to give reliable estimates of the energy of the radical pair, even though they are enthalpies, because the entropy correction is usually small. These are listed in Table 3 for some of the more common leaving groups for PhCH$_2$—LG. The bond can be photoreactive to homolytic cleavage and formation of the radical (RP) only if the excitation energy of the singlet excited state ($E_{S1}$) is higher than the BDE. For benzene derivatives, this, in principle, would only rule out CN and $^+$OH$_2$, but F and $^+$NH$_3$ also seem unlikely to react. In practice, the bond being broken is often considerably weaker than the available excitation energy. For instance, benzyl alcohol (BDE = 81 kcal/mol) and benzylmethyl ether (BDE = 71 kcal/mol) are quite unreactive in methanol even though, with $E_{S1}$ = 106 kcal/mol, the excited state has 20–30 kcal/mol excess energy. In contrast, benzyl acetate (BDE = 62 kcal/mol) reacts quite efficiently [25]. As the excitation energy drops on going to naphthalene and anthracene chromophores, more of these leaving groups will become unreactive. Because the excited states of these chromophores are π,π*, the energy gap between $E_{S1}$ and $E_{T1}$ is large. Therefore, the excited

**Table 3** Bond Dissociation Energies for Formation of Radical Pairs and Ion Pairs from PhCH$_2$-LG Substrates

| LG[a] (PhCH$_2$—LG) | BDE (RP)[b] (kcal/mol) | E$_{ox}$ (LG)[c] (V, SCE) CH$_3$CN | ΔG$_{ET}$[d] (kcal/mol) CH$_3$CN | BDE (IP) (kcal/mol) CH$_3$CN | E$_{ox}$ (LG)[c] (V, SCE) H$_2$O | ΔG$_{ET}$[d] (kcal/mol) H$_2$O | BDE (IP) (kcal/mol) H$_2$O |
|---|---|---|---|---|---|---|---|
| $^+$OH$_2$ | 120[e] | > +3[f] | >50 | <70 | | | |
| CN | >100 | +1.3 | −13 | >87 | +1.7 | −22 | >78 |
| $^+$NH(CH$_3$)$_2$ | 101[e] | +1.1[f] | −9 | 92 | | | |
| F | 97[g] | +2.6 | −43 | 54 | +3.4 | −62 | 35 |
| H | 88 | +0.36 | +25 | 113 | | | |
| OH | 81 | +1.0 | −6 | 75 | +1.7 | −22 | 59 |
| Cl | 72 | +1.8 | −24 | 48 | +2.3 | −36 | |
| N(CH$_3$)$_2$ | 71 | h | | | | | |
| OCH$_3$ | 71 | +0.5 | +6 | 77 | +1.2 | −10 | 61 |
| C(CH$_3$)$_3$ | 67 | −2.1 | +65 | 132 | | | |
| CH$_2$Ph | 66[i] | −1.0 | +40 | 104 | | | |
| Ac | 62 | +1.5 | −18 | 44 | +2.2 | −34 | 28 |
| NH$_2$ | 62 | h | | | | | |
| SCH$_3$ | 61 | +0.5 | +6 | 67 | +0.7 | +1 | 62 |
| Br | 58 | +1.4 | −15 | 43 | +1.8 | −25 | 33 |
| I | 48 | +0.9 | −4 | 44 | +1.2 | −11 | 37 |

[a] The leaving groups are written with the charge as it would be when attached to the carbon.

[b] Ref. 35.

[c] Ref. 37.

[d] Assuming the oxidation potential of the benzyl radical is 0.73 V versus SCE in acetonitrile, Ref. 36.

[e] Assuming the increases in BDE (RP) for protonation of PhCH$_2$OH and PhCH$_2$NH$_2$ are 41 and 31 kcal/mol, respectively. Boyd, R.J.; Glover, J.N.M.; Pincock, J.A. J. Am. Chem. Soc. **1989**, *111*, 5152–5155.

[f] Mann, C.K. Anal. Chem. **1964**, *36*, 2424.

[g] Assuming a value of 13 kcal/mol less than that for CH$_3$—F.

[h] Oxidation potentials of these anions are not available because they are not stable in acetonitrile or water.

[i] Assuming a value of 13 kcal/mol less than that for PhCH$_2$—CH$_3$.

triplet states have considerably lower excitation energies than the singlet states and will be thermodynamically even less likely to react. For instance, 1-naphthylmethyl acetate (BDE ~ 60 kcal/mol) is reactive to photocleavage from the singlet state ($E_{S1}$ = 90 kcal/mol) but not the triplet ($E_{T1}$ = 61 kcal/mol) [31]. Some potential leaving groups [C(CH$_3$)$_3$, N(CH$_3$)$_2$] are unreactive despite having fairly weak bonds. As will be discussed in Sec. III, this may result from the low polarity and polarizability of these bonds. In contrast, the very strong bonds in positively charged groups like $^+$NH(CH$_3$)$_2$ and $^+$OH$_2$ do react.

Estimates of the energy of the ion pair, BDE(IP), can be obtained from the oxidation potential of the benzyl radical [36] and the reduction potential of the leaving group [37], in the same solvent (see Table 3). Some caution is required here because the coulombic term for the ion pair, which will be distance- and solvent-dependent, has been ignored. Moreover, the oxidation potential for the benzyl radical has been measured only in acetonitrile. Estimation of the $\Delta G_{ET}$ for other solvents, particularly protic ones like water and alcohols in which these reactions have mostly been studied, can only be made by assuming that the oxidation potential of the benzyl radical is not very solvent-dependent—perhaps a safe assumption for a delocalized cation. Using this assumption and the reduction potential of some of the leaving groups in water allows estimation of the energy of the ion pair in that solvent. These are also given in Table 3.

The most important conclusion reached from these calculations is that, as indicated by the negative $\Delta G_{ET}$ values, the ion pair in polar solvents is more stable than the radical pair for almost all of the common photochemically reactive leaving groups. The exception is SCH$_3$ and, in fact, no ion-derived products are detected in the photolysis of arylmethyl thioethers [38]. Therefore, in the other cases where the singlet excitation energy, $E_{S1}$, is sufficient to induce homolytic cleavage, the ion pair will also be thermodynamically accessible. This also means that, although radical pairs can be converted to ion pairs ($k_{etri}$ in Scheme 1) by electron transfer, the conversion of ion pairs to radical pairs ($k_{etir}$) can be ignored. Much of the debate on the mechanism of these photocleavages from the excited singlet state has centered on the question of the relative yields of ion-derived products, from two of the possible pathways: $k_{het}^{S1}$, direct heterolytic cleavage from the excited state, or $k_{etri}$, electron transfer from an initially formed radical pair. For most leaving groups, except sulfonium salts (Sec. IV.H.2) [39], benzoates [40,41], and sulfonates (Sec. IV.D.7) [5], little evidence has been presented for the pathway beginning with electron transfer in the excited singlet state ($k_{et}^{S1}$) resulting in an intramolecular charge transfer complex. In fact, bond cleavage reactions do not occur efficiently from the charge transfer complex for the benzoates.

From the data assembled in Tables 1 and 3, fairly reliable reaction coordinate diagrams can be drawn. An example is shown in Fig. 1 for 1-naphthylmethyl acetate. In direct irradiations, S$_1$ is the excited state first formed. After excitation the aromatic ring will rapidly undergo vibrational relaxation to the equilibrium geometry for S$_1$; the singlet lifetime for this ester is 41 ns [31]. The observation of 0,0 band overlap between the absorption and fluorescence spectra indicates that minimum on the $S_1$ surface cannot be very different in geometry from that of $S_0$. Intersystem crossing gives $T_1$ but this state is unreactive because homolytic cleavage to the radical pair is endergonic and heterolytic cleavage to the ion pair is spin-forbidden. High-level molecular orbital calculations [42] for benzene itself show that the $S_1$ minimum still has $D_{6h}$ symmetry but with all bonds slightly longer at 1.43 Å than at 1.39 Å for $S_0$. These calculations also predict a barrier on the $S_1$ surface of 23 kcal/mol to a conical intersection that leads to the ground state surface of a "prefulvene" biradical, analogous to **10**. Possibly, the barrier height to these nonplanar singlet state biradicals/zwitterions may be decreased or even disappear for other aromatic rings or substituted benzenes. Certainly, the electron density distribution will vary as a function of substituents. Bond cleavage from $S_1$ can form, at least thermodynamically, either the radical pair or the ion pair. The question mark in Fig. 1 is to highlight the questions of whether nonpla-

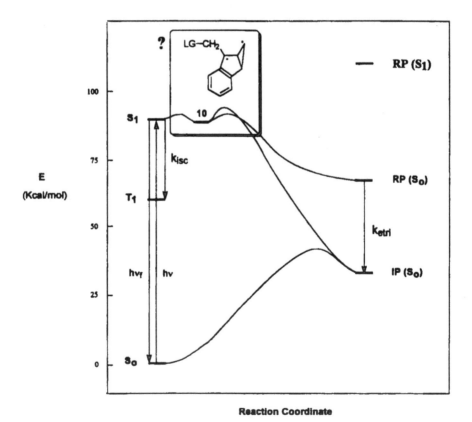

**Figure 1**   Energy versus reaction coordinate for the cleavage of the C—O bond in 1-naphthylmethyl acetate.

nar intermediates are important and whether the barrier for the second possibility may be higher than that for the first because the available evidence indicates that homolytic cleavage is preferred for this and other arylmethyl esters [5]. The RP($S_0$) to IP($S_0$) process represents the electron transfer step. The energy of the first excited state of the radical pair, RP($S_1$), has been estimated from the absorption spectrum of the 1-naphthymethyl radical [43]. Therefore, adiabatic conversion of $S_1$ to RP($S_1$), the excited state of the 1-naphthylmethyl radical, is energetically impossible. Finally, the difference in energy between $S_0$ and IP($S_0$) provides an estimate for the activation energy for ground state solvolysis; the large endergonic barrier explains the slowness of this process.

## E.  Substituent Effects

Substituents on the aromatic ring play an important role in controlling the relative yield of ion- and radical-derived products as well as the efficiency of the photochemical reaction. These effects could operate at many places in Scheme 1. A seminal paper on substituent effects, published by Zimmerman and Sandel in 1964, reported on the photochemistry of methoxy-substituted benzyl acetates. The results are summarized in Scheme 2 [44]. Their important observation was that the yield of ion-derived product, the substituted benzyl alcohol in each case for the aqueous dioxane solvent, was higher for the 3- (or *meta*-) methoxy-substituted esters, 8 and 9, than for 4-methoxy compound, 7. Moreover, the quantum efficiency was also higher. This effect, which is contrary to expectations for the corresponding ground state solvolysis reactions, was attributed to changes in electron density that occur upon excitation to $S_1$. Essentially, the idea is that the methoxy group activates benzylic reactivity in the excited state by being a better electron-donating group (EDG) by its resonance effect from the meta position. Resonance pictures like 11 were drawn to approximate the excited singlet state.

This argument was supported by electron density calculations at the simple Hückel MO level and more recently by higher lever calculations estimating the energy of the ion pairs and radical pairs as a function of the position of methoxy substituents [4]. This impact of substituents on excited state reactions has frequently been called the "meta effect" although, in fact, both ortho and meta activation are predicted. For instance, 1,3-dimethoxybenzene undergoes photoprotonation of its excited singlet state preferentially at $C_2$ (ortho to both methoxy groups) and more slowly at $C_5$ (meta to both methoxy groups). No reaction is observed at $C_4$ (ortho to one methoxy group and para to the other) [45]. The calculations also indicate that electron density decreases ortho and meta to electron-withdrawing groups (EWGs) on excitation to $S_1$.

Using the above argument, the substituent is thought to control the relative magnitude of the two rate constants $k_{het}^{S1}$ and $k_{hom}^{S1}$ and hence the yield of ion pairs and radical pairs. Possibly, the planar resonance form depicted as 11 in Scheme 2

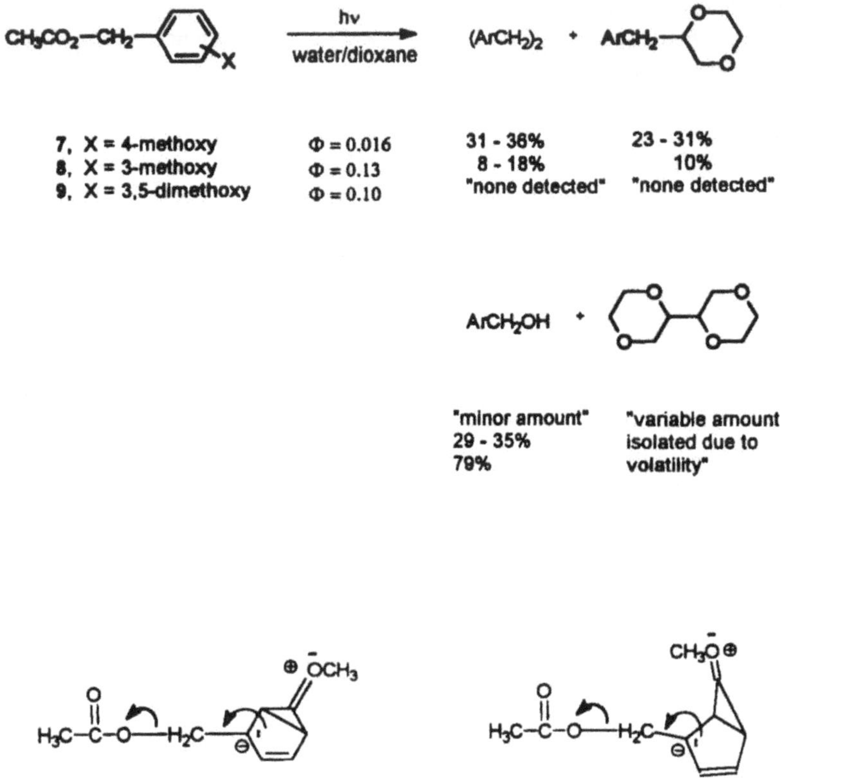

**Scheme 2**  Photolysis of methoxy-substituted benzyl acetates in 80% aqueous dioxane.

is rationalizing rapid geometry changes by low-energy barriers to nonplanar structures like **12**, i.e., like **10** in Fig. 1. The two barriers for radical pair and ion pair formation from **12** would be expected to be substituent-dependent. However, other rate processes of the radical pair or the ion pair must also vary with substituents and these variations could alter the yield of the critical intermediates after they are formed by photocleavage. For example, recent results from the laboratories of Peters [6–8] and Pincock [5] have assessed the importance of the electron transfer step, $k_{etri}$. Substituents on the ring are known to alter the oxidation potential of the arylmethyl radical by their electron donating (stabilization) and electron withdrawing (destabilization) ability [36,46]. These potentials correlate with the substituent $\sigma^+$ parameters resulting in $\rho^+ = -9.3$ for benzyl radicals

[36] and $\rho^+ = -8.4$ for 1-naphthylmethyl radicals [46], both in acetonitrile. This means that the $\Delta G_{ET}$ values for the oxidation of the radicals to their corresponding cations becomes more favorable by 19 kcal/mol for the benzyl radicals and by 16 kcal/mol for the 1-naphthylmethyl radicals as the substituent changes from 4-CN (EWG) to 4-OCH$_3$ (EDG). For 1-naphthylmethyl acetate and other aryl-methyl esters, good evidence has been presented that this electron transfer process is critical in controlling the yield of ion pairs and that very little of the excited state chemistry proceeds through direct homolytic cleavage of the sigma bond [5].

## III. QUALITATIVE THEORETICAL APPROACHES

Michl and Bonačić-Koutecký have written an excellent book on the theory of organic photochemical reactions [47] and its terminology, examples, and diagrams will be used as the basis for this section. The method involves first constructing a molecular orbital (MO) correlation diagram by symmetry. Next, electronic configurations are constructed by the possible occupancies of the MOs, which then leads to a configuration correlation diagram. Finally, consideration of configuration mixing, again as permitted by symmetry, results in a state correlation diagram. For sigma bond cleavage reactions, the first example used is a simple, ditopic (one active orbital per atom) molecule, like H$_2$, which is instructive but not particularly relevant to organic photochemistry.

In H$_A$-H$_B$ the only two active MOs are $\sigma(1s_A + 1s_B)$ and $\sigma(1s_A - 1s_B)$ and, with respect to a plane perpendicular to the breaking sigma bond, the former is symmetric and the latter is antisymmetric. As a function of bond length, $\sigma$ and $\sigma^*$ are degenerate at long distances but have a large difference in energy (the bond energy) at short distances. At short distances, this leads to four states: G, the ground state has the $\sigma$ orbital doubly occupied; S$_1$, the first singlet excited state, with one electron in each of $\sigma$ and $\sigma^*$; S$_2$, the doubly excited singlet state, with both electrons in the $\sigma^*$ orbital; and T, the triplet state, with the same orbital occupancy as S$_1$ but with the electrons triplet paired. At long distances, where there is no interaction between 1s$_A$ and 1s$_B$, there are also four configurations: $^1$D, a singlet diradical, with one electron in each atomic orbital; $^3$D, the corresponding triplet diradical; and two zwitterionic states, Z$_1$ and Z$_2$, with double occupancies in either 1s$_A$ or 1s$_B$, respectively. For the symmetric molecule, H$_2$, at long distances, the two ionic configurations are degenerate and at higher energy than the two radical ones, which are also degenerate. Moreover, the ionic MOs are not symmetry-correct and linear combinations must be taken; a symmetric combination, Z$_1$ + Z$_2$, and a corresponding antisymmetric one, Z$_1$ − Z$_2$; however, because the bond distance is long, they remain degenerate. The state correlations are obvious but the shape of the potential energy surfaces is not. Only by including CI in this MO approach do the minima in the S$_1$ and S$_2$ surface become apparent (Fig. 2). The conclusions drawn from these arguments are as follows: the ground state, G, is bound and correlates with the singlet diradical, $^1$D; the triplet state, T, is dissociative and correlates with the triplet diradical, $^1$T; both $S_1$ and $S_2$ are weakly bound, although at longer bond lengths than G, and correlate with the two high-energy zwitterion states, $Z_1 + Z_2$ and $Z_1 - Z_2$.

Extension of these ideas to those more relevant to organic chemistry requires other examples: 1.) introduction of polarity into the bond, as in $CH_3$—$NH_3^+$, which is still ditopic; 2.) introduction of the aromatic chromophore, as in $PhCH_2$—$NH_3^+$, which is still ditopic at the reactive sigma bond but provides an "interactive subunit," the aromatic ring; and 3.) a system of higher topicity, e.g., $CH_3$—$NH_2$, with the lone pair on the nitrogen making it tritopic. Systems of higher topicity like $CH_3$—$OH$ (tetratopic) and $CH_3$—$Br$(pentatopic) are then logical extensions.

For example 1, the obvious major change is that there is now a large electronegativity difference between the carbon and nitrogen and therefore the energy separation between "$Z_1$" and "$Z_2$" [48] at long distances is large. Michl and Bonačić-Koutecký [49] provide a helpful introduction to the effects of bond polarity in sigma bond cleavage reactions by first considering an artificially polarized $H_2$ case where the nuclear charges vary from their natural values of 1.0 at both $H_A$ and $H_B$ to values of 0.8/1.2 and 0.6/1.4, respectively. In the case of

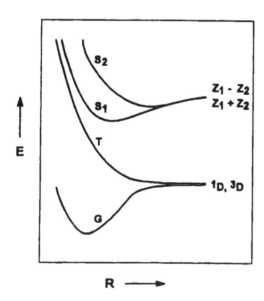

**Figure 2**   Potential energy curves for the dissociation of $H_2$. (Modified from Ref. 47.)

$CH_3$—$NH_3^+$, the structure formed by heterolytic cleavage with the electron pair going to nitrogen, $CH_3^+:NH_3$, "$Z_1$" is energetically much preferred to the alternative possibility with the pair going to carbon, $CH_3^-\ NH_3^{2+}$, "$Z_2$." In fact, "$Z_1$" is more stable even than $CH_3:NH_3$, the species formed by homolytic cleavage, because the ionization potential of $CH_3\cdot$ is lower than that of $NH_3$. This case is more clearly relevant to the photochemical examples where the ion pair in polar solvents is more stable than the radical pair. Now, as the bond stretches, an avoided crossing between $S_1$ and G results. If the difference in polarity ($\delta$) is great enough and the configuration mixing strong enough, $S_1$ becomes dissociative (Fig. 3). The triplet state, T, is unaffected by this increase in the polarity of the bond and is still dissociative, $^3D$.

For example 2, $PhCH_2$—$NH_3^+$, an interaction between the sigma bond and the aromatic chromophore is required. This example is clearly relevant to the photochemical reactions that are the subject of this review. Although Michl and Bonačić-Koutecký use the interaction of the locally excited triplet state of the phenyl ring with the $\sigma,\sigma^*$ triplet state of the sigma bond, the interaction between the locally excited singlet state of the phenyl ring with the $\sigma,\sigma^*$ singlet state (Fig. 4) will result in the same conclusions [50]. Specifically, they state: "The conditions that favor the dissociation of a $\sigma$ bond in a molecule upon local excita-

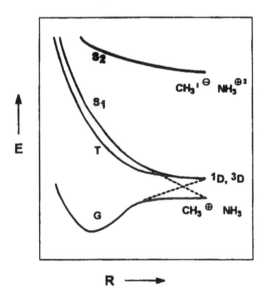

**Figure 3**  Potential energy curves for the dissociation of $CH_3NH_3^+$. The dotted lines indicate an avoided crossing.

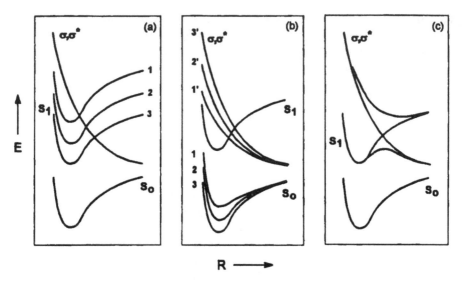

**Figure 4** Potential energy curves for dissociation in the singlet excited state, $S_1$, of a polar, covalent, benzylic σ bond. (Modified from Ref. 47.) (a) The lowering of the excitation energy of $S_1$ (1 to 2 to 3) increases the barrier from the minimum in $S_1$ to the crossing point with the σ,σ* state. (b) The increasing σ bond strength (1 to 2 to 3) increases the energy of the σ,σ* state (1′ to 2′ to 3′) and increases the barrier from the minimum in $S_1$ to the crossing point with the σ,σ* state. (c) Mixing between $S_1$ and the σ,σ* state lowers the barrier to fragmentation from $S_1$.

tion of a π chromophore by minimizing a potential barrier along the way can be summarized as illustrated in (Fig. 4):

1.  The local π excitation energy is large (Fig. 4a).
2.  The sigma bond energy is weak (Fig. 4b).
3.  The excitation is into the triplet state, or intersystem crossing is efficient.
4.  If the excitation is into the singlet state, the sigma bond is polar (large $\delta_{AB}$) and the solvent solvates ions well.
5.  The bond is allylic (benzylic) with respect to the excited π system (Fig. 4c).
6.  The bond is lined up as close to perpendicular to the plane of the π system as possible (Fig. 4c).
7.  The coefficients of the MOs involved in the π excitation are large in the position of attachment (Fig. 4c).

Conditions 1–4 optimize the position of the zero-order curve crossing, and conditions 5–7 optimize the degree to which crossing is avoided.''

For $PhCH_2$—$NH_3^+$, the last factor (vii) is outlined in Fig. 5. The $S_1$ excited state, $L_b$ in simple aromatics, is formed from $S_0$ by a symmetry forbidden electronic transition with low probability (low ε value in Table 1). It also has the wrong symmetry to interact with the sigma bond. Formation of the $S_2$ excited state, $L_a$, is an allowed transition ($\lambda_{max}$ = 204 nm, ε = 7,900, hexane solvent). As shown in Fig. 5b, this state has the correct symmetry and a large coefficient at $C_1$ to interact with the C—N sigma bond. Therefore, mixing is strong and the barrier to formation of the (radical/radical ion) pair is low. In contrast, the barrier on the $L_b$ surface will be high and the natural correlation is to an excited state of the arylmethyl cation. This latter adiabatic process is therefore symmetry allowed but endergonic. Moreover, in solution, rapid internal conversion from upper excited states to lower ones (i.e., $L_a$ to $L_b$) is normally faster than other processes of the excited state so that even if the initially excited state is $S_2$ (i.e., $L_a$), the reactive excited state should still be $L_b$. Therefore, the final prediction is that there should be significant symmetry-created barriers to excited state reactivity in arylmethyl photochemistry. Any perturbation of this symmetry (the meta effect, perhaps) might lower these barriers by allowing mixing between $L_b$ and σ,σ*

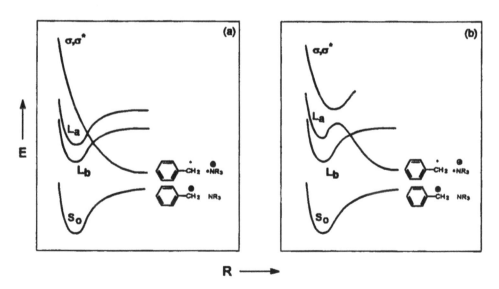

**Figure 5** Potential energy curves for the bond dissociation from $S_1$ of the benzylammonium ion: (a) before interaction between the aromatic ring and the σ bond and (b) after interaction. (Modified from Ref. 47.)

and result in increased efficiency of conversion of $S_1$ by homolytic bond cleavage to the radical pair (in the case of $PhCH_2$—$NH_3^+$, the radical/radical ion pair).

For the last example, (3), $CH_3$—$NH_2$, the additional lone pair on nitrogen provides another active orbital and therefore this case is tritopic. Again, without the aromatic chromophore, the connection to organic photochemistry is tenuous. The ground state and the first excited singlet, n,σ*, are symmetric and antisymmetric, respectively, relative to a plane of symmetry containing the C—$NH_2$ bonds (Fig. 6). If this plane of symmetry is maintained, mixing between these states is forbidden; if nonlinear geometries of bond cleavage occur, the symmetry is lowered and mixing between these two singlet states results in an avoided crossing. Both the excited singlet and triplet n,π* states now have exothermic dissociative pathways available. In a bitopic bond cleavage, as in $H_2$ (Fig. 2), there is a barrier on the $S_1$ surface but this barrier will be lowered, and perhaps disappear, if there is a large difference in electronegativity, as in $CH_3$—$NH_3^+$ (Fig. 3) or if there is strong mixing of the antibonding sigma orbital with a benzylic chromophore, as in $PhCH_2$—$NH_3^+$ (Fig. 4). The inclusion of the lone pair (increased topicity) also removes the barrier that was present on the $S_1$ surface for a ditopic case. The reason for this is the presence of the low-energy n,π* state (absent in the ditopic case) and also the low-lying excited singlet state in the radical-pair product. Finally, the fact that the increased topicity does not change the symmetry of the σ,σ* state should be noted; the triplet σ,σ* still

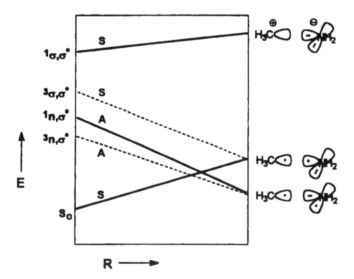

**Figure 6**  State correlation diagram for the dissociation of the C—N bond in methylamine as an example of higher than ditopic case. (Modified from Ref. 47.)

correlates with the triplet radical pair and the singlet $\sigma,\sigma^*$ state with the very high-energy ion pair.

The effect of adding the aryl chromophore in PhCH$_2$—NH$_2$ can now be qualitatively assessed. First, the BDE (RP) is high (84 kcal/mol) [35] so that the interaction between the aromatic chromophore and the sigma bond will be weak. Moreover, the symmetry of the $\sigma,\sigma^*$ state is still wrong for interaction with the lower energy L$_b$ state of the aromatic chromophore. Therefore, no reaction is expected or observed. In analogy to ground state chemistry, the amino group is a poor photochemical leaving group. In order for the photochemical reaction to be efficient, the energy of the radical pair must be lower than it is for PhCH$_2$—NH$_2$, i.e., the bond must be weaker. Moreover, interaction of the L$_b$ state with the $\sigma,\sigma^*$ orbital must be increased by symmetry breaking in order to allow the mixing that produces an avoided crossing and lowers the activation barrier to bond cleavage.

## IV. SURVEY BY LEAVING GROUP

### A. Hydrogen

To our knowledge, photochemical homolytic cleavages of benzylic C—H bonds do not occur. For instance, laser excitation of triphenylmethane at 266 nm in CH$_3$CN gave a transient species with a lifetime of ~1 ns that was assigned to the excited singlet state. No evidence for the triphenylmethyl radical was obtained although it, and the corresponding cation, was observed with other leaving groups [33].

Heterolytic cleavages of C—H bonds have been observed, but with the opposite polarity to that which is the topic of this article, i.e., the compounds behave as photo acids in the excited state, not as hydride donors [51–54]. This topic has been reviewed recently by Wan and Shukla [55]. These reactions are particularly efficient when the conjugate base of the acid is a delocalized, Hückel's rule, 4n $\pi$-electron carbanion (known as antiaromatic). Thus, the compounds, 13, 14, and 15, all exchange hydrogen for deuterium at the methylene carbon when irradiated in deuterated aqueous media. Compound 13 is the most reactive one and undergoes exchange in 1 : 1 D$_2$O : CH$_3$CN in the absence of a base whereas 14 and 15 require base catalysis [1 : 1 (1 M NaOD/D$_2$O) : EtOD or ethanolamine in CH$_3$CN]. Coumpounds like 16 and 17, which cannot form 4n $\pi$-electron carbanions, are inert to photochemical H/D exchange.

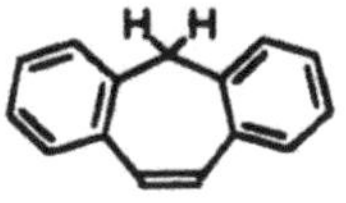

13 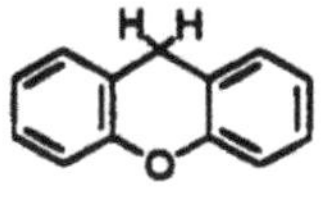       14 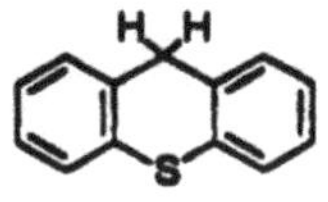       15

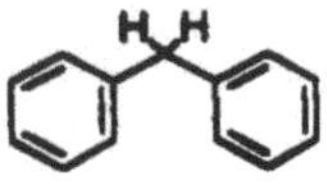

**16** 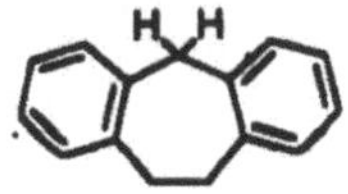                    **17**

## B.  Carbon

Although carbon is not a heteroatom and strictly does not fall into the category of the leaving groups covered by this chapter, suitably substituted carbons do undergo photochemical benzylic cleavage reactions, cyanide ion being perhaps the most obvious example. It and the six others summarized below are included for completeness; they are arranged in order of decreasing oxidation state of the carbon.

## 1.  Nitrile Carbon: —C≡N

Cyanide ion is apparently not a very good leaving group because the only known examples of photoreactive substrates are those that yield intermediates of high stability like the triarylmethane derivatives, malachite green leucocyanide, **18** [56], and Victoria blue leucocyanide **19** [57]. Both were shown to undergo photo-fragmentation from their S₁ state to give the highly colored triaryl cation/cyanide anion ion pair. In both of these compounds, the electron-donating substituent is para to the reactive benzylic-leaving group bond. This ionization occurs in competition with fluorescence, intersystem crossing and, for **19**, intramolecular charge transfer. No evidence was found for homolytic cleavage of the carbon–carbon bond. For **18**, LFP studies showed that the heterolytic cleavage was very dependent on the dielectric constant of the solvent with the rate constant (2 × 10⁸ s⁻¹ to 10 × 10⁹ s⁻¹) and quantum yield (0.40 to 0.97) of ion pair formation increasing and the quantum yield of fluorescence (0.07 to 0.0035) dropping as the solvent changed from ethyl acetate ($\varepsilon = 6.0$) to methanol ($\varepsilon = 33$). The focus of this work was on the picosecond dynamics of ion pairs.

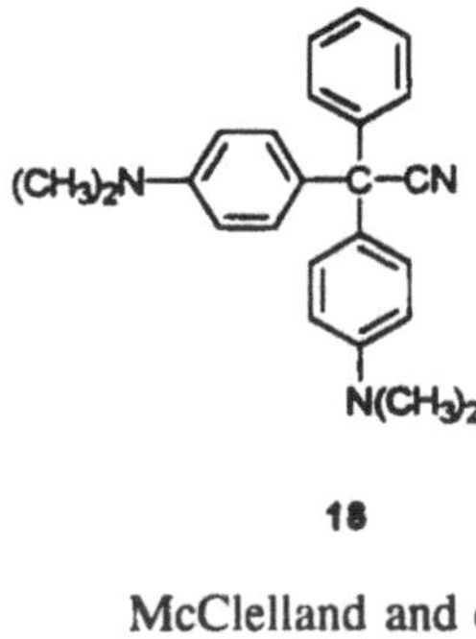

**18** 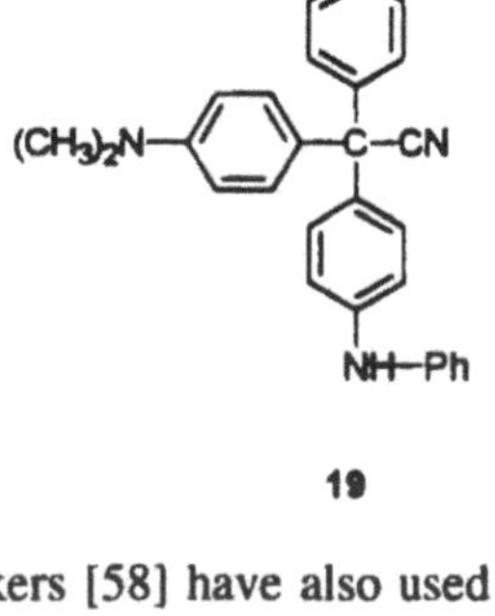                              **19**

McClelland and co-workers [58] have also used cyanide ion as a leaving group to generate carbocations for nanosecond LFP studies from methoxy-substituted triarylmethane derivatives. No cations were observed for the analogous diarylacetonitriles. Similarly, in related vinylogous benzylic derivatives **20**, Pischel and co-workers [59] have shown that tropylium ions are not formed on photolysis in acetonitrile or methanol if the potential leaving group is cyanide ion but are if it is methoxide ion.

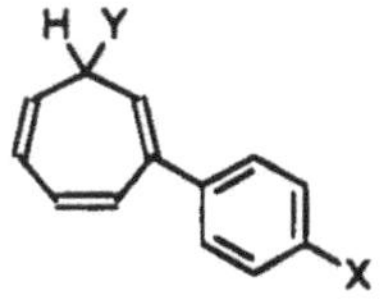

**20**: Y = methoxy, cyanide
X = H, methyl, dimethylamino

### 2. Acyloxy Carbon: —(C=O)—OH

Decarboxylation reactions at benzylic carbon are quite common and the subject has been reviewed recently [60]. For instance, irradiation of phenylacetic acid, Eq. (3), in methanol gives bibenzyl, clearly indicating the formation of benzyl radicals. Moreover, flash photolysis in methanol gives a transient species that was assigned to the benzyl radical [61].

Recently, Wan and co-workers [54,62] have published a number of papers on the photodecarboxylation reactions of arylmethyl carboxylates as in Eqs. (4) and (5). The heterolytic cleavage is in the direction expected forming benzylic carbanions rather than cations.

$$PhCH_2CO_2H \xrightarrow[CH_3OH]{h\nu} [PhCH_2\cdot \ \cdot CO_2H] \longrightarrow PhCH_2CH_2Ph \qquad (3)$$

(4)

(5)

### 3. Acyl Carbon: —(C=O)—R

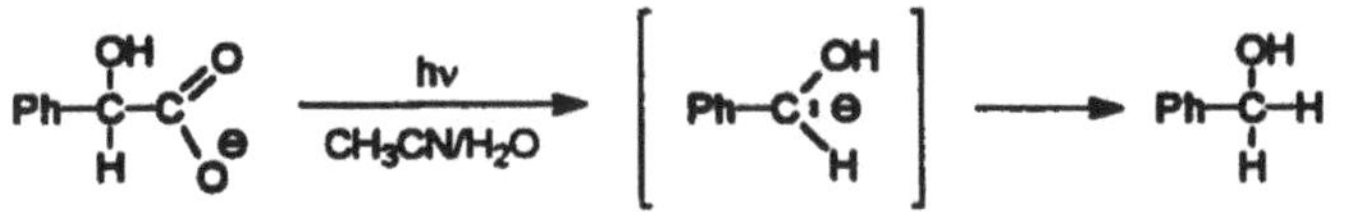

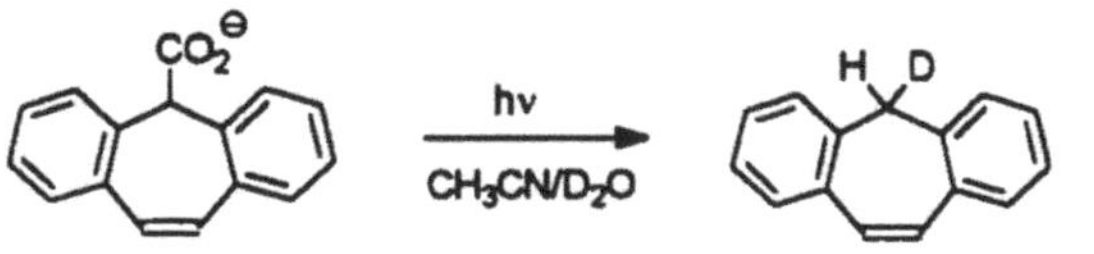

Arylmethylketones, like dibenzylketone, undergo excited state arylmethyl carbon–carbon homolytic bond cleavage but the reactions occur from the n,π* triplet excited state of the carbonyl functional group (Norrish type I reaction) and are not induced by the aromatic $\pi,\pi^*$ state [63].

## 4. Ketal Carbon: —C(OR)₂—R

Photoheterolysis of nitrobenzyl ketals, **21**, Eq. (6), has been shown to generate benzylic cations and nitrobenzyl anions, both by nanosecond LFP and by characterization of the products obtained [64,65]. In this case, both of the carbons are benzylic but the lower energy chromophore should be the nitro-substituted one. Moreover, normally nitro groups induce rapid intersystem crossing and formation of $n,\pi^*$ triplet states. More information is therefore needed to clarify the surprising observation of the generation of an ion pair by direct heterolytic cleavage.

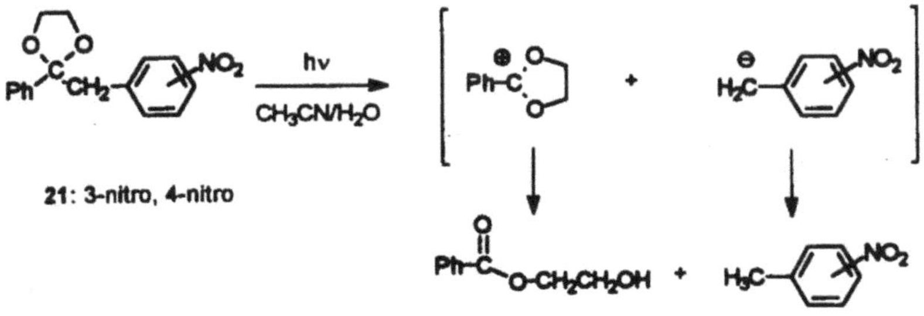

**21**: 3-nitro, 4-nitro

## 5. Hydroxymethyl Carbon: —C(OH)R₂

Photochemical retroaldol reactions at benzylic carbon have also been observed, Eq. (7), the high pH required indicating that these reactions are base-catalyzed [64]. The product ratio of toluene to bibenzyl is also pH-dependent with the latter compound favored at higher pH. The toluene is formed by protonation of the intermediate benzylic carbanion. At higher pH, this protonation step becomes slower and the carbanion forms the dimer. The mechanism for this oxidative dimerization, which occurs in the absence (argon purging) of oxygen, is not yet clearly understood.

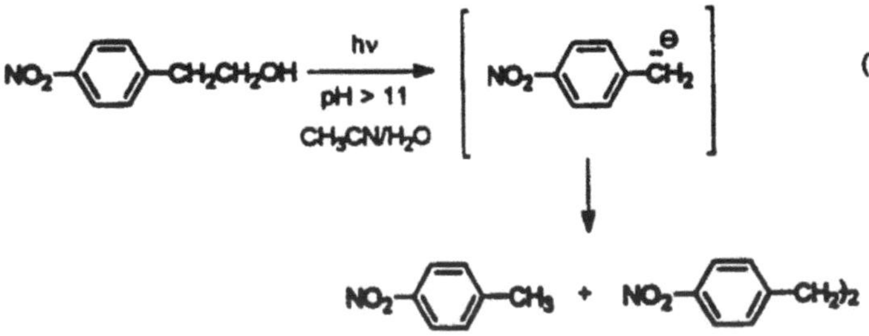

In contrast, irradiation of **22**, Eq. (8), at 254 nm in acetonitrile/water at pH 7 gives bibenzyl and dibenzosuberenone, clearly indicating homolytic cleavage of the benzylic bond [52].

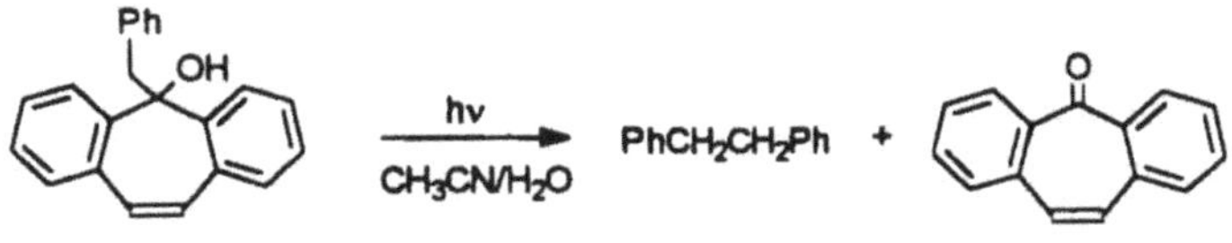

$$\text{\textbf{22}} \quad \xrightarrow[\text{CH}_3\text{CN/H}_2\text{O}]{hv} \quad \text{PhCH}_2\text{CH}_2\text{Ph} \;+\; \qquad\qquad (8)$$

## 6. Cyano-Substituted Carbon: —C(C≡N)R₂

The reactivity of compounds **23**, **24**, and **25**, which are benzylic at both ends of the reactive carbon–carbon bond, has been studied by both picosecond and nanosecond LFP after excitation at 355 nm [66]. The bond cleaves heterolytically to give the triphenylcyclopropenyl cation and the corresponding substituted carbanion. UV absorption spectra demonstrate that the two chromophores overlap so that excitation occurs in both the cyclopropene and the substituted aromatic ring of these molecules. The mechanism for cleavage begins with intramolecular electron transfer in the excited singlet state to give the radical cation of cyclopropene and the radical anion of the substituted aromatic. In competition with back electron transfer, the sigma bond then cleaves to give the triphenylcyclopropenium cation and the substituted benzylic carbanion. For **23c**, only back electron transfer occurs, and for **23a** and **24** only the cation, but not the anion, was observed. No evidence for homolytic cleavage to radicals was obtained. Redox potentials indicate that the ion pair is more stable than the radical pair.

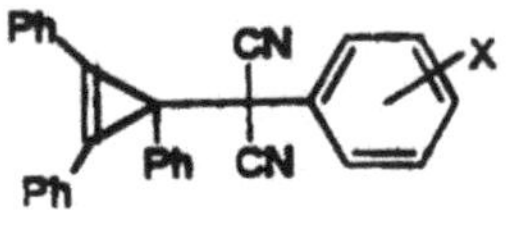

**23a**, X = 4-cyano
**23b**, X = 4-nitro
**23c**, X = 3-nitro

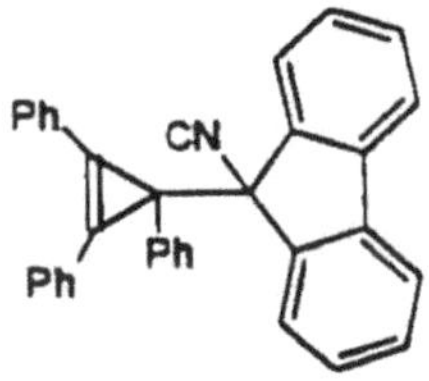

**24**

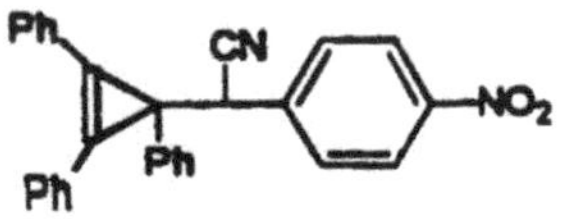

**25**

## 7.  β-Keto Carbon: —CH₂(C=O)—R

The only known examples of this case are the xanthylketones, **26**, which can be
used to generate xanthyl cations, **27**, Eq. (9), by LFP [67]. However, this reaction
occurs by β cleavage of the n,π* triplet of the ketone and the xanthyl radical
formed is oxidized to the cation by oxygen. In the absence (nitrogen purging)
of oxygen, the xanthyl radical persists.

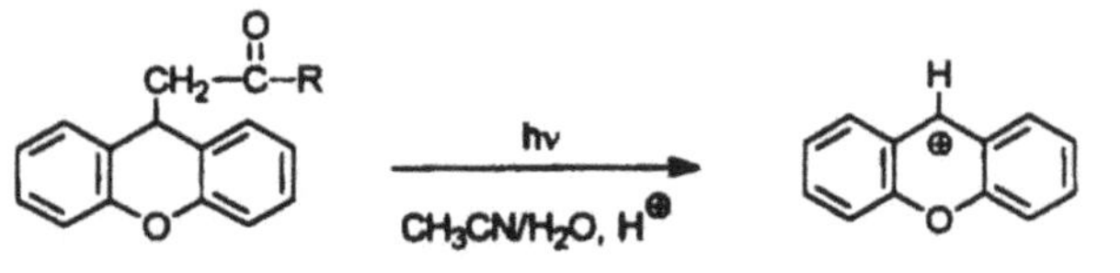

**26**: R = methyl, phenyl                                    **27**

## 8.  Cyclopropyl Carbon

An interesting example of carbon behaving as a leaving group occurs for phenyl-
substituted cyclopropanes, **28** [68]. As shown in Eq. (10), photolysis in acetonitrile results in trans-to-cis isomerization along with formation of the rearranged
acetate. Evidence for ion pair intermediates was provided by results from photolysis in nucleophilic media like methanol and water-acetonitrile where ethers and
alcohols, respectively, were formed. Quantum yields of reaction in acetonitrile
along with singlet lifetimes were used to obtain the rate constants for these excited
singlet state reactions. Surprisingly, the most reactive compounds were the 3-
cyano–($k > 157 \times 10^6$ s$^{-1}$) and 3-trifluoromethyl–($140 \times 10^6$ s$^{-1}$) substituted
ones and the least reactive one was the 3-methoxy compound ($<0.25 \times 10^6$ s$^{-1}$).
The transition state for this heterolytic cleavage apparently has an electron demand
opposite to that of the usual benzylic carbon–oxygen bond cleavage reactions.

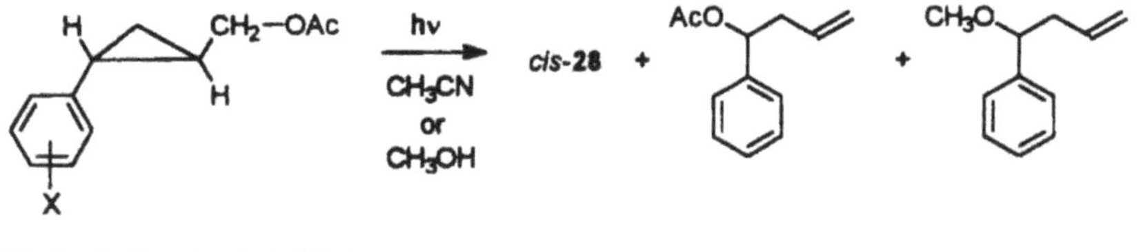

**28**: X = hydrogen, 4-methoxy,
3-methoxy, 4-methyl, 3-methyl,
4-cyano, 3-cyano, 3-trifluoromethyl

## C.  Nitrogen

To our knowledge, no unambiguous results on the direct photochemical cleavage
reactions of benzylic amines or amides have been reported. Presumably both of
these leaving groups are very unreactive; e.g., N-(1-naphthylmethyl)aniline [69]
and the naphthylmethylamides, **29** and **30** [70], are photochemically inert in

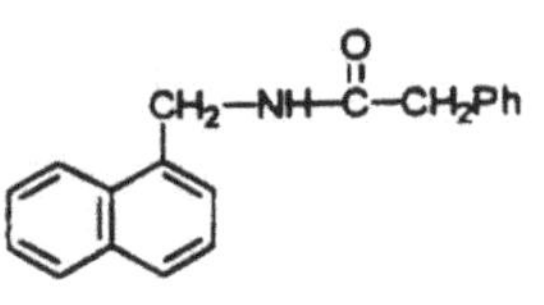

**29** 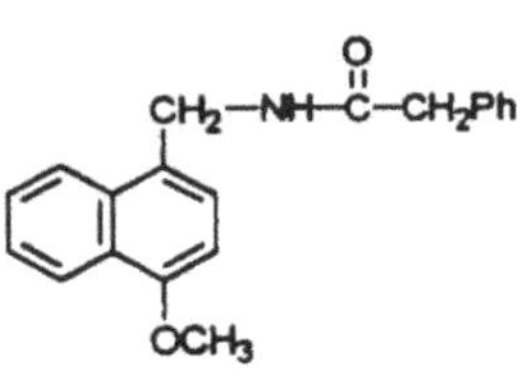                              **30**

methanol. Photolysis at 254 nm of benzylic adenine derivatives **31** in water, Eq. (11), results in efficient formation of the free amine and the benzyl alcohol [71]. However, the mechanism of these reactions is not clear because both the adenine and the benzylic protecting group are chromophores. Moreover, the reactions are most efficient for the 3-bromo and 3-chloro compounds, which suggests intervention of triplet states. However, both the 3-methoxy and 3,5-dimethoxy compound also react faster than the unsubstituted compound, suggesting meta activation.

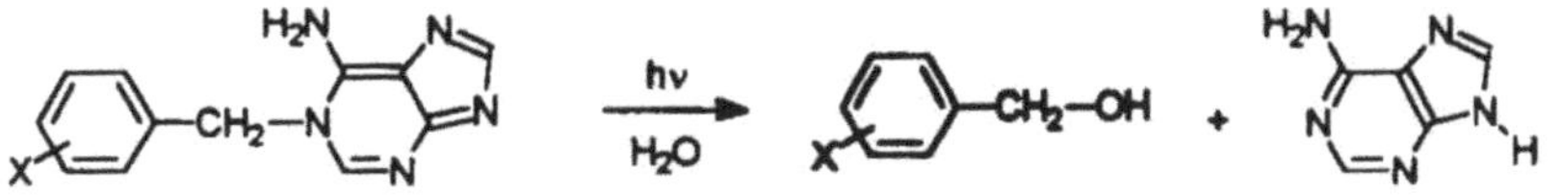

$$\text{(11)}$$

**31**: X = hydrogen, 3-methoxy, 3-fluoro,
3-chloro, 3-bromo, 3-amino, 3,5-dimethoxy

A recent LFP, electron spin resonance, and product study on the photodissociation of *N*-(triphenylmethyl)anilines concludes that the primary photochemical event is homolytic cleavage [72]. However, the chromophore in these molecules is the aromatic amine. Triphenylmethyl cations were observed when the excitation wavelength was 248 nm but these were apparently formed by a biphotonic process of electron photoejection from the triphenylmethyl radical.

## 1. Amide Nitrogen: —NH(C=O)R

There is considerable interest in the photochemical release of biologically active amines and amino acids from biologically inactive "caged" precursors for the study of fast kinetics, time-resolved crystallography, and combinatorial chemistry [73]. To this end, the *ortho*-nitrobenzyl photolabile precursor has received considerable attention. A recent study [73] on structural effects on the rates of the photochemical cleavage has concluded with the finding that the half-life for cleavage of the acetamide bond in the model compound, **32**, Eq. (12), is under a minute on irradiation at 350–400 nm in aqueous buffers, methanol, and dioxane. The four-carbon amide side chain in **32** can be used to couple the model compound to solid supports and photoreactivity is still maintained. These reactions do not, however, result from excited state benzylic cleavage of the carbon–nitrogen bond. Instead, the excited state process is hydrogen atom abstraction by the nitro group of the benzylic hydrogen followed by oxidation at the benzylic carbon and reduction of the nitro group to a nitroso group, **33**. The amide is then released by ground state hydrolysis. Similar *ortho*-nitrobenzyl protection [74] of NAD(P)$^+$ and other biologically active amides [75,76] has also been reported.

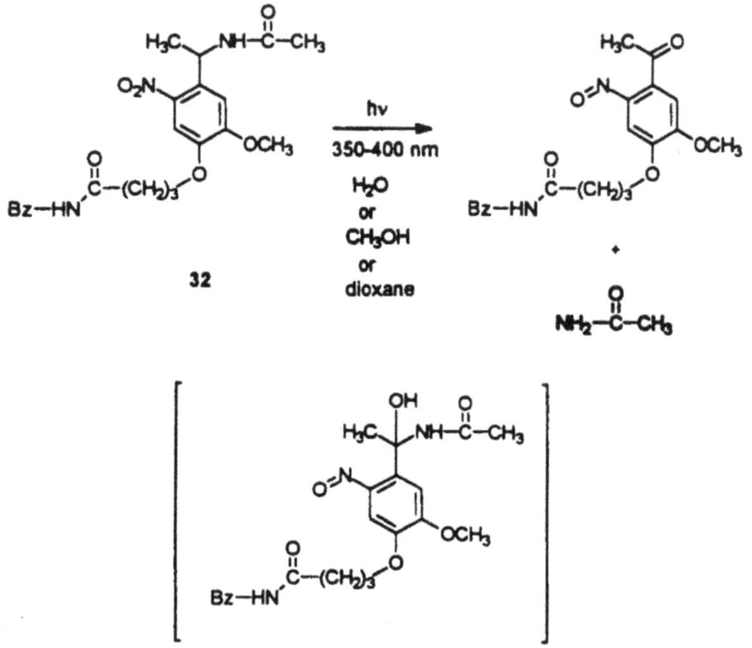

(12)

**32**

**33**

## 2.  Ammonium Salts: —N$^+$R$_3$

The early work on the photochemistry of benzylammonium salts has been reviewed by Cristol and Bindel [1]. Two examples are given in Eqs. (13) and (14) [21]. These reactions exhibit the usual mechanistic dilemma that arises in explaining observations on this class of reactions. The ether product in each case is clearly derived from trapping of an intermediate benzylic carbocation by the solvent, methanol (photosolvolysis); the other products come from an intermediate benzyl radical. Acetone sensitization of the unsubstituted bromide in water was used to indicate that the triplet state gave, as expected, only radical-derived products, but in considerably reduced yield, particularly in the case of toluene. Therefore most of the toluene was singlet state–derived. Quenching studies with piperylene for the unsubstituted bromide in aqueous *t*-butyl alcohol indicated that the yield of ion-derived products was unaffected by the piperylene. In contrast, the radical-derived products were quenched, although in two different ways. The Stern-Volmer plot for bibenzyl was linear and gave a triplet lifetime of 4.1 × 10$^8$ s$^{-1}$; the analogous plot for toluene was nonlinear, indicating that approximately 40% of the toluene was singlet-derived. The unsubstituted bromide showed both fluorescence ($S_1 = 112$ kcal/mol) in solution and phosphorescence ($T_1 = 82$ kcal/mol) in low-temperature glasses.

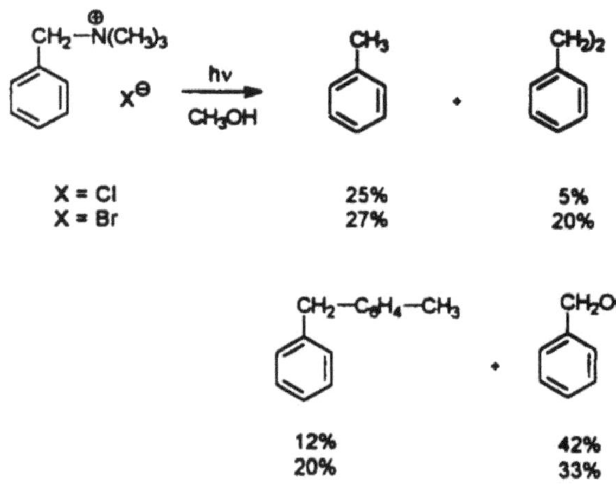

|              |       |        |
|--------------|-------|--------|
| X = Cl       | 25%   | 5%     |
| X = Br       | 27%   | 20%    |

|              |       |        |
|--------------|-------|--------|
|              | 12%   | 42%    |
|              | 20%   | 33%    |

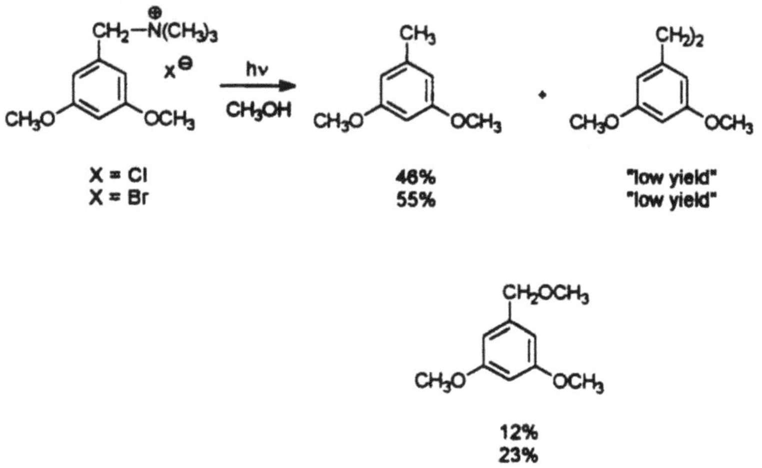

|              |           |             |
|--------------|-----------|-------------|
| X = Cl       | 46%       | "low yield" |
| X = Br       | 55%       | "low yield" |

|              |       |
|--------------|-------|
|              | 12%   |
|              | 23%   |

The authors proposed a mechanism similar to the one in Scheme 1 (see p. 213). The excited singlet state reacts to give both the in-cage ion pair, 34, and the radical pair, 35, but the question of whether these are two separate species or two resonance contributors was left open. Toluene and the methylether are derived from this singlet pair; the quantum yields are 0.19 and 0.11, respectively, in 50:50 t-butyl alcohol/water for the unsubstituted bromide. The triplet radical pair is formed from the excited triplet state with a quantum yield of 0.2 in 50:

50 t-butyl alcohol/water, again for the unsubstituted bromide. The triplet radical pair gives mostly bibenzyl by out-of-cage radical coupling. A surprising observation of this study was that the 3,5-dimethoxy substrates, Eq. (14), gave a lower yield of the ion-derived products than did the unsubstituted ones, Eq. (13). This is not consistent with the meta effect arguments summarized in the Introduction and Scheme 2.

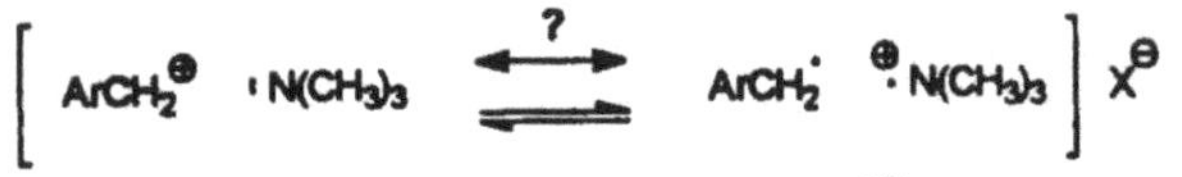

$$\left[ \text{ArCH}_2^{\ominus} \quad \cdot \text{N(CH}_3)_3 \underset{\phantom{?}}{\overset{?}{\rightleftarrows}} \text{ArCH}_2^{\cdot} \quad \overset{\oplus}{\cdot} \text{N(CH}_3)_3 \right] X^{\ominus}$$

$$\phantom{xxxxxxxxx}\textbf{34}\phantom{xxxxxxxxxxxxxxxxx}\textbf{35}$$

In order to explore substituent effects for ammonium salts in more detail, Pincock and co-workers examined the series of substituted 1-naphthylmethylammonium salts, 36, in Eq. (15) [22]. The fluoroborate salts were chosen to avoid the complication of nucleophilic quenching of carbocation intermediates by the counterion. As discussed in the Introduction, the naphthalene chromophore is a simpler one to study because singlet lifetimes are longer and, additionally, both sensitization and quenching experiments are easier to interpret. In this case, both xanthone sensitization and diene quenching indicated that only the excited singlet state was reactive. This is probably because the triplet excited state, at approximately 60 kcal/mol (Table 1), does not have sufficient energy to induce cleavage of the strong carbon–ammonium nitrogen bond (Table 3). As shown in Eq. (15), the reactions are very clean with excellent mass balance for the two products. The methylnaphthalene is formed by hydrogen atom transfer for the in-cage radical pair, Eq. (16), and the methyl ether by nucleophilic trapping of the ion pair by the solvent, methanol. As for the benzylammonium salts discussed above, the yield of ion-derived product is lower for the meta-methoxy compound than for the unsubstituted one.

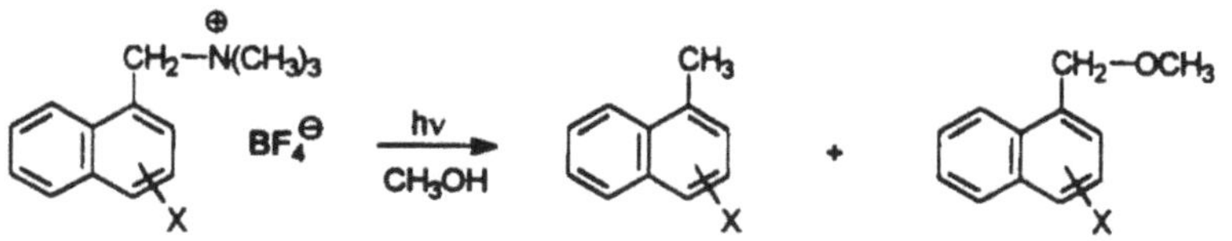

| | Yield | $k^{S_1}{}_{hom}$ $\times 10^{-7}$ (s$^{-1}$) | Yield | $k^{S_1}{}_{het}$ $\times 10^{-7}$ (s$^{-1}$) |
|---|---|---|---|---|
| **36a**: X = H | 21% | 1.4 | 72% | 4.9 |
| **36b**: X = 4-methoxy | <5% | <0.1 | 90% | 18.7 |
| **36c**: X = 3-methoxy | 35% | 3.6 | 61% | 6.0 |
| **36d**: X = 4-cyano | 62% | 2.6 | 34% | 1.2 |
| **36e**: X = 3-cyano | 46% | 0.5 | 48% | 0.7 |

$$(16)$$

By measuring excited singlet lifetimes and quantum yields of product formation, rate constants for both homolytic ($k_{hom}^{S1}$, Scheme 1) and heterolytic ($k_{het}^{S1}$) cleavage from $S_1$ can be estimated. These are also given in Eq. (15). Obviously, several assumptions are being made in calculating these values, the major one being that the radical pair is not converted to the ion pair even though the ion pair is expected to be more stable than the radical pair in methanol (Table 3). In other words, the yield of radical-derived and ion-derived products is assumed to be determined by only the partitioning between homolytic and heterolytic cleavage in the excited state. If this mechanism is correct, these valves reveal that the ion-derived product is formed fastest from the 4-methoxy substrate and the radical-derived product fastest from the 3-methoxy one, an order that is again contrary to expectations based on the meta effect.

As will be discussed in Sec. IV.D.3, the decarboxylation rate of acyloxy radicals can be used as a radical clock to probe the reactivity of the radical pairs formed in benzylic ester photochemistry. This is not straightforward for the radical pair from ammonium salts because the hydrogen atom transfer rate will be dependent on the substituents, X, Eq. (16). However, this rate effect is likely to be small. If so, and if the assumption is made that the initial excited state process is only homolytic cleavage to the radical pair, then the yield of ion pairs will be controlled by the rate of electron transfer and therefore the change in the oxidation potential of the 1-naphthylmethyl radical as a function of substituents. Figure 7 shows a plot of the log $\{[\% \text{ ether}]/[100 - \% \text{ ether}]\} = \log (k_{etn}/k_{ht})$ for the radicals derived from compounds **36** versus the free energy of electron transfer. These latter values were calculated from the oxidation potential of the 1-naphthylmethyl radical [46] and the oxidation potential of trimethylamine (0.82 V versus $Ag^+$) [77]. The hydrogen atom transfer rate, $k_{ht}$, is assumed to be constant. The yield of ion-derived product is well correlated to the change in oxidation potential, with all points falling in the "normal" region of the Marcus equation [78], i.e., the more favorable the electron transfer is thermodynamically, the faster the rate.

In order to increase the range of substituent effects, Pincock and co-workers [79] have also studied the photochemistry of two anilinium salts, **37**, Eq. (17). Synthetic difficulties have prevented an extension of this work. The radical-derived products are now more complicated because radical coupling to give photo-Fries-type products competes with formation of 1-methylnaphthalene.

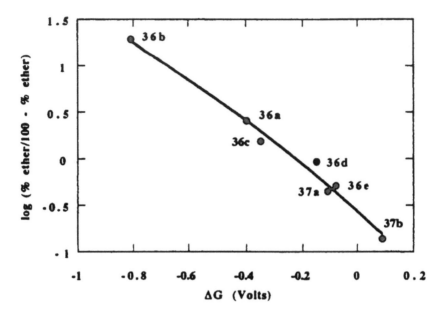

**Figure 7** Plot of log [% ether/(100 − % ether)] versus $\Delta G$ (volts) for conversion of the radical pair to the ion pair formed in the photolysis of ammonium salts **35** and **37**.

The yield data for the two compounds can be added to the plot for the results from compounds **36** (Fig. 7) using the oxidation potential for aniline (0.53 V versus Ag⁺) and 4-methoxyaniline (0.33 V versus Ag⁺). The correlation is still very good, providing further support for a mechanism involving homolytic cleavage from $S_1$ followed by electron transfer in the radical pair to form the ion pair.

$$\text{(17)}$$

**37a**, X = H                    31%
**37b**, X = 4-methoxy            12%

## D.  Oxygen

### 1.  Alcohols: —OH

The photocleavage of arylmethyl alcohols has been studied extensively, mainly by Wan and co-workers, and the results have been recently reviewed as part of an article on acid–base behavior of excited states [55].

For substrates that are precursors to stable cations these photoheterolysis reactions occur in neutral media. For instance, leucohydroxides like **38** are photochromic and give increased pH values when irradiated in aqueous solutions, Eq. (18) [80]. As well, these reactions are at least partially adiabatic and this topic was recently reviewed [81]. For instance [82], irradiation of 9-phenylxanthen-9-ol, **39**: X = H, at 270 nm in aqueous buffer at pH 7, gives observable emission with a $\lambda_{max}$ of 507 nm. This emission, which was only observed in aqueous media and not in ethanol or acetonitrile, was assigned to fluorescence from the excited state of the corresponding cation. The cation was also trapped by methanol giving the corresponding methyl ether. These observations were rationalized by a proton-assisted adiabatic heterolytic carbon–oxygen bond cleavage with water being the proton source. Of interest is the observation that the heterolytic ring oxygen is ortho to the reactive benzylic center perhaps providing, as predicted by the ortho/meta effect, the driving force for ion pair formation. Later, 248-nm LFP experiments in aqueous acetonitrile demonstrated that the quantum yield of ion formation was 0.4 and that only 1% of this yield was by the adiabatic route [83]. The 9-phenylxanthenyl radical was also shown to be the major transient produced in nonpolar hydrocarbon solvents. A series of aryl-substituted xanthylium cations have also been generated by LFP at 248 nm of the xanthenols, **39**, and the rate constants for the reaction of the ions with various nucleophiles ($H_2O$, $OH^-$, $CN^-$, $SO_3^{-2}$, $N_3^-$, $R$-$NH_2$) have been measured [84]. The Hammet slopes (versus $\sigma$) give low positive $\rho$ values of approximately 0.5, although higher slopes were observed for cyanide (1.0) and azide (0.8) ions. The nucleophilicities were also shown to follow the Ritchie equation for $N^+$ with a slope near unity. For the xanthylium ion itself, **40**, the slope was considerably smaller (0.65), suggesting that as the reactivity of the cation increases, the rate constants are being controlled more by diffusion or perhaps desolvation. Other examples of adiabatic dehydroxylations have been observed for dibenzosuberenol, **41**: R = H[85] and the thioxanthenols, **42** [86].

(18)

**38**

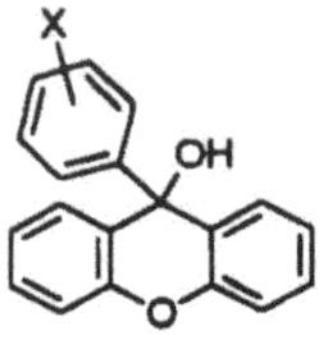

**39**: X = H, 4-methoxy, 4-methyl,
3-methyl, 4-fluoro, 3-methoxy,
4-chloro, 4-trifluoromethyl

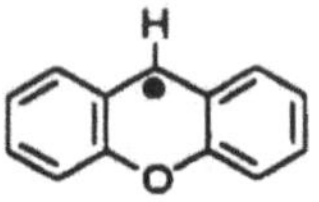

**40**

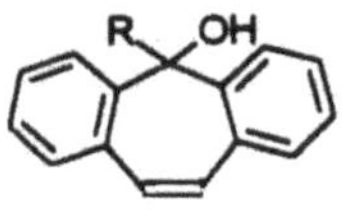

**41**: R = H, methyl

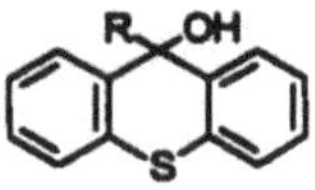

**42**: R = H, methyl,
phenyl, benzyl

These photodehydroxylations are particularly efficient when the cation formed is a cyclically conjugated $4\pi$ electron system as in the case of fluorenyl substrates, **43**, Eq. (19), as compared to the 5-suberenol derivatives, **41**, which were less reactive [87]. Photolysis of **43**: R = H in pure methanol also gave the ion-derived ether, although in lower yield and now accompanied by radical-derived products. Photolysis in acetonitrile gave mostly radical-derived products and only a low yield (10%) of the amide derived from trapping of the cation by the solvent. Picosecond LFP results by Mecklenburg and Hilinski [88] for **43**: R = H and R = Ph in 9:1 water/methanol demonstrate prompt (<20 ps) formation of the corresponding fluorenyl carbocations. The signals for the two cations decayed with lifetimes of <20 ps and 275 ps, respectively. In the more weakly ionizing solvent, methanol, excited singlet states, $S_1$, of both substrates were observed and they decayed to give the fluorenyl radical. No evidence was obtained for formation of the excited states of carbocations.

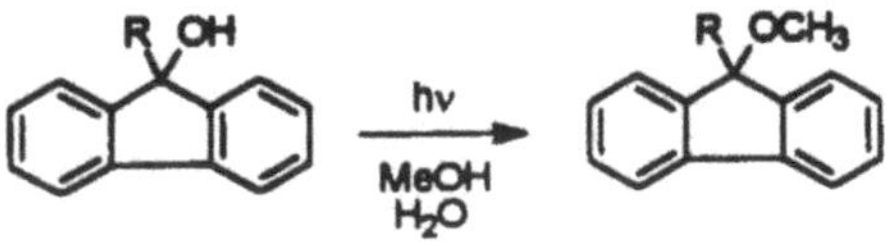

(19)

**43**: R = H, methyl, phenyl

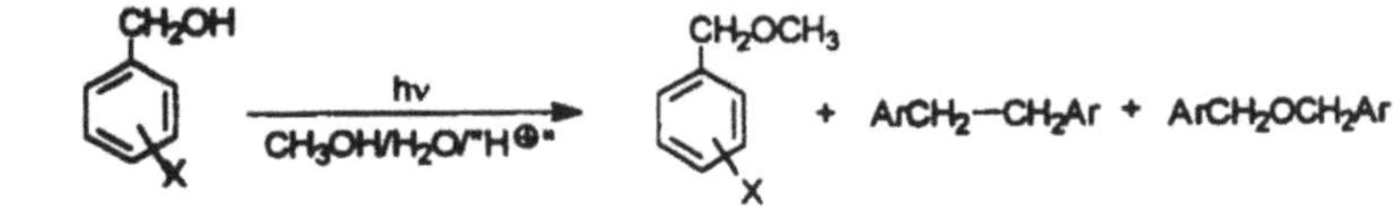

$$\text{(CH}_2\text{OH structure)} \xrightarrow[\text{CH}_3\text{OH/H}_2\text{O/"H}^{\oplus\text{"}}]{hv} \text{(CH}_2\text{OCH}_3\text{ structure)} + ArCH_2\!-\!CH_2Ar + ArCH_2OCH_2Ar \quad (20)$$

**44**: X = 2-methoxy, 3-methoxy,        >90%                    <5%
4-methoxy, 3,5-dimethoxy

Much of this work on non-acid-catalyzed photodehydroxylations was stimulated by the early study of **44**, Eq. (20), of Turro and Wan [89] and later studies by Wan and co-workers on similar acid-catalyzed reactions of substituted benzyl alcohols [90–93]. As indicated in Eq. (20), the ion-derived ether products are clearly the dominant ones formed under these conditions. Trapping of the carbocations by other alcohols, acetic acid, and cyanide ion has also been observed. Moreover, both steady-state and time-resolved fluorescence quenching as a function of increased acidity showed that the products were formed in parallel with the decrease in fluorescence. These sigmoid quenching curves showed half quenching in the pH range 1.5–0.5. Therefore, a mechanistic scheme was proposed in which the fluorescence quenching rate constant, $k_Q$, is equivalent to the acid-catalyzed photodehydroxylation rate constant, $k_H$, from the excited singlet state. Results from other substituted benzyl alcohols soon followed and the $k_Q = k_H$ values are given in Table 4. All of these alcohols also undergo photodehydroxylation at a much slower rate at pH 7; quantum yields, where available, are also given in Table 4. These rate constants and quantum yields clearly show the order expected by the ortho/meta effect with the 2,6-dimethoxy compound being the most reactive, followed by other 2- and 3-substituted cases, with the 4-methoxy being the least reactive by far.

Finally, this catalysis of photodehydroxylation can be intramolecular as in Eq. (21). Excitation increases the acidity of the phenolic oxygen of **45** and,

**Table 4**   Rate Constants ($k_Q$) and Quantum Yields ($\Phi$) for the Photodehydroxylation of Substituted Benzyl Alcohols

| Substituent | $k_Q = k_H \times 10^{-9}$ (M$^{-1}$ s$^{-1}$) | $\Phi$ in 50% CH$_3$OH/H$_2$O |
|---|---|---|
| 2,6-Dimethoxy | — | 0.31 |
| 2,5-Dimethoxy | — | 0.18 |
| 2-Methoxy | 12 | 0.058 |
| 3,5-Dimethoxy | 14 | 0.025 |
| 3-Methoxy | 3.7 | 0.01 |
| 3-Methyl | 0.35 | — |
| 3-Fluoro | 0.16 | — |
| 4-Methoxy | — | 0.0 |

presumably, also the basicity of the benzylic oxygen, so that proton transfer to form the intermediate but unstable *ortho*-quinonemethide is efficient.

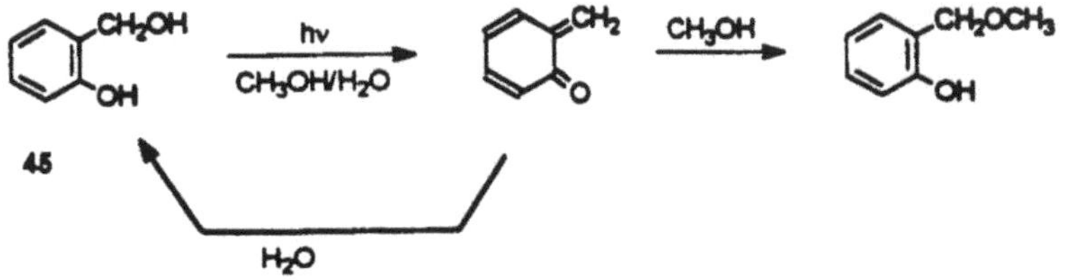

(21)

## 2. Ethers: —OR

The alkoxy group, like hydroxide, is not a particularly good leaving group in benzylic photochemistry and examples of reactions seem to be known only in cases where the benzylic fragment is highly stabilized as a radical or a cation. Thus, in a pioneering study on substituent effects on benzylic reactivity, Zimmerman and Somasekhara examined nitro- and cyano-substituted aryltriphenylmethyl ethers, **46**, Eq. (22) [94]. The yield of the ion-derived products, the phenol and the triarylcarbinol, are considerably higher than those for 9-phenylfluorene, which is radical-derived. In terms of substituent effects, the quantum yields (in brackets) for formation of the carbinol are interesting: the efficiency is higher for 3-nitro and 3-cyano when compared to the corresponding 4-substituted compounds. Again, this observation was rationalized by excited state meta effects but, in these cases, with the opposite polarity to that of the electron-donating groups. Structures like **47** were drawn to rationalize the enhanced reactivity. Although these are photochemical benzylic cleavage reactions, the excitation energy is now mainly localized in the leaving group, not in the benzylic fragment.

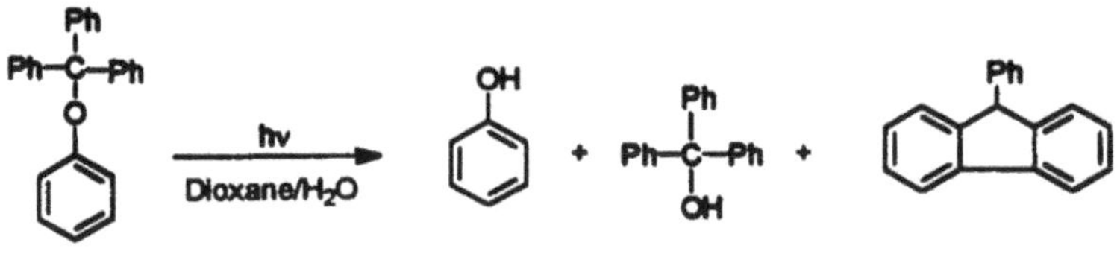

(22)

| | | | |
|---|---|---|---|
| **46**: X = 3-nitro | 81% | 65%<br>(0.034) | 20% |
| **46**: X = 4-nitro | — | —<br>(<0.007) | — |
| **46**: X = 3-cyano | 80% | 66%<br>(0.31) | 14% |
| **46**: X = 4-cyano | 72% | 65%<br>(0.15) | 19% |

Similarly, the 4-cyanophenyl group has been used as a leaving group to generate carbocations in 1:2 acetonitrile/water by nanosecond LFP as in **48** where a large range of substituted aryl rings was examined [95]. In most cases, aryl-substituted methyl radicals were also observed, with the ratio of cation to radical signal varying from as high as 10:1 for methyl- and methoxy-substituted examples of **48**: R = Ar to as low as 1:10 for unsubstituted **48**: R = H. The cation formation (with one exception) was complete within the 20-ns laser pulse, but with compounds **48**: R = H the radical signal continued to grow after the flash. These experiments were aimed at measuring the reactivity of the cations generated.

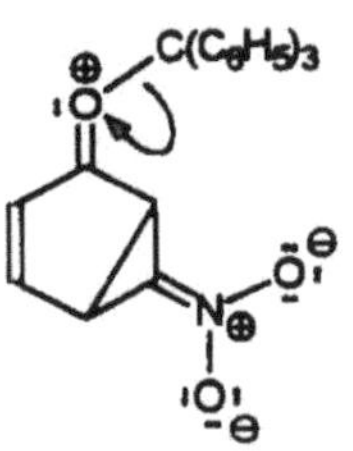

47 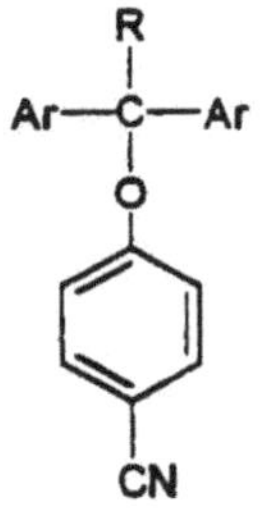                    48: R = H, Ar

The photochemistry of benzylic ethers of phenols has also been examined, Eqs. (23) (the numbers in brackets are quantum yields of formation) [96] and (24) [97]. For **49**, the variation in yield on photolysis in cyclodextrins was also examined. The products are all radical-derived as is typical of photo-Fries chemistry. Analogous products are also observed for 1-naphthylmethylphenyl ether [69]. Again, although these are benzylic cleavage reactions, the excitation energy will be mainly localized in the aryloxy and not the benzylic chromophore.

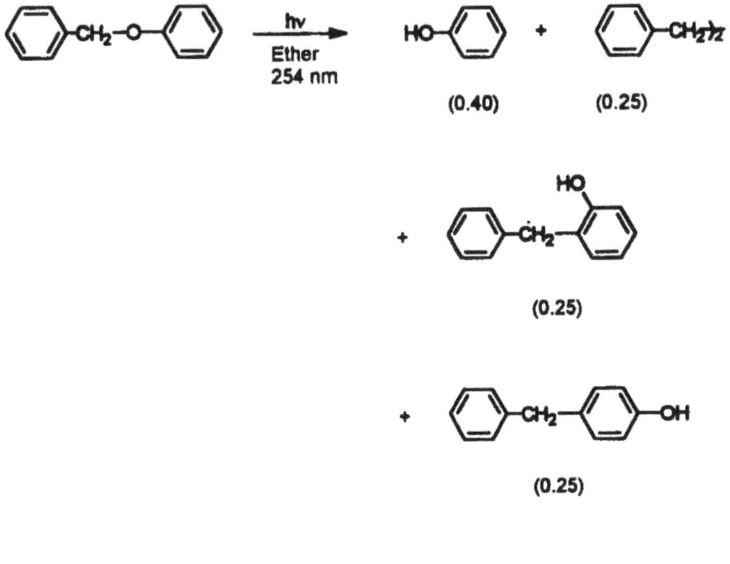

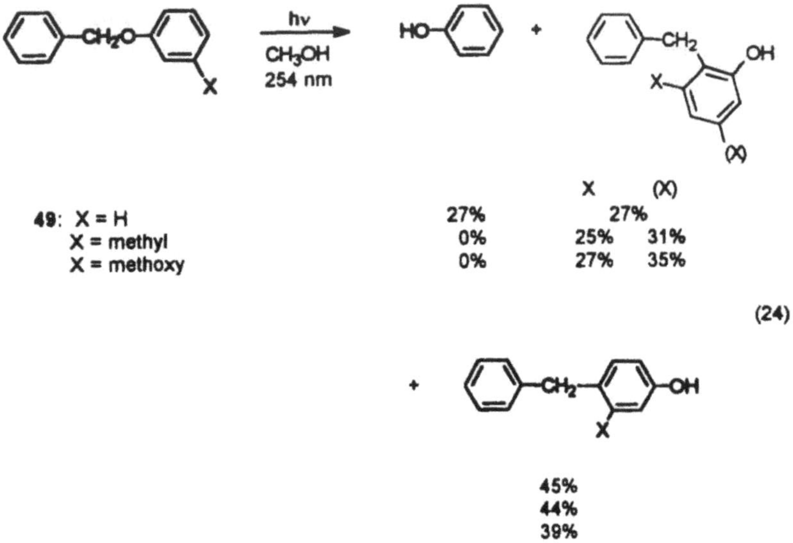

49: X = H
    X = methyl
    X = methoxy

|   | X | (X) |
|---|---|---|
| 27% | 27% |  |
| 0% | 25% | 31% |
| 0% | 27% | 35% |

(24)

45%
44%
39%

Only a few examples are known where the chromophore is definitely benzylic. For instance, picosecond LFP experiments for **50** in both acetonitrile and cyclohexane gave only a short-lived transient assigned to the excited singlet state [33]. No triphenylmethyl cation or radical was detected. Malachite green leucomethoxide, **51**, in cyclohexane behaved similarly. In contrast, **51** showed rapid growth of the triaryl cation from excited $S_1$ with a rise time of 1 ns in 9:1 acetonitrile/water, and an even faster rise time in methanol. No evidence for the triarylmethyl radical was obtained. Also, the dithioketal, **52**, has been used to generate the corresponding relatively long-lived cations by photoheterolysis of the carbon–oxygen ether bond; the half-life for cation decay in 1:1 ethanol/water is 0.17 s. Rate constants for the reactions of this cation with a number of amine and anionic nucleophiles were measured.

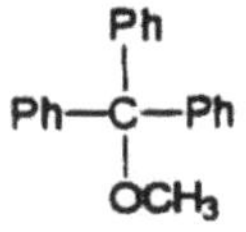

50 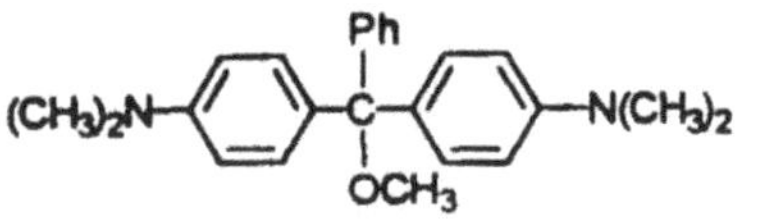                               51 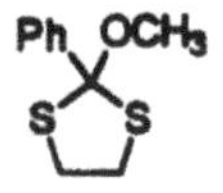                               52

The formation of the deuterated hydrocarbon, **54**, on irradiation of the methyl ether, **53**, in acetonitrile/water, Eq. (25), suggests homolytic cleavage of the benzylic–ether bond followed by disproportionation of the in-cage radical pair [87]. Finally, *ortho*-nitrobenzylic chromophores have been used to photorelease carbohydrates from the corresponding ethers [76].

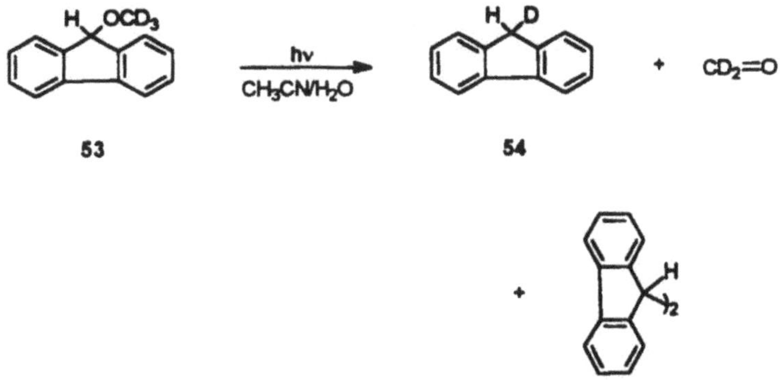

$$53 \xrightarrow[\text{CH}_3\text{CN/H}_2\text{O}]{hv} 54 \quad + \quad CD_2{=}O$$

(25)

## 3.  Esters: —O—(CO)—R

Benzylic esters have been studied in considerable detail often as a continuation of the pioneering work by Zimmerman and co-workers (Scheme 2) in 1963 [44]. There are several reasons for this. First, the synthesis of compounds with the structural variables required to probe specific mechanistic questions is often straightforward. Second, products are usually formed from both ion pairs and radical pairs and, therefore, the structural variables that control this partitioning can be systematically studied. Third, the radical pair ($ARCH_2\cdots O$—(CO)—R) incorporates a built-in radical clock, the decarboxylation of the acyloxy radical, which serves as a useful probe for the reactivity of the radical pair. If the carbon of the acyloxy radical is $sp^3$ hybridized, this decarboxylation rate is on the 1- to 1000-ps time scale, depending on R, so that decarboxylation will often occur within the solvent cage before diffusional escape. The topic of benzylic ester photochemistry has been recently reviewed twice by Pincock [5,98] and therefore only a brief summary will be given here.

Scheme 3 summarizes the mechanistic conclusions reached from the study of the product distribution, quantum yields, and the photophysical properties of nine structurally different sets of substituted arylmethyl esters, 55 [99,100], 56 [101], 57 [101], 58 [41], 59 [100], 60 [25,102], 61 [103], 62 [104], and 63 [105] (73 compounds in all), mostly in the solvent methanol. Quenching and sensitization studies for those cases that have been examined [55: R = $CH_3$, 60: R = $CH_3$, X = 4-CN, 4-$CH_3$, 4-$OCH_3$, 3-$OCH_3$, and 60: R = $C(CH_3)_3$, X = 4-$OCH_3$, 3-$OCH_3$] demonstrate that these reactions occur from the excited singlet state, $S_1$. With the exception of some of the benzoates (58, X = H, Y = CN; X = $OCH_3$, Y = H, CN), where the dominant process is intramolecular electron transfer in $S_1$ from the naphthalene chromophore to the benzoate ring, all of these substrates react from $S_1$ by carbon–oxygen bond cleavage to give products that result from radical and/or ion pairs. The debate on the mechanism in Scheme 3 has been focused on the importance of the competition between homolytic ($k_{hom}^{S1}$) versus heterolytic ($k_{het}^{S1}$) cleavage from $S_1$ in determining the final yields of radical- and ion-derived products, respectively.

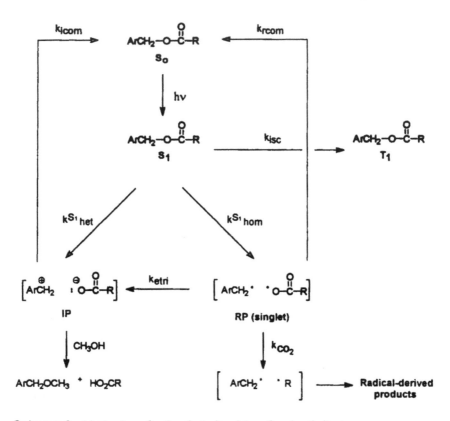

**Scheme 3**   Mechanisms for the photochemistry of arylmethyl esters.

In the "Pincock mechanism" the dominant pathway is thought to be homo-lytic cleavage. Ion pairs are then formed by electron transfer ($k_{etri}$) and the rates of electron transfer follow Marcus's theory in both the normal and "inverted" regions. The rate of decarboxylation of the acyloxy radical was used as a radical clock to determine the values of $k_{etri}$. The meta effect operates only kinetically in that a 3-methoxy group increases the rate constant for bond cleavage from $S_1$ but does not significantly increase the yield of ion pairs formed directly from $S_1$. The arguments supporting this mechanism have been reviewed in detail [5,98].

In contrast to this proposal, in the "Zimmerman mechanism" the competi-tion that controls the yield of radical pairs and ion pairs occurs between $k_{hom}^{S_1}$ and $k_{het}^{S_1}$. In other words, both homolytic and heterolytic cleavage occur from $S_1$ and the partitioning between these two pathways is controlled by the electron distribu-tion of the excited state of the benzylic chromophore and, in particular, by substit-uents on the aromatic ring. Thus, the meta effect operates by, for instance, a 3-methoxy substituent being an electron donor in $S_1$ and, therefore, enhancing heterolytic cleavage. The arguments supporting this mechanism have also been recently discussed [4].

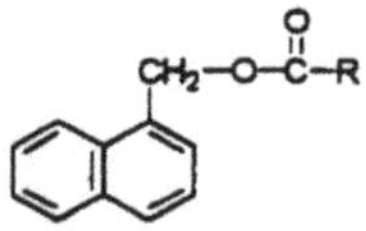

**55**: R = methyl, ethyl, 2-propyl, t-butyl, allyl, cyanomethyl, methoxymethyl

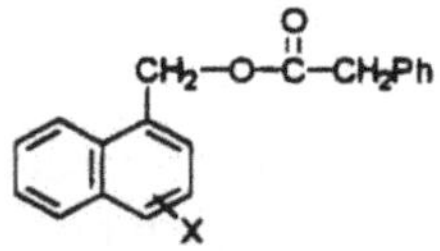

**56**: X = 4-cyano, 4-carbomethoxy, 4-methoxy, 4-ethoxy, 3-methoxy, 4-methyl, hydrogen, 4-fluoro, 4,6-dimethoxy, 4,5-dimethoxy, 4,7-dimethoxy

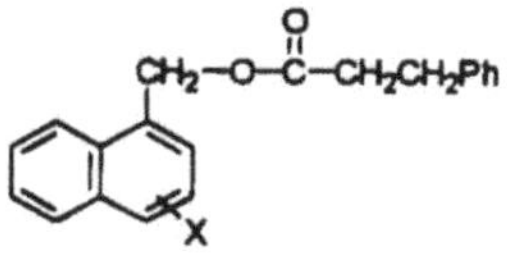

**57**: X = 4-methoxy, 3-methoxy, hydrogen

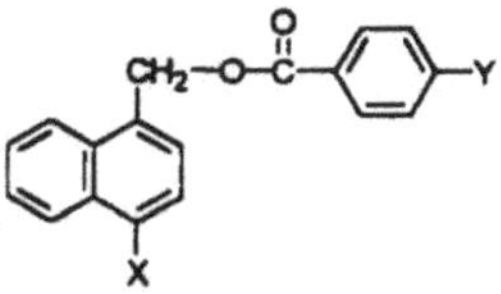

**58**: X = hydrogen, Y = hydrogen, cyano, methoxy
X = methoxy, Y = hydrogen, cyano
X = cyano, Y = hydrogen, methoxy

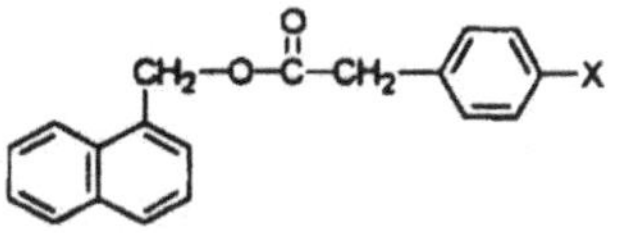

**59**: X = 4-cyano, 3-cyano, 4-trifluoromethyl, 3-trifluoromethyl, hydrogen, 4-fluoro, 3-fluoro, 4-methyl, 3-methyl, 4-methoxy, 3-methoxy

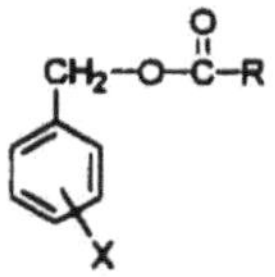

**60**: R = methyl, t-butyl
X = 4-cyano, 4-trifluoromethyl, hydrogen, 4-methyl, 3-methyl, 4-methoxy, 3-methoxy, 3,4-dimethoxy, 3,5-dimethoxy, 3,4,5-trimethoxy

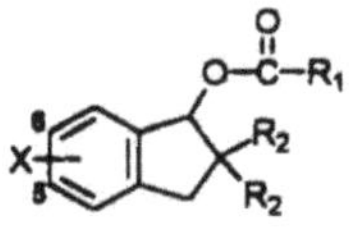

**61**: R$_1$ = methyl, R$_2$ = hydrogen, X = hydrogen
R$_1$ = t-butyl, R$_2$ = hydrogen, X = hydrogen
R$_1$ = methyl, t-butyl, R$_2$ = methyl, X = hydrogen, 5-methoxy, 6-methoxy

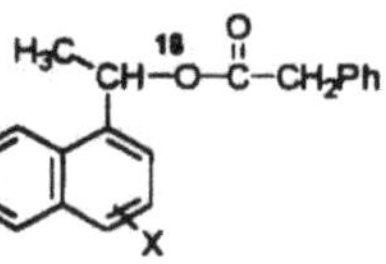

**62**: X = 4-cyano, hydrogen, 4-methoxy, 4,7-dimethoxy

**63**: R = CH$_2$Br, CH$_2$-(SO)—CH$_3$, CH$_2$—(SO)—C$_6$H$_5$

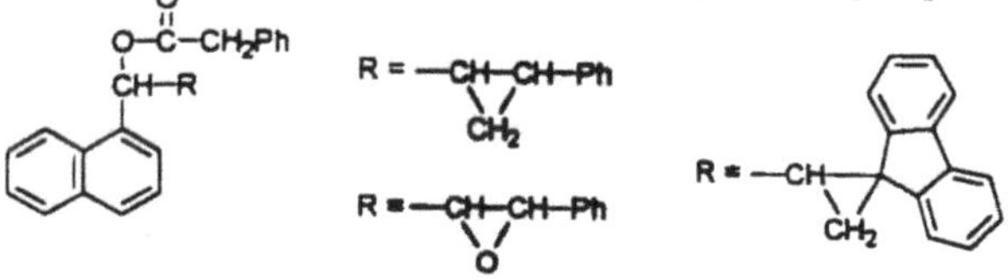

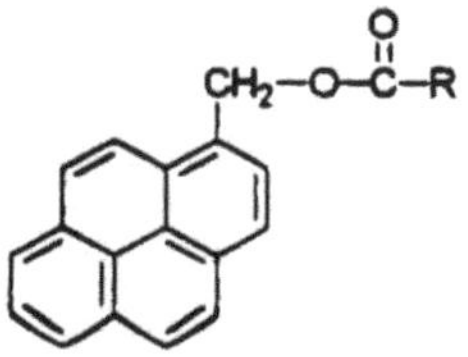

R = —CH—CH—Ph
         |
        CH$_2$

R = —CH—CH—Ph
         \ /
          O

R = —CH
        |
       CH$_2$

**64**: R = cyclohexyl, 4-toluyl, benzyl, N-CBZ-phenylalanine, N-CBZ-leucine

It is likely that both of these mechanisms are acting simultaneously for all benzylic substrates and depending on reaction conditions, the chromophore, and the leaving group, either one or both may be the major pathway(s). For instance, irradiation of 1-pyrenylmethyl esters of alkyl- and arylalkylcarboxylic acids, 64, through Pyrex in methanol gave only ion-derived products in high yield. The lack of decarboxylation products, even for substrates where the acyloxy radical will decarboxylate rapidly, suggests that radical intermediates, if formed at all, are very short-lived [106]. Direct observation of the decay of the reactive excited state ($S_1$) along with the growth and decay of the radical and ion pairs by picosecond LFP could clarify many of these questions. However, these experiments will not be possible for most esters because their $S_1$ lifetimes are usually in the nanosecond time domain. Therefore, the rise times of the initially formed in-cage radical and ion pairs will be slower than the conversion of one to the other. Nanosecond LFP studies on diphenylmethyl acetate and trifluoroacetate in acetonitrile give transient absorption bands for both the diphenylmethyl radical and cation in a ratio of $>12:1$ and $7:1$, respectively [34]. The quantum yield for formation of the radical from the trifluoroacetate is 0.31. Both species were present immediately after the pulse so that the time evolution of their growth was not observable. For more reactive substrates, as will be discussed below for the diphenylmethyl halides (Sec. IV.I), picosecond LFP experiments have been successfully used to monitor the disappearance of $S_1$ and the growth and interconversion of radical and ion pairs.

Nevill and Pincock [105] attempted to study the reactivity of the radical pair by incorporating substrates, 63, with a second radical clock that would operate on the picosecond time scale as shown in Eq. (26). However, none of the fragmenta-

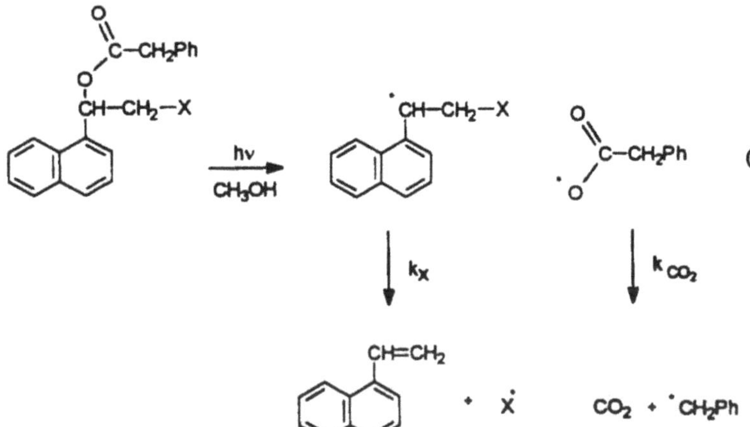

$$(26)$$

tion reactions ($k_r$) are fast enough to compete with the normal chemistry of the 1-naphthylmethyl radical. For the fastest radical clock studied (63, R = cyclopropylfluorenyl), the excited state of the ester reacts by a different pathway that does not include carbon–oxygen bond cleavage.

Recently, Cozens and co-workers [107] have shown that photolysis of 3,5-dimethoxybenzyl acetate in methanol gives the nonaromatic isomer, **65**, Eq. (27). This triene derivative reacts by ground state solvolysis in methanol (half-life, 2.7 min) to give the methyl ether expected from the ion pair. This pathway provides an alternate photochemical (plus thermal) route to ion-derived products. The implications that this observation has on benzylic ester photochemistry are not yet clear but an interesting speculation is that this pathway could be the major one from $S_1$. Secondary photolysis, combined with ground state solvolysis, of **65** could then lead to the radical and ion-derived products normally observed. All previous mechanistic conclusions would then require reconsideration!

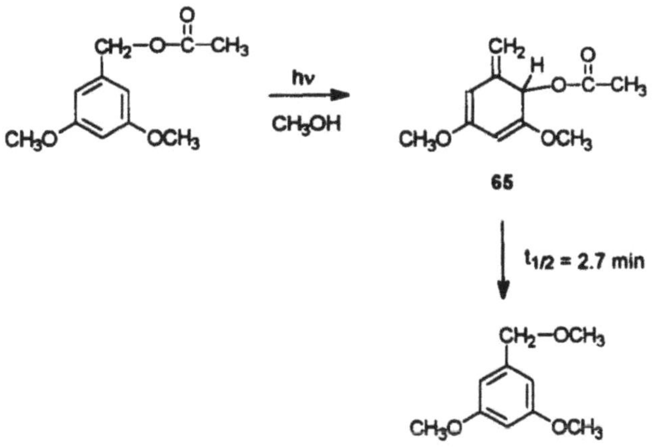

One other class of benzylic compounds has received considerable attention in the design of photocages or phototriggers. These are the benzoin derivatives like **66**, Eq. (28) [108,109]. However, these reactions occur from the

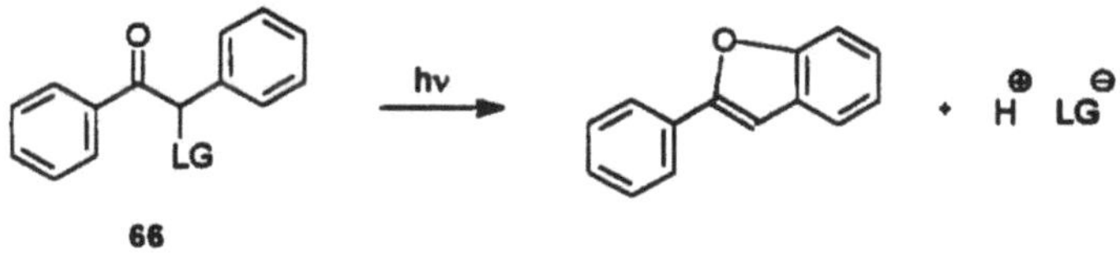

n, $\pi^*$ triplet state of the aromatic ketone, not the benzylic chromophore. Moreover, Wan and co-workers [110] have recently shown that when the LG is an ester [—O—(CO)—R] as in **67**, Scheme 4, the mechanism begins with electron transfer from the 3,5-dimethoxybenzene ring to the ketone chromophore. This intramolecular exciplex then cyclizes to **68** which rearomatizes by loss

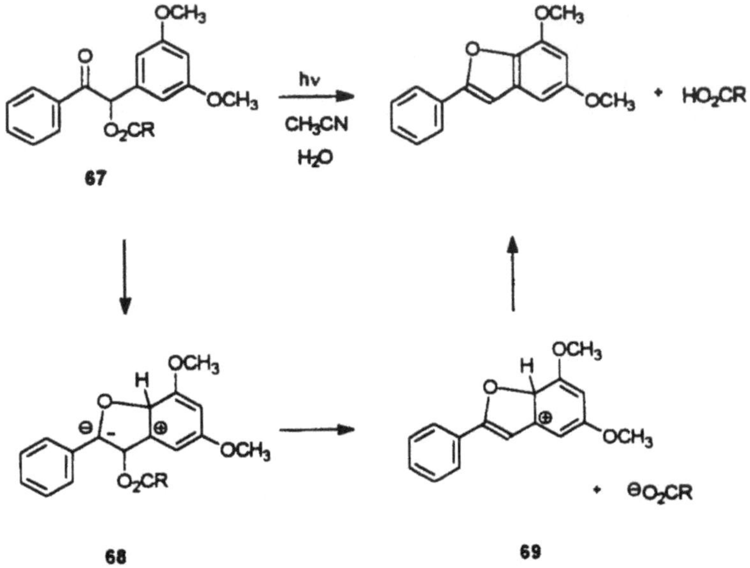

OCH₃ ... hv / CH₃CN / H₂O ... OCH₃ ... OCH₃ + HO₂CR

**67**

**68**                                        **69**

**Scheme 4**  Mechanism for the release of carboxylic acids from esters **67**.

of the carboxylic acid. A transient species with a lifetime of 1 µs, which was generated by nanosecond LFP of **67** at 355 nm, was assigned to the cation, **69**. Superficially, these reactions look like heterolytic cleavage of the benzylic carbon–oxygen bond activated by 3,5-dimethoxy substitution. In fact, the methoxy groups have two other roles. The first is to lower the oxidation potential of the ring so that it is a better electron donor for intramolecular exciplex formation. Second, the methoxy groups are correctly placed to stabilize the intermediates, **68** and **69**.

Finally, *ortho*-nitrobenzyl chromophores have been used to photorelease carboxylic acids from esters [111], but as in the case of amides [Eq. (12), Sec. IV.C.1], carbamates (Sec. IV.D.5), phosphates (Sec. IV.D.6), and sulfonates (Sec. IV.D.7), these reactions are induced by hydrogen atom abstraction by the nitro group from the benzylic carbon.

## 4.  Carbonates: —O—(CO)—OR

The photochemistry of benzylic carbonates has not been extensively studied. However, as shown in Eq. (29), the limited results available parallel those for esters [69]. The reactive excited state is probably $S_1$ because both **70a** and **70b** have very similar quantum yields of fluorescence (0.11 and 0.14, respectively) and singlet lifetimes (31 and 36 ns, respectively), which are significantly lower than the corresponding values for naphthalene itself (0.20, 92 ns) even though the absorption bands ($\lambda_{max}$ and $\varepsilon_{max}$) for **70a** and **70b** are similar to those for naphthalene.

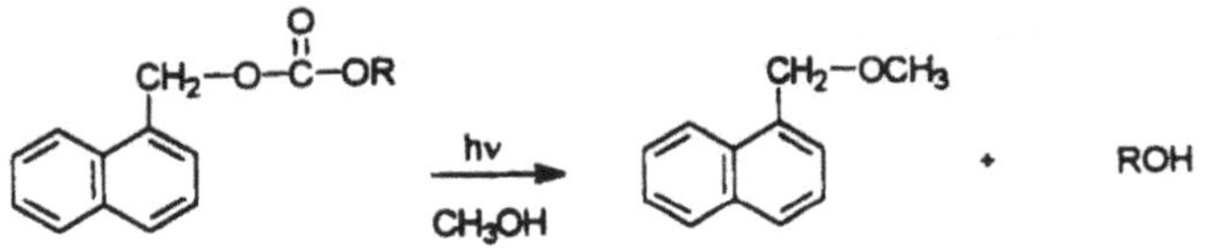

The structures for Eq. (29) are shown here:

$$70a: R = Ph \qquad 43\% \qquad 43\%$$
$$70b: R = CH_2Ph \qquad 93\% \qquad 97\%$$

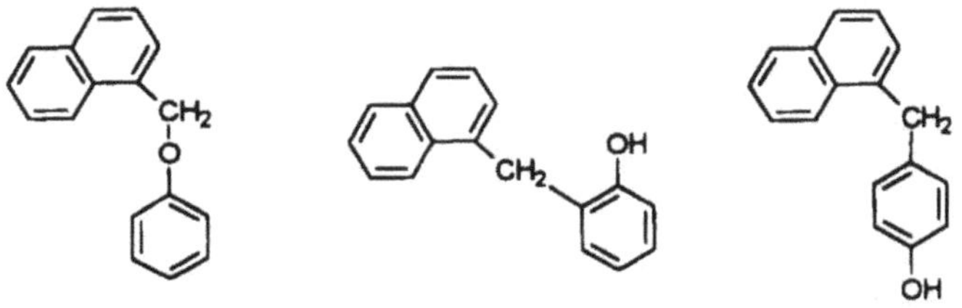

(with naphthalenyl compounds: $CH_2$-O-$\overset{O}{\overset{||}{C}}$-OR converting via $hv$ / $CH_3OH$ to $CH_2$-$OCH_3$ + ROH, Eq. (29))

74    2%         31%         24%

The obvious major difference between the results for esters and those for the carbonates is that none of the products from the carbonates contain $CO_2$. For esters, the carboxylic acids formed are stable but for carbonates this is not so. From literature results on the kinetics of the decarboxylation of alkylcarbonates, the half-life for $PhCH_2O$—$CO_2$—H in water can be estimated at about 10 min [112]. For the benzylcarbonate, **70b**, the equivalent and essentially quantitative yield for the methyl ether and benzyl alcohol indicates that both products are derived from the ion pair, **71b**. This is reasonable because the rate of decarboxylation of the $PhCH_2O$—$CO_2$ radical has been estimated at $10^6$ s$^{-1}$ [113]. Therefore, any radical pair, **72b**, formed by initial homolytic cleavage of S$_1$ will be converted to the ion pair before cage escape.

For the radical pair, **72a**, derived from **70a**, decarboxylation to the more stable phenoxyl radical should be much more rapid, probably in the nanosecond time domain. Therefore, any initially formed radical pair will be rapidly converted to either the ion pair, **71a**, by electron transfer or to the new radical pair, **73a**, by decarboxylation. Coupling of this radical pair then gives the other three products; the yield of the product **74**, Eq. (29), does not necessarily reflect its primary photochemical yield because it is rapidly converted to the other two coupling products by irradiation in methanol (Sec. IV.D.2).

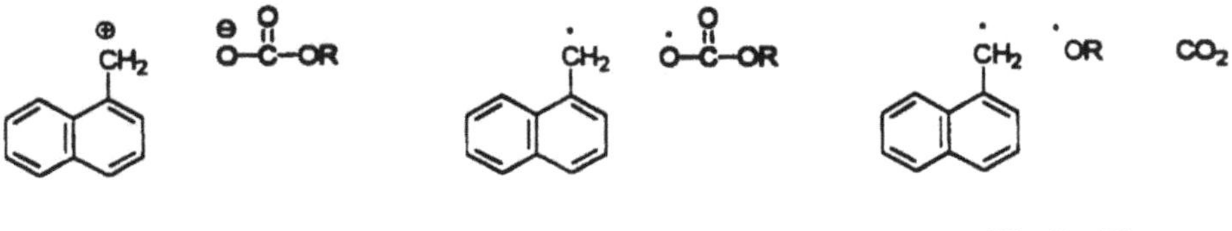

**71a**: R = Ph
**71b**: R = CH₂Ph

**72a**: R = Ph
**72b**: R = CH₂Ph

**73a**: R = Ph
**73b**: R = CH₂Ph

## 5.   Carbamates: —O—(CO)—NR$_2$

The photochemistry of benzylic carbamates has recently attracted considerable attention both for their utility in the photogeneration of bases (amines) and as photocages for biologically active amines.

Fréchet, Cameron, and co-workers have published a series of papers on photoamine generation. The initial compound studied was the urethane, **75**, Eq. (30) [114]. The choice of the α,α-dimethyl-3,5-dimethoxybenzyl group (Ddz)

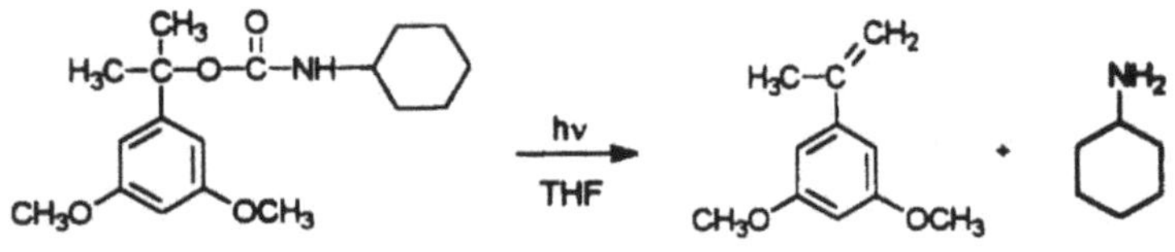

(30)

**75**

was made on the basis of earlier work on amine protection in peptide chemistry. The authors propose a mechanism of initial heterolytic cleavage of the carbon–oxygen bond to form an ion pair which, on subsequent elimination, forms the substituted styrene, $CO_2$, and the amine. Although this is a reasonable mechanistic possibility, exclusive homolytic cleavage from S$_1$ to form a radical pair could also account for the same products. An example to support this possibility is provided by the photolysis in methanol of the carbamate, **76** [69]. The only products observed are derived from the ion pair. In contrast, **77** gives products derived from both the ion pair and the radical pair (photo-Fries-type coupling products). Because **76** and **77** have essentially the same photophysical properties (fluorescence quantum yield, S$_1$ lifetime, S$_1$ energy) their primary photochemical processes are also likely to be the same. Therefore, the observation of only ion-derived products from **76**, and by analogy **75**, does not prove that the initial excited state process is only heterolytic cleavage.

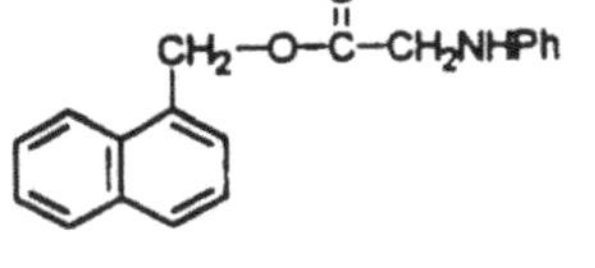

**76** 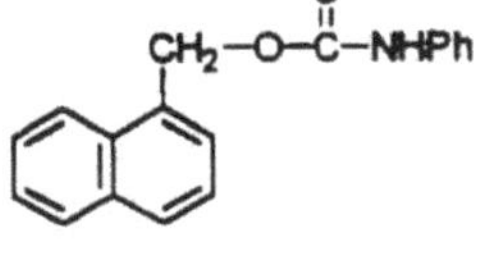                                              **77**

Fréchet, Cameron, and co-workers have also examined a variety of 2-nitrophenylcarbamate derivatives, like **78**, which release amines on photolysis [115,116]. As in the case of 2-nitrobenzylic amides discussed above, Eq. (12), these reactions proceed by hydrogen atom abstraction in the excited state by the nitro group from the benzylic carbon. The dimethoxynitro derivatives, **79**, have also been found to photorelease amines [117]. More recently, Cameron and co-workers have also studied α-ketocarbamates, **80**, but as in the case for benzoin carboxylates discussed above, Eq. (28) and Scheme 4, these reactions begin with the n, $\pi^*$ excited state of the aromatic carbonyl group and not by the $\pi$, $\pi^*$ excited state of the benzylic chromophore [118,119].

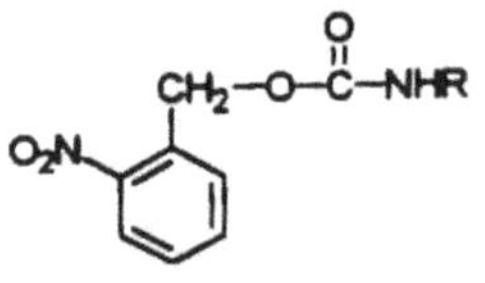

**78** 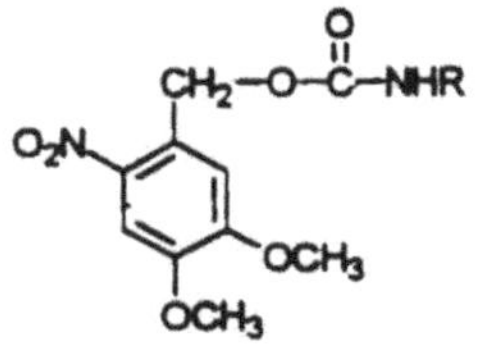                    **79** 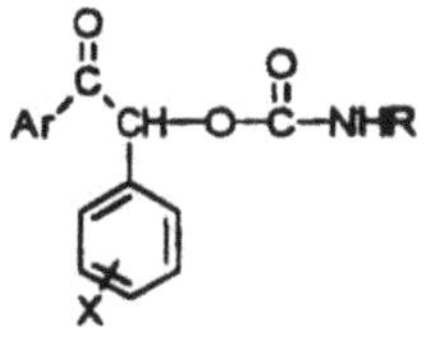                    **80**

Finally, 1-pyrenylmethylcarbamates have been shown to release "caged" amino acids, in yields varying from 38% to 98%, on photolysis in aqueous dioxane, Eq. (31) [120]. The advantage that the pyrenyl chromophore provides is a very large extinction coefficient ($\sim$36,000 M$^{-1}$ cm$^{-1}$) and long wavelength of absorption ($\lambda_{max}$ at 340 nm). The compounds are strongly fluorescent ($\Phi_F \sim 0.5$) and the quantum yields of reaction, where measured, are somewhat low (0.01 for alanine and 0.005 for leucine). This may be a result of the rather low $E_{S1}$ ($\sim$76 kcal/mol). The mechanism proposed begins with homolytic cleavage from $S_1$ to form the radical pair followed by electron transfer to form the ion pair. Products are clearly derived from both species, although the ion-derived alcohol is the major one.

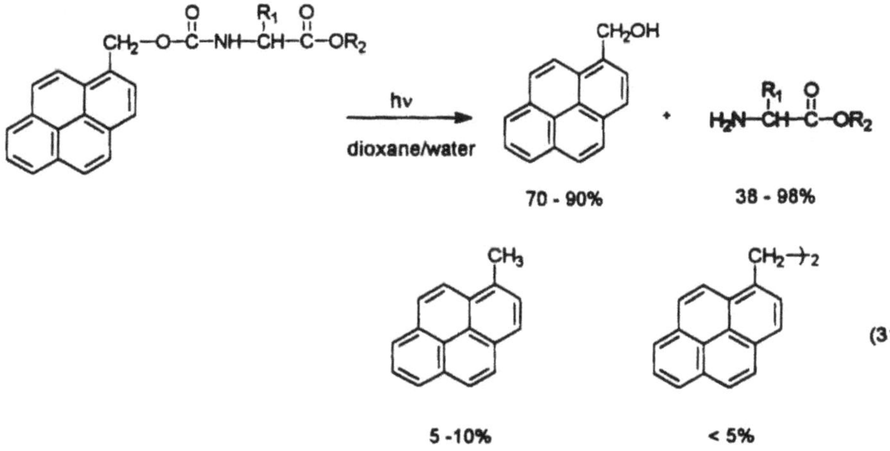

70 - 90%          38 - 98%

5 -10%          < 5%          (31)

## 6.   Phosphates: —O—(PO)—(OR)$_2$

The photochemistry of phosphate esters, including benzylic phosphates, has recently been reviewed by Givens and Kueper [121]. Much of this research has

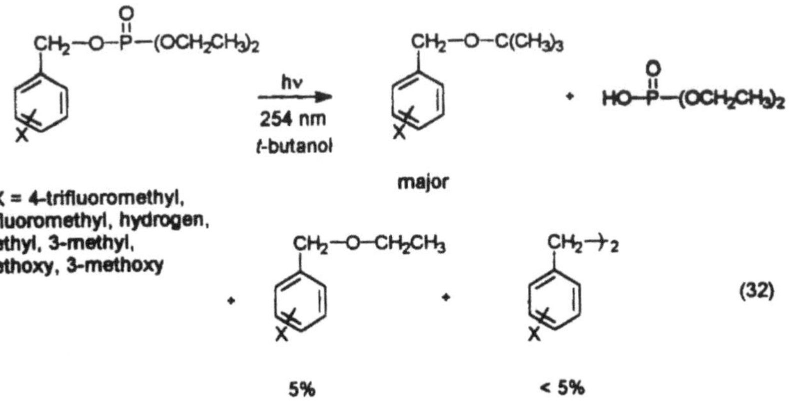

81: X = 4-trifluoromethyl,
3-trifluoromethyl, hydrogen,
4-methyl, 3-methyl,
4-methoxy, 3-methoxy

(32)

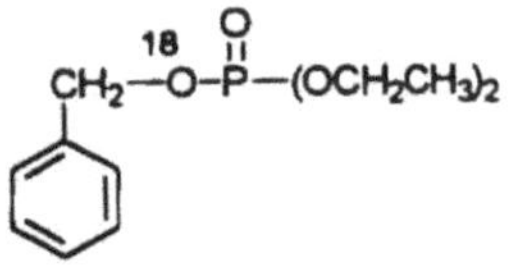

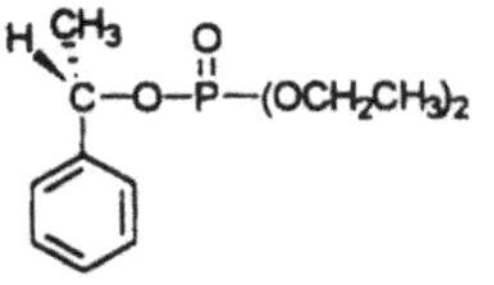

been aimed at the photorelease of biologically active phosphates [13,122]. The results obtained from the compounds, 81, Eq. (32) [123,124], indicate that the major products, the *t*-butyl ether, and phosphoric acid are derived from the ion pair. The singlet state $S_1$, not the triplet $T_1$, is the reactive one. Internal return has also been studied by examining $^{18}O$ exchange in the selectively labeled compound, 82, and the chiral phosphate, 83. For photolysis of 82 in *n*-butanol, carbon–oxygen bond cleavage (92%) and fluorescence (5%) account for essentially all of the excited states formed. The quantum yield for disappearance of 82 is 0.26, the other 66% of bond cleavage resulting in internal return, probably from the ion pair. Therefore, 83 racemizes efficiently. Moreover, the ether product from 83 is almost racemic, having a 5.5% ee for the product of retention of configuration. By measuring singlet lifetimes and quantum yields of disappearance of starting material and by assuming that the efficiency of internal return is constant and independent of substituent, rate constants for the benzylic carbon–oxygen bond cleavage process can be estimated. These values correlate reasonably well with Hammett σ values with ρ = −0.90 indicating an electron-deficient carbon at the reactive benzylic carbon during fragmentation. This means that the most reactive compound is 81: X = 4-methoxy ($k = 4.0 \times 10^8$ s$^{-1}$) and that the unsubstituted ($2.7 \times 10^8$ s$^{-1}$) and the 3-methoxy ($1.5 \times 10^8$ s$^{-1}$) ones are less reactive. This observation does not agree with the meta effect proposal. The final mechanism proposed is direct heterolytic cleavage from $S_1$ to give the ion pair. An attempt was made to probe for the intervention of radical pairs which could be precursors to the ion pairs by incorporating a radical clock into the benzylic radical, 84, Eq. (33) [125]. However, this clock reaction has a rate constant for cyclization of only $10^5$ s$^{-1}$ and is therefore not fast enough to effectively compete with other rapid processes of the benzylic radical. In agreement with this fact, the observed yield of products from **85** was less than 2%.

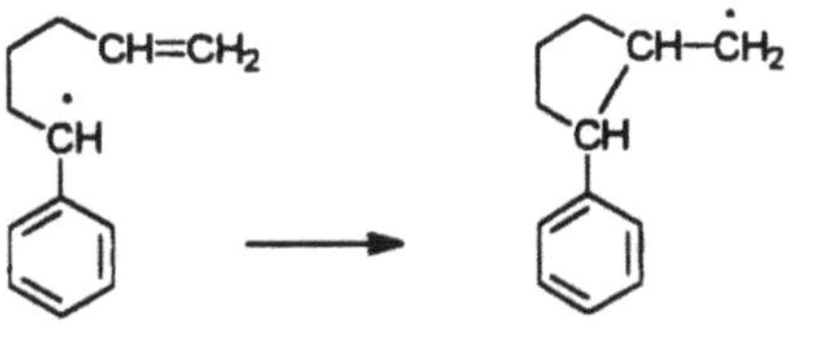

(33)

**84**                                   **85**

Phosphates, including cAMP derivatives [126], have also been released from other benzylic substrates, **86** [127], **87** [126], and **88** [31]. The pyrenyl case, **86**, as in the case of pyrenyl esters (Sec. IV.D.3) and carbamates (Sec. IV.D.5), are particularly efficient because of the high extinction coefficient.

The *ortho*-nitrobenzyl chromophore has also been employed for phosphate deprotection but, as in Eq. (12) for amides, this photochemistry again does not result from excited state carbon–oxygen bond cleavage [126,128]. Similarly, desyl phosphates like **89** release phosphate photochemically but, as discussed in Sec. IV.D.3, by a pathway beginning with n,π* excited state of the carbonyl group [13,126,129].

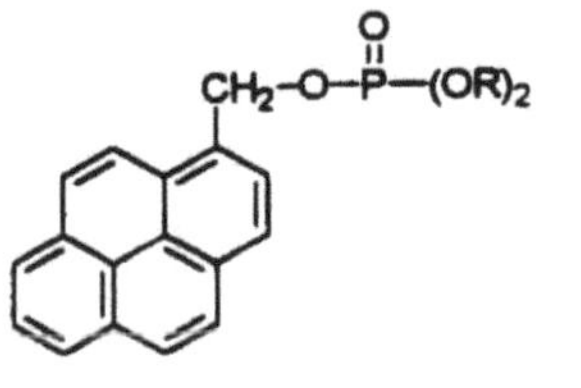

**86**

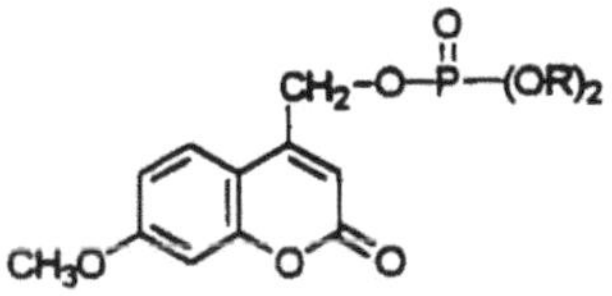

**87**

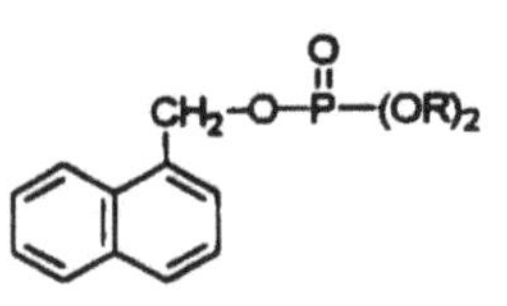

**88**

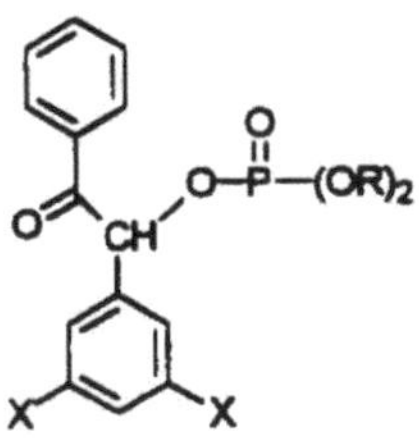

**89**: X = H, methoxy

## 7. Phosphites: O—P(OR₂)

Benzylic phosphites like **90**, **91**, and **92** have been studied recently by Bentrude and co-workers. For **90**, irradiation in benzene gave, as the major product, the photo-Arbuzov phosphonate, **93** (85–95% yield) [130]. The very low yield of the out-of-cage dimer, 1,2-diphenylethane, the absence of formation of the ion-derived methyl ether in the presence of added methanol, and the lack of time-resolved ESR signals [131] all provide good evidence that this reaction proceeds by homolytic cleavage from S₁ to an in-cage singlet radical pair or by a concerted 1,2-sigmatropic shift. In agreement with this proposal, the stereochemistry of the process, using individual stereoisomers of specific phosphites, has been deter-mined as retention of configuration at both phosphorous [132] and the migratory carbon [130].

In contrast, for the acetyl-substituted compound, **91**, where intersystem crossing from the excited singlet state to $T_1$ is expected to be efficient, homolytic cleavage to triplet radical pairs was proposed on the basis of formation of the out-of-cage coupling product and by CIDEP [131] and CIDNP [133] studies. Finally, **92** reacted from S₁ on direct irradiation again to give the phosphonate as the major product, but from $T_1$ on sensitization with benzophenone [131].

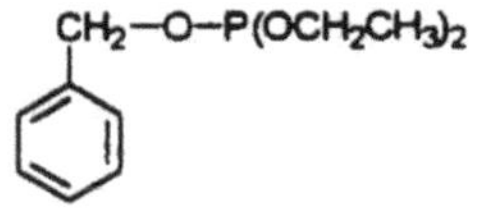

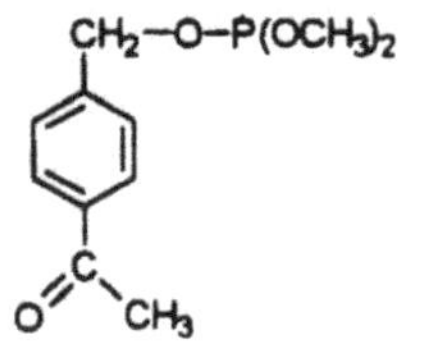

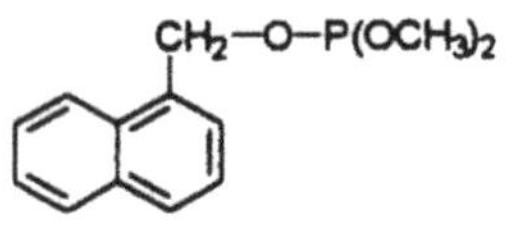

## 8.  Sulfonates: O—(SO₂)—R

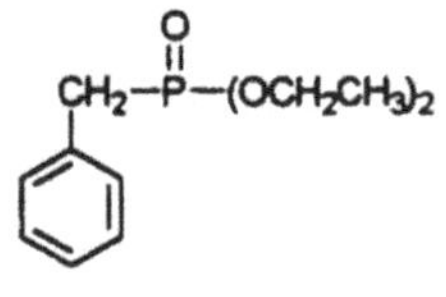

Benzylic sulfonates, because of their high solvolytic reactivity, have been studied extensively in the ground state. These substrates show the classic change from $S_N1$ to $S_N2$ reactivity as the substituents change from electron-donating to elec-tron-withdrawing groups [134]. However, very little is known about their photo-chemical reactivity, probably because many are too reactive to study conve-niently. For instance, the approximate half-lives at 25°C for solvolysis of benzyltosylates in 80% aqueous acetone, a solvent with similar ionizing power to that of methanol (Table 2) as a function of substituents, are 4-methoxy, 1 s;

4-methyl, 40 min; H, 97 min; and 4-nitro, 193 h. Clearly, only benzylic sulfonates with electron-withdrawing groups, like 4-nitro, will be stable enough for their photochemical reactivity to be studied in the absence of competing ground state solvolysis unless solvents with lower ionizing power are used. A good choice would be $t$-butanol, which would lower these rates by about three orders of magnitude (Table 2).

There are a few examples of benzylic sulfonate photochemistry that have been studied as potential photoacid generators [15]. All have electron-withdrawing groups on the benzylic ring. For instance, photolysis of the sulfonate, **94**, Eq. (34), gives the sulfonic acid (quantum yield for sulfonic acid formation in brack-

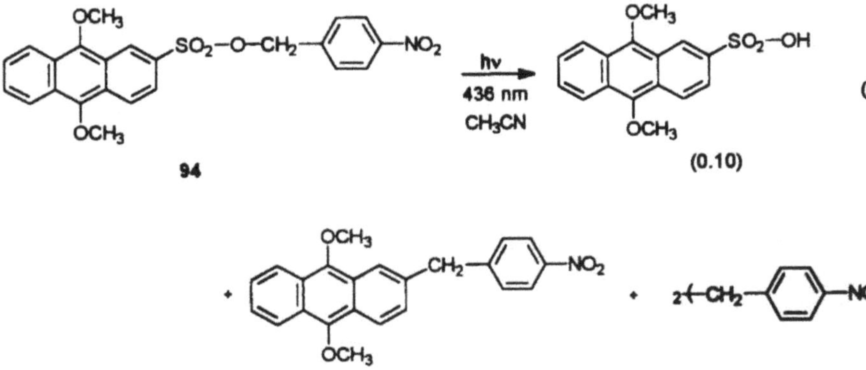

ets) [135]. LFP studies demonstrated that the mechanism of this reaction is intramolecular electron transfer ($k_{set}^{S1}$ in Fig. 1) from the excited state of the electron-rich anthracene ring to the nitrophenyl group. Heterolytic bond cleavage of the carbon–oxygen bond then gives the radical pair. The source of the hydrogen atom for sulfonic acid formation is not obvious. Other similar sulfonate derivatives have also been examined [15]. In the absence of the possibility of intramolecular electron transfer, bond cleavage to generate 4-nitrobenzenesulfonic acid can be induced in compounds like **95** by intermolecular electron transfer from an added sensitizer. Several substituted aryl derivatives of pentafluorobenzylsulfonates, **96**, have also been studied [15]. The deprotection of sugar mesylates and tosylates has been performed by irradiation in methanol in the presence of iodide ion as the electron donor; cleavage is again from the radical anion of the sulfonate [136].

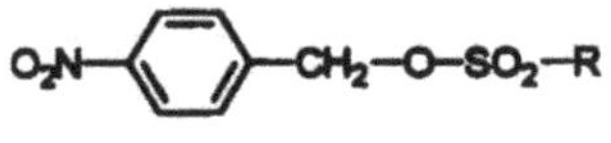

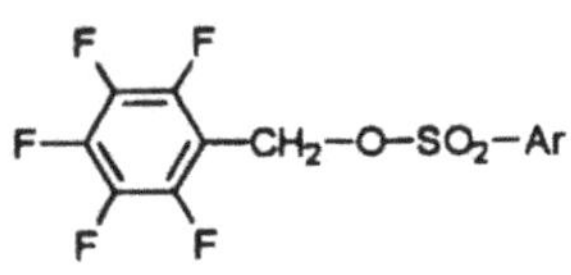

## E.  Fluorine

Because of the strong carbon–fluorine bond, fluorine, as in ground state solvolysis reactions, is not a very good leaving group in photocleavage reactions.

An unusual but interesting example comes from the photolysis of trifluoro-
methylnaphthols, 97 and 98, and phenols, 99, in aqueous base [137]. The mecha-

97: 5-, 6-, 7-, 8-CF₃        98: 3-, 4-, 5-, 6-CF₃        99: 2-, 3-, 4-CF₃

nism proposed for the conversion to the phenolic carboxylic acids, Eq. (35), begins with heterolytic cleavage from the excited singlet state of the naphthoate or
phenolate; the high electron donating ability of the negatively charged phenolic
oxygens is expected to provide large rate accelerations. Quantum yields of reaction for the *ortho*- and *meta*-trifluoromethylphenols (0.5 and 0.8, respectively)
were considerably higher than for the para isomer (0.01), in agreement with meta
(or ortho/meta) effect predictions. Moreover, for the naphthol derivatives, the
rates for the photoreactions (determined from quantum yields of reaction and
singlet excited state lifetimes) were found to correlate with calculated changes
in electron density that occur upon excitation to $S_1$. Because these reactions provide a clear observation of this predicted effect, these and other related examples
deserve further study.

$$ HO-Ar-CF_3 \xrightarrow[\text{H}_2\text{O, basic}]{h\nu} HO-Ar-CO_2H , \qquad (35) $$

As part of an extensive nanosecond LFP study of diarylmethyl derivatives
[34], 248-nm excitation of diphenylmethyl fluoride in acetonitrile produced transient absorptions for both the diphenylmethyl cation ($\Phi$ = 0.05) and radical
($\Phi$ = 0.20). At this time scale, no information on their growth was obtained.

A recent brief study of the photochemistry at 254 nm in tetrahydrofuran
(THF) of the fluorinated 1,2-diphenylethane derivatives, 100, reported very low
and inefficient conversions to the corresponding stilbene derivatives, Eq. (36)
[138]. The addition of a radical inhibitor, 2,6-di-*t*-butyl-4-methylphenol, had no
effect. The mechanism for the loss of hydrogen fluoride is therefore not known.
Irradiation in the presence of LiAlH₄ gave reduction products and the mechanism
proposed begins with electron transfer from the hydride to the excited state of
the substrate to form a radical ion pair.

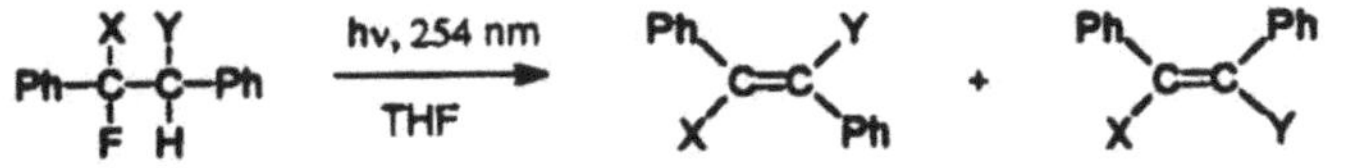

$$
\underset{\underset{F\;\;H}{|\;\;\;|}}{Ph-C-C-Ph} \quad\xrightarrow[\text{THF}]{h\nu,\ 254\ nm}\quad \underset{X}{\overset{Ph}{\diagdown}}C=C\underset{Ph}{\overset{Y}{\diagup}} \quad + \quad \underset{X}{\overset{Ph}{\diagdown}}C=C\underset{Y}{\overset{Ph}{\diagup}} \qquad (36)
$$

| | | |
|---|---|---|
| 100: X = Y = H | 14% | 4% |
| X = F, Y = H | 3% | 5% |

## F. Silicon

The photochemistry of $ArCH_2SiR_3$ [139–142] derivatives has been reported. Each of the studies was carried out under electron transfer conditions, which are not covered in this chapter. Whether there would be photocleavage under direct conditions is not clear.

## G. Phosphorus

### 1. Phosphine: —PR₂

We are not aware of any examples in the literature that describe the photocleavage of a benzylic phosphine. This is consistent with the previously noted poor photoreactivity of the related benzylamines.

### 2. Phosphonium: —P⁺R₃

The first report of benzylphosphonium photochemistry appears to be the early work by Griffin and Kaufman [143]. Their findings were rationalized by an initial photoinduced electron transfer from the chloride counterion to give the benzyltriphenylphosphoranyl radical. Radical cleavage of the benzyl bond was proposed as the mechanism for formation of bibenzyl, Eq. (37). Biphenyl was obtained in this study and phosphorus–phenyl bond cleavage of the phosphoranyl radical was proposed as its source. Diphenylmethane was also observed and the suggested paths for its formation were coupling of benzyl and phenyl radicals or benzyl radical attack on the solvent (benzene/ethanol) followed by rearomatization. No toluene was reported in agreement with the fact that abstraction of a hydrogen atom from ethanol by a benzyl radical is endothermic by 6 kcal/mol [35]. No mention was made of other counterions being studied or substituent effects. If heterolytic cleavage had occurred, then benzyl ethyl ether would have been observed. Although no mention was made of the experimental procedure with regard to oxygen exclusion, oxidized phosphorus products were observed.

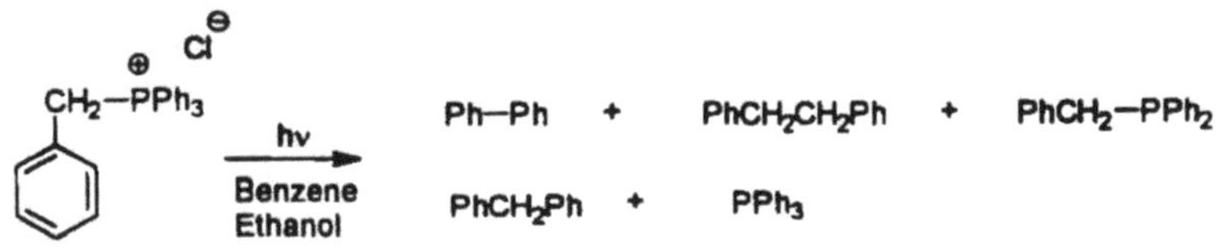

$$
\underset{\underset{\text{Benzene}}{\text{Ethanol}}}{\xrightarrow{h\nu}}\ \ Ph–Ph \ + \ PhCH_2CH_2Ph \ + \ PhCH_2–PPh_2
$$

$$
PhCH_2Ph \ + \ PPh_3 \qquad (37)
$$

Oliveira and co-workers have reported that irradiation in alcohol solvents of 1-(pyrenylmethyl)triphenylphosphonium bromide and chloride gives high yields (>90%) of ethers and this reactivity parallels a decrease in the fluorescence quantum yield [144]. Moreover, the corresponding tributylphosphonium chloride had a very high fluorescence quantum yield and was photochemically inert. The authors suggest a mechanism of heterolytic cleavage from $S_1$ but the fluorescence behavior is indicative of intramolecular electron transfer from the pyrene chromophore to the triphenylphosphonium moiety followed by homolytic cleavage to an ion pair.

Saeva and Breslin [145] reported the photocleavage of triphenyl (p-cyanobenzyl)phosphonium tetrafluoroborate in 1988. The products from this reaction are clearly indicative of a radical pathway, Eq. (38); 4,4'-dicyanobibenzyl was the only nonphosphorus containing product obtained. Interestingly, the phosphorus group was recovered as the phosphine oxide. Apparently, oxygen was not excluded in this study. The only counterion used was $BF_4^-$ and its inability to readily act as an electron donor probably accounts for the different products when comparing this study with the work by Griffin and Kaufman. No other substituents on the benzyl group were included in the work by Saeva and Breslin and there is no indication that the para-cyano group had any effect on the reaction, although it certainly would stabilize the benzyl radical intermediate. No products from heterolytic cleavage were reported in these experiments which were performed in acetonitrile.

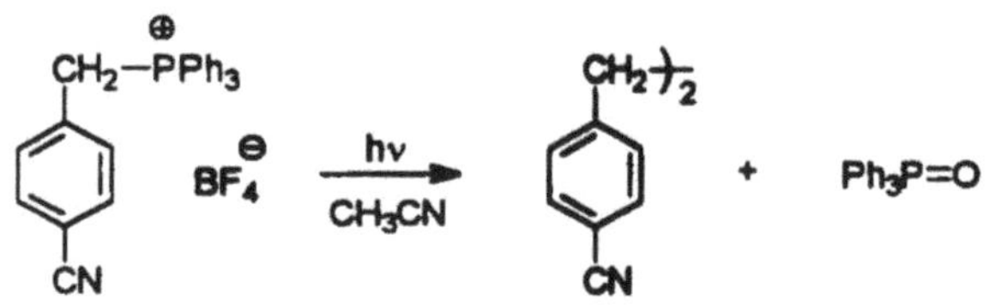

Alonsono and co-workers [146] have used substituted 1-naphthyl and phenylphosphonium chlorides as precursors for the generation of the corresponding arylmethyl radicals and cations in both nanosecond LFP and product studies. For instance, the salt 101 has a quantum yield for cation formation of 0.71 in methanol and the sole product observed was the corresponding methyl ether. No transient radical was observed in this solvent. In contrast, in 5% acetonitrile in dioxane, the radical was observed but now the cation was absent. No fluorescence was observed in either solvent suggesting that $S_1$ is very reactive. Redox potentials indicate that the conversion of the radical/radical ion pair to the cation/triphenylphosphine pair would be exothermic by some 25 kcal/mol. Therefore, both heterolytic cleavage from $S_1$ or homolytic cleavage followed by electron transfer were suggested as possible pathways for cation formation.

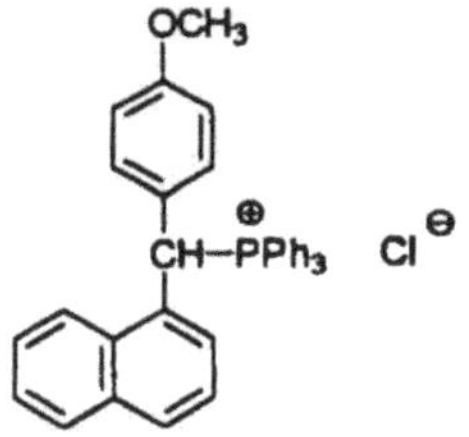

**101**

Modro and co-workers [147] recently reported a thorough study of the photochemistry of 3-methoxy and several 4-substituted triphenylbenzylphosphonium salts, **102**. Irradiation in acetonitrile, Eq. (39), or methanol resulted in homolytic and heterolytic photocleavage products. According to the authors, **103** was obtained from homolytic cleavage of the arylmethyl–phosphorus bond followed by hydrogen abstraction from acetonitrile, which would also produce cyanomethyl radicals. The arylpropanonitrile, **104**, would then logically result from coupling of arylmethyl radical and cyanomethyl radicals. This proposed pathway is surprising because abstraction of a hydrogen atom from acetonitrile by a benzyl radical is endothermic by 5 kcal/mol [35]. In contrast, the same reactions in methanol do not give the corresponding 2-arylethanol. Other radical-derived products include the commonly observed dimers, **105**, which result from coupling of the arylmethyl radicals.

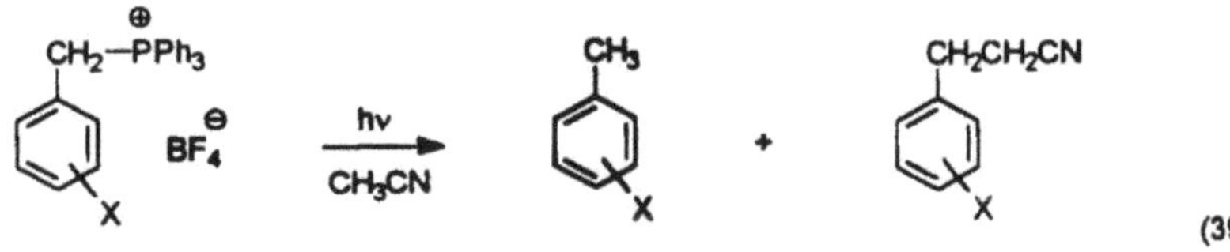

(39)

**102**: X = H, 4-methyl, 4-t-butyl, 4-methoxy, 3-methoxy, 4-fluoro, 4-chloro, 4-cyano, 4-nitro

**103**
9% (X = H)
78% (X = 4-OCH₃)
79% (X = 3-OCH₃)

**104**
8% (X = H)
- (X = 4-OCH₃)
- (X = 3-OCH₃)

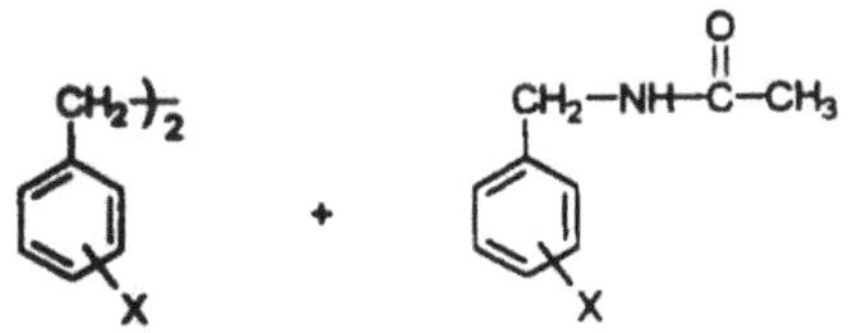

**105**
3% (X = H)
14% (X = 4-OCH₃)
9% (X = 3-OCH₃)

**106**
79% (X = H)
8% (X = 4-OCH₃)
12% (X = 3-OCH₃)

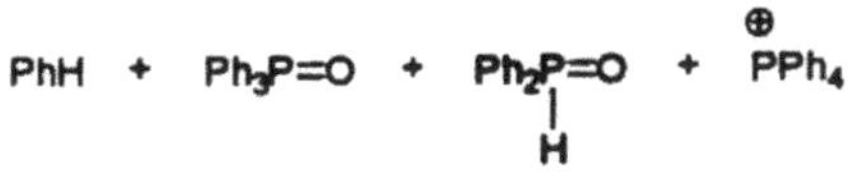

PhH + Ph₃P=O + Ph₂P=O + PPh₄
                          |
                          H

The major product in these irradiations when tetrafluoroborate was the counterion was the amide **106**, which is obtained by acetonitrile capture of the arylmethyl cation followed by hydrolysis. Interesting exceptions to this trend were the 4-methoxy- and 3-methoxy-substituted compounds. Both gave radical-derived products preferentially over the solvent-trapped, ion-derived compounds (amide **106** in acetonitrile and arylmethylmethyl ether in methanol). The yields are given with Eq. (39). The authors argued that the ionic pathway was a result of initial homolytic cleavage followed by single-electron transfer (SET) ($k_{etri}$ in Scheme 1). The low yield of ion-derived products for the 4-methoxy compound (which would have the highest thermodynamic driving force for electron transfer) was justified as a result of the Marcus "inverted region" where the electron transfer becomes slow because it is prohibitively exothermic. No explanation was offered for the ineffective SET of the postulated 3-methoxybenzyl radical.

Sensitized irradiations resulted in exclusive formation of products from radical cleavage (**103–105**). Consistent with this observation is the increased selectivity for radical-derived products when bromide and iodide were the counterions. The 4-cyanobenzylphosphonium substrate also demonstrated considerable selectivity for the radical pathway. This was explained as a result of unfavorable oxidation potential of the 4-cyanobenzyl radical, but could also be a function of the cyano group enhancing ISC to yield the triplet excited state [148]. Oxygen was not rigorously excluded in these studies; the solutions were simply degassed with argon prior to photolysis. This may or may not have any impact on the singlet/triplet population for the excited state. Oxygen did play a role in product formation; three oxidized phosphorus products were formed in acetonitrile and two were formed when methanol was the solvent.

Benzene, triphenylphosphine oxide, and diphenylphosphine oxide were observed in this work indicating that the phosphorus–phenyl bond is also photocleaved. There was no evidence, however, for cleavage of the $CH_2$—Ar bond.

## H.  Sulfur

### 1.  Thioether: —SR

Fleming and co-workers previously reported that the photocleavage of benzyl-sulfur bonds proceeds via a radical mechanism [14,149,150]. As part of that study they irradiated benzylphenyl sulfide and several of its derivatives, **107**, which allowed a comparison of the substituent effects on the photocleavage of benzyl sulfides. On the basis of this work and transient absorption studies [151], benzyl sulfides apparently undergo almost exclusive radical photocleavage. No products resulting from benzyl cations were detected. These irradiations were performed in *t*-butyl alcohol and acetonitrile.

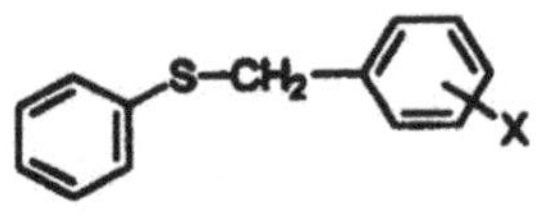

**107**: X = H, 4-methoxy, 3-methoxy
4-methyl, 3-methyl, 4-trifluoromethyl,
3-trifluoromethyl, 4-cyano, 3-cyano,
4-nitro, 3-nitro

Although there was no indication of heterolytic cleavage, there was a trend to decreased quantum yields for loss of starting material for the meta-substituted compounds when compared to the para ones (Table 5). The mechanistic implication is that the excited state, prior to cleavage, has a sense of the substituent in the meta position. In fact, calculations indicated that the excited state has an out-of-plane deformation of the ipso position that would potentially allow a meta substituent to stabilize the species prior to cleavage (see Fig. 1).

Oxygen was also found to complicate these studies. Under freeze-pump-thaw (FPT) conditions there were fewer side reactions and a consistent (albeit minor) $\sigma^+$ behavior for the meta substituents. Simple degassing was not sufficient to avoid oxidized products that clouded the substituent effects. This observation may imply that intersystem crossing has a significant impact on the product distribution of benzyl photocleavages.

Sucholeiki [152] has recently suggested the use of benzyl sulfide photocleavage for detachment from a solid support. His results are consistent with Fleming's work. The *para*-phenyl-substituted benzyl sulfide, **108**: R = Ph, has enhanced benzylic cleavage compared to benzyl sulfide, **108**: R = H. The increase in benzyl cleavage precludes the competitive Norrish type II reaction pathway which is shown in Eq. (40). However, the wavelength of excitation was

**Table 5**   $\Phi$ Values for Disappearance of 103

| 103: X = | N$_2$ bubbling $\Phi$ | F—P—T $\Phi$ |
|---|---|---|
| H | 0.12 ± 0.045 | 0.033 ± 0.015 |
| 4-Methoxy | 0.31 ± 0.021 | 0.071 ± 0.026 |
| 3-Methoxy | 0.054 ± 0.028 | <0.020 |
| 4-Methyl | 0.18 ± 0.073 | 0.18 ± 0.031 |
| 3-Methyl | 0.11 ± 0.009 | 0.047 ± 0.003 |
| 4-Trifluoromethyl | 0.10 ± 0.059 | 0.078 ± 0.037 |
| 3-Trifluoromethyl | 0.096 ± 0.022 | 0.053 ± 0.002 |
| 4-Cyano | 0.068 ± 0.015 | 0.044 ± 0.011 |
| 3-Cyano | 0.139 ± 0.078 | 0.39 ± 0.01 |
| 4-Nitro | 0.063 | 0.023 |
| 3-Nitro | 0.081 ± 0.017 | 0.071 ± 0.016 |

350 nm and therefore the excited chromophore should be the phenone group. It is not clear that the substituted benzyl group would be responsible for the photochemistry unless the mechanism involves energy transfer from the triplet acetophenone moiety ($E_T$ = 74 kcal/mol) to the biphenyl group ($E_T$ = 65 kcal/mol) [19]. Irradiation in the presence of oxygen led to aldehydic products, presumably from oxygen trapping of the intermediate benzyl radicals.

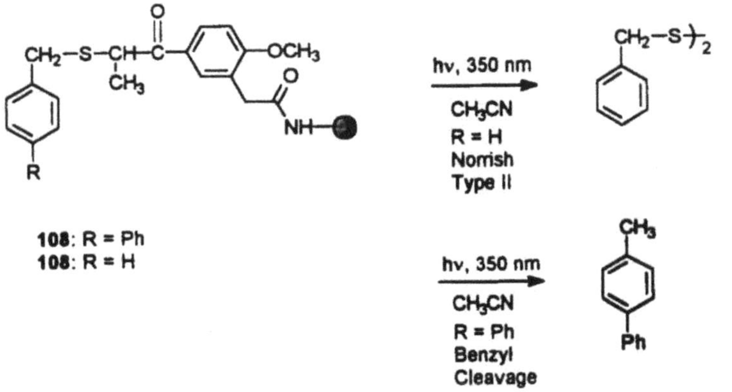

(40)

108: R = Ph
108: R = H

A related naphthylmethyl sulfide cleavage was reported by Ouchi, Yabe, and Adam [153]. They irradiated (248 nm) a degassed cyclohexane solution of the bisarylmethyl sulfide, 109. A two-photon process allowed them to detect the diradical intermediate, 110, which then coupled to give acenaphthene. The efficiency of this reaction was attributed to the possible absorption of energy by the phenyl groups, sulfur atoms, participation of the $S_2$ state of the naphthalene, or some combination of these chromophores.

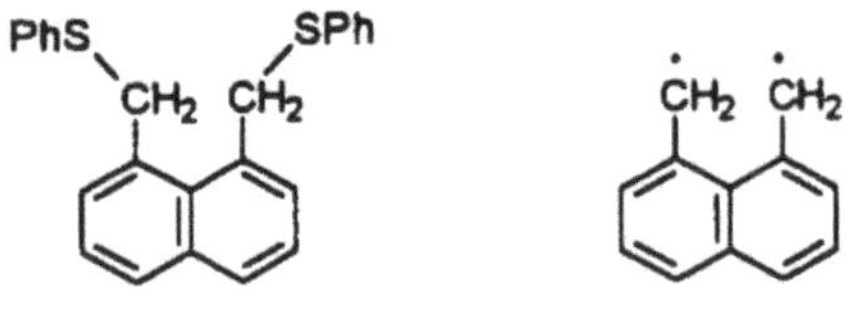

109                                110

## 2.  Sulfonium: —S⁺R₂

Maycock and Berchtold [154] reported the photochemical cleavage of a variety of benzylsulfonium tetrafluoroborates in 1970. In their work, they examined the photochemistry of the benzyl derivatives **111–115**. Their irradiations were carried out in methanol, *t*-butyl alcohol, or acetonitrile. Both heterolytic and homolytic cleavage were proposed to occur on the basis of the observed solvent-captured benzyl cations and the coupled benzyl radicals, respectively. The ratio of radical-derived to ion-derived products was nearly 1:1 for photocleavage of benzylsulfonium tetrafluoroborate, **111**, in methanol. The *meta*-methoxybenzyl species, **112**, had a 1:5 ratio favoring the cationic pathway in the same solvent. The authors argued that the difference in product selectivity was likely due to the increased absorption of the methoxy compound rather than any reflection of a substituent effect. The mechanistic explanation they offered was that the solvolytic products arise from direct heterolytic cleavage and the radical-coupled products resulted from homolytic cleavage.

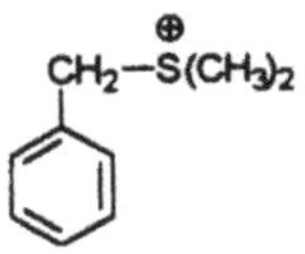

**111** 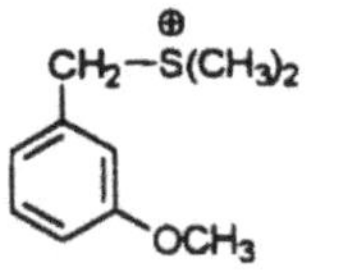        **112** 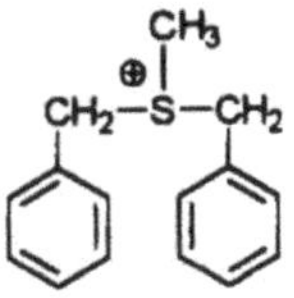        **113**

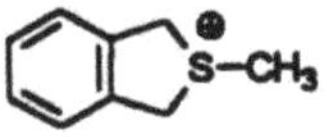

**114** 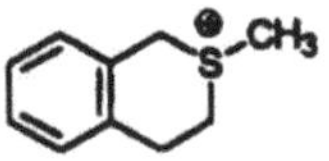        **115**

Saeva and co-workers [155] have reported the photochemistry of an anthracenyl-substituted benzylsulfonium trifluoromethanesulfonate salt, **116**, Eq. (41). Unlike most of the benzyl cleavage reactions, this study involves initial excitation of a moiety other than the benzyl group. The mechanistic picture proposed involves SET from the excited anthracenyl group to the sulfonium species, which leads to homolytic cleavage of the benzylic–sulfur bond as shown in Eq. (42). The arylmethyl radical couples with the radical cation of the anthracenyl moiety which rearomatizes to produce the phenyl sulfide **117**. An intriguing observation is that the anthracenyl-substituted benzyl group does not photocleave although it is the chromophore. This study also revealed solvent capture of the *para*-cyanobenzylic cation to yield amide **118**. This product is apparently the result of internal return by in-cage back electron transfer to reform the neutral anthracenyl compound and the benzylic cation. Their irradiations were carried out in acetonitrile-water and oxygen outgassed with argon.

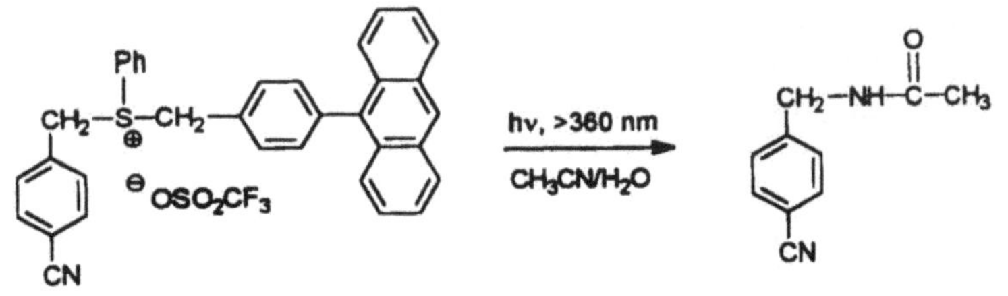

(41)

116                                                       118

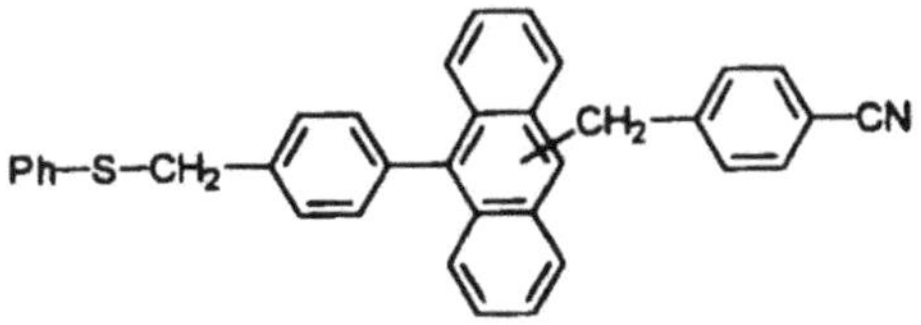

117 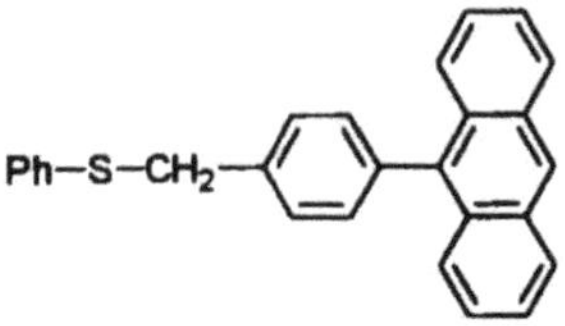                                                       119

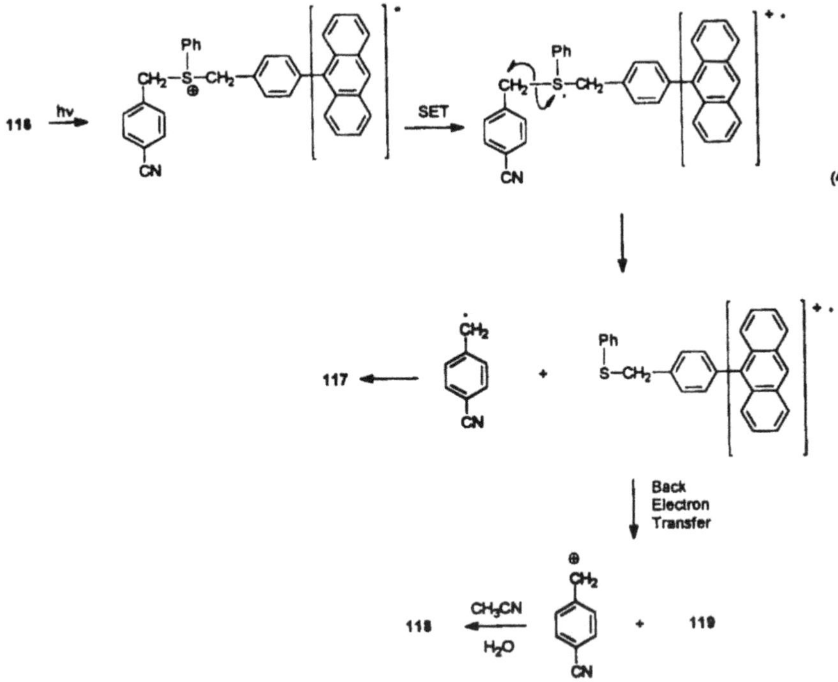

(42)

Pincock and co-workers [31] found that 1-naphthylmethylsulfonium tetrafluoroborate photocleaves in nitrogen-purged methanol to give radical-trapped and
nucleophilic-trapped products in a 12:40 ratio, respectively. Two mechanisms for
formation of the cationic intermediates (which is trapped by the solvent to form
naphthylmethyl methyl ether, **120**) were considered; direct heterolytic cleavage of
the benzyl–sulfur bond or homolytic cleavage followed by electron transfer within
the radical pair. There were insufficient data for the authors to distinguish between
these pathways. The quantum yield of formation of the coupled 1-naphthylmethyl
radical product, 1,2-dinaphthylethane, **121**, was 0.019 and the quantum yield for
naphthylmethyl methyl ether, **120**, formation was 0.30. Less than 5% 1-methyl-
naphthalene was observed but no other photoproducts were noted. Lifetime studies
for the 1-naphthylmethylsulfonium excited state indicated that the singlet lifetime
was too short (<0.5 ns) to be quenched. The photochemical pathway appears to be
through the singlet manifold for this sulfonium example.

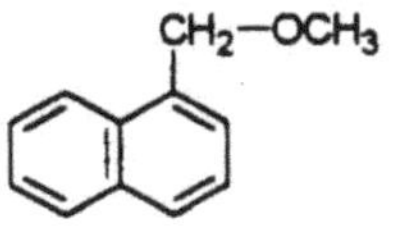

**120** 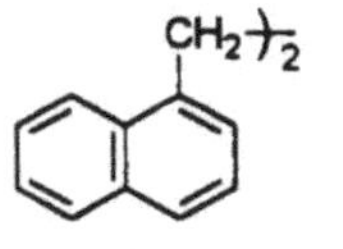          **121**

## 3. Sulfoxide: —S(O)R

A thorough examination of the photochemistry of benzylphenylsulfoxides has been
reported by Gao and Jenks [156]. Although this benzyl cleavage is a result of initial
excitation of the sulfoxide group [157], followed by an α cleavage from the singlet
state, it is included in this report due to the similarity with other benzyl cleavage
reactions. Irradiations of benzylphenylsulfoxide in argon-flushed *t*-butyl alcohol,
acetonitrile, or acetone gave the singlet radical pair as shown in Eq. (43). The radical
pair partitioned between recombination and coupling to give the sulfenic ester,
**122**. Cage escape gave trace amounts of bibenzyl- and sulfur-coupled products. No
substituent studies were reported for the benzyl group and only homolytic cleavage
was observed. The efficiency of the cleavage for the singlet process was estimated
at 0.42 with substantial recombination. The back reaction was measured by moni-
toring the loss of optical rotation of the optically pure starting sulfoxide.

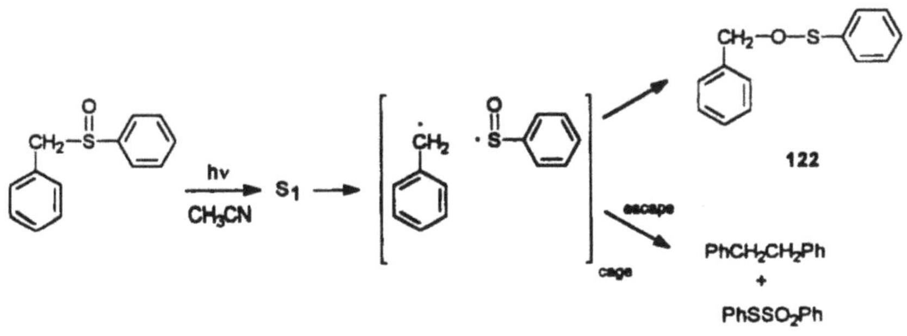

**122**                    (43)

PhCH$_2$CH$_2$Ph
+
PhSSO$_2$Ph

Other benzylsulfoxides have been studied, but their cleavage also appears to be the result of sulfoxide excitation [158].

## 4.  Sulfone: —SO$_2$R

Both benzylsulfone and naphthylmethylsulfone photochemistry have been re- ported by several groups. Givens and co-workers [159,160] and Amiri and Mellor [161] described the homolytic cleavage of dibenzylsulfone, 1-naphthylmethyl- benzylsulfone, and 2-naphthylmethylbenzylsulfone. Internal return for these reac- tions, measured by employing the optically active methyl analogs for each of the sulfones, was at least 20%. The singlet pathway predominated for the benzyl and 2-naphthylmethyl cleavages. The 1-naphthylmethyl compound reacts exclusively from the triplet manifold. These studies were performed in nitrogen-degassed, benzene solutions.

The naphthylmethylbenzylsulfones undergo homolytic cleavage on the naphthyl side preferentially as shown in Eq. (44). Both the singlet and triplet pathways were reported to lead to ultimiate desulfonylation. The resulting radical pairs can couple in-cage or after escaping the solvent cage to give the 1,2-diaryl- ethanes.

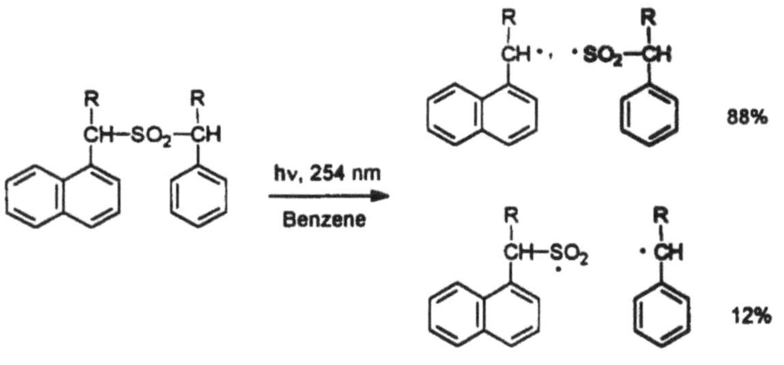

(44)

No mention is made of the potential for simple α cleavage of the sulfone as is observed in sulfoxide photochemistry, but aryl absorption is clearly involved in the naphthylmethyl cases as shown by emission and transient absorption stud- ies [162]. Since the dibenzylsulfone studies were performed at 254 nm and dial- kylsulfones (unlike the sulfoxide analogs) are transparent in the near-UV region [163], one can conclude that the benzyl group is the chromophore.

Pincock and co-workers [31] found similar results in their work with 1- naphthylmethylmethylsulfone, Eq. (45). No solvent capture was observed in these nitrogen-purged methanol studies. Only radical coupled products 1,2-dinaph- thylethane, **115**, and 1-methylnaphthalene were obtained and those in rather low quantum efficiency (0.005 and 0.001, respectively). Homolytic cleavage was clearly demonstrated. Consistent with Givens's sulfone work, the triplet state was implicated for the cleavage process based on quenching studies which resulted in a significant slowing of the photocleavage process.

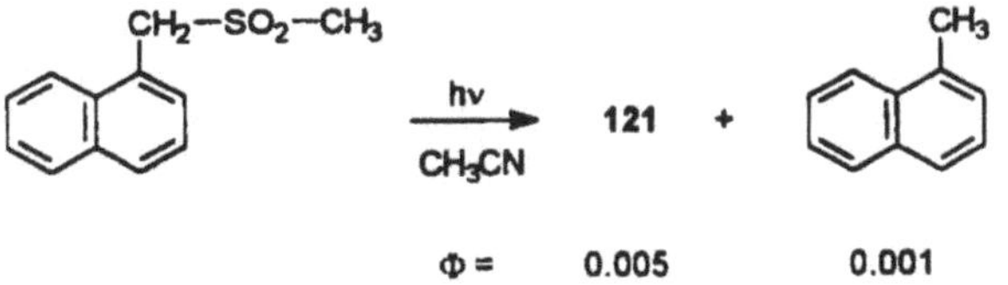

$$\Phi = \quad 0.005 \qquad\qquad 0.001 \qquad\qquad (45)$$

Pitchumani and co-workers [164] recently reported the photocleavage of benzylphenylsulfone in β-cyclodextrin in the absence of solvent. Radical coupling between the ortho position of the benzyl radical and the benzenesulfonyl radical to yield o-methyldiphenylsulfone, **123**, was reported. No products resulting from heterolytic cleavage were observed. Interestingly, none of the typical radical coupled products (biphenyl, diphenylmethane, or dibenzyl) were formed in these conditions. There was no discussion concerning the nature of the excited state.

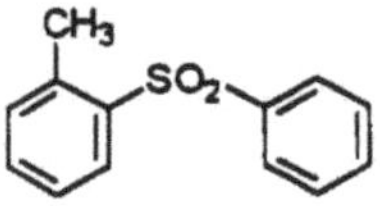

**123**

## I. Halogen

Early contributions by Cristol and Bindel [1] set the stage for benzyl chloride and benzyl bromide photochemistry. More recently, Bartl, Steenken, Mayr, and McClelland [34] have implicated both heterolytic and homolytic cleavage from nanosecond LFP of diphenylmethyl halides. Transient absorption spectroscopy in the 70-ns time frame allowed for observation of both cationic and radical intermediates. The ratio of the two pathways (heterolytic/ homolytic) was essentially the same for the chloride and the bromide.

Moreover, they found that para substitution offered little or no stabilization of the radical intermediate species in homolytic cleavage. The conditions used in this study were 248 nm LFP of the benzyl species, Eq. [46], in acetonitrile that was deoxygenated by bubbling with argon. The ratio of observed products was dependent on the presence of oxygen. The cationic-derived products, **125** and **126**, were unaffected by oxygen, whereas the total yield of radical products, **127–130**, was diminished and benzophenone formed instead. One of the major conclusions from this work was that the homolytic and heterolytic cleavage can occur from either the singlet or triplet excited state. They also pointed out that solvation probably plays a significant role in the reaction pathway. Thus, if solvation occurs on the same time scale as cleavage, then the heterolytic pathway would be favored. In cyclohexane, ion solvation would not be possible and homolytic cleavage would dominate.

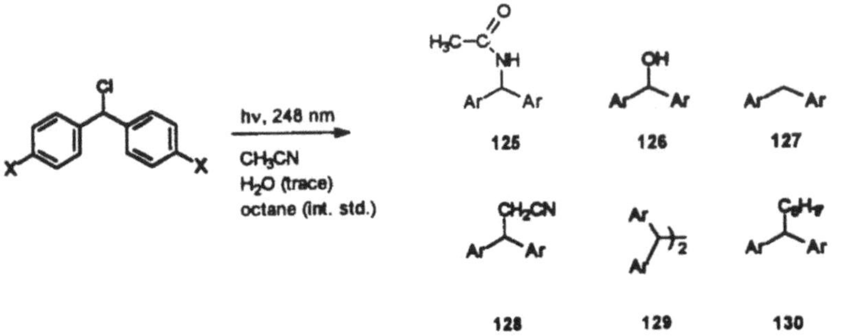

The 1-naphthylmethyl halides have been examined in some detail both by product studies [28,31] and by nanosecond LFP [28]. In methanol solvent, the reported ratio of radical- to ion-derived products in direct irradiations was 61: 39 [28] and 56:31 [31] for the chloride, 12:88 [28] and 16:74 [31] for the bromide, and close to 0:100 [28] for the iodide. The quantum efficiencies reported for disappearance of starting material were 0.24 [28] and 0.15 [31] for the chloride, 0.20 [28] and 0.08 [31] for the bromide, and 0.10 [28] for the iodide. Quenching experiments for the chloride and bromide established that the direct irradiations resulted in reaction from $S_1$. The question of whether the ion-derived methyl ether was formed by direct heterolytic cleavage or by homolytic cleavage followed by electron transfer was left open. The iodide reacted by a completely different mechanism [28] involving homolytic cleavage from $S_1$ followed by reaction of the iodine atom with the substrate. Methanol then reacts with this adduct to give the ether and the radical anion of molecular iodine. As expected, direct irradiations in cyclohexane gave only radical-derived products for all three leaving groups, although the iodide reacted very inefficiently.

Surprisingly, sensitization with xanthone [31] and a variety of other sensitizers [28] initiated photocleavage reactions for the chloride and bromide that gave very similar product distributions as in the direct irradiations. The mechanism proposed [28] for these cases, however, was not triplet–triplet energy transfer but rather exciplex formation followed by reaction of the exciplex either with methanol or by homolytic cleavage.

In a recent series of important papers, Peters and co-workers [6–8] have reported detailed kinetics by picosecond LFP of the photocleavage of a series of substituted diphenylmethyl chlorides, 131, in acetonitrile. These substrates are reactive enough from $S_1$ that measurable concentrations of ion pairs and radical pairs can be generated and their growth and decay monitored. Both geminate pairs are formed within the 20-ps time resolution of the instrument; the quantum yields ($\Phi_{hom}$ and $\Phi_{het}$) are given by structure 131. The radical pair does not combine but rather decays by both cage escape and return to the ground state surface as the contact ion pair. The quantum yields, $\Phi_{etri}$, for formation of ion pairs from radical pairs indicate that this process becomes more favorable as it becomes thermodynamically more favorable. An important aspect of this work is the consideration of the complex question as to how solvent polarity and polarization dynamics control the trajectory of the excited state on the pathway toward the radical pair (less stable) and the avoided crossing to the ion pair (more stable).

McGowan and Hilinski [3] have looked at the picosecond photodissociation of 9-chloro- and 9-bromofluorene, **132**. These analogs of a benzyl halide were found to undergo homolytic cleavage from the singlet excited state. Intersystem crossing to the triplet competes with cleavage particularly for the bromo compound. There was no observed cleavage from the triplet state up to 20 ns. These studies were performed in argon-bubbled cyclohexane. Involvement of a higher, dissociative triplet state could not be excluded in this report. Cristol and Bindel [165] proposed such a pathway in their work with benzyl halides.

A recent application of the homolytic photocleavage was reported by Gajewski and Paul [166]. They irradiated dichlorodifluorenylbenzene, **133**, in a 2-methyltetrahydrofuran glass through a Pyrex filter and monitored the benzylic cleavage by ESR. The spectral evidence indicated that the diradical species was a triplet state.

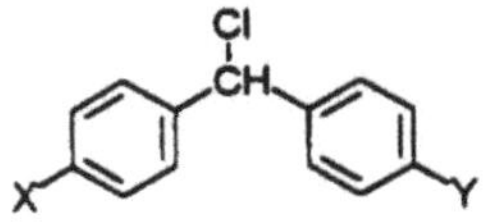

| | $\Phi_{hom}$ | $\Phi_{het}$ | $\Phi_{tri}$ |
|---|---|---|---|
| **131:** X = Y = H | 0.37 | 0.28 | 0.00 |
| X = H, Y = methyl | 0.35 | 0.18 | 0.05 |
| X = Y = methyl | 0.52 | 0.28 | 0.07 |
| X = H, Y = methoxy | 0.62 | 0.43 | 0.19 |
| X = Y = methoxy | 0.54 | 0.34 | 0.23 |

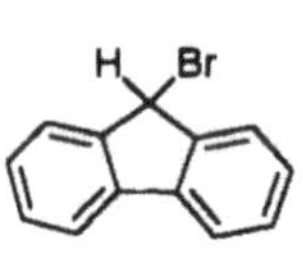

**132** 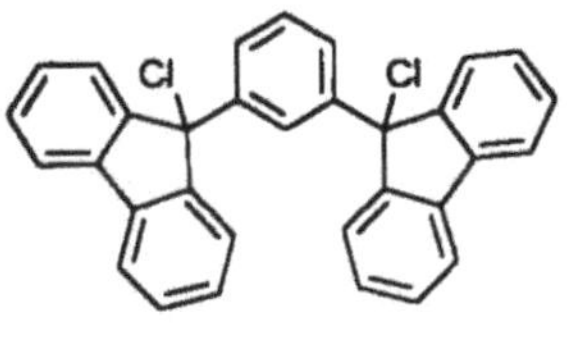 **133**

## J. Arsenic

Saeva has also reported the photocleavage of the triphenyl(*p*-cyanobenzyl)arsonium cation [145]. The counterion in this study was tetrafluoroborate. Irradiation in acetonitrile gave both radical- and solvent-trapped cationic products, Eq. (47). The mechanistic rationale for this chemistry was based on the same argument as Saeva's phosphonium salt results [see Eq. (38)]. In contrast to the phosphonium leaving group, however, the arsonium one gave the ion-derived amide product, **118** [see Eq. (41)]. The radical coupled products in this study were a mixture of diarylmethane, **134**, from in-cage coupling and diarylethane, **135**, from out-of-cage coupling. The initially formed singlet radical pair can undergo intersystem crossing ($k_{-isc}$, in Scheme 1) to the triplet radical pair more readily due to the magnetic properties (increased spin-orbit coupling) of the arsonic-centered radical. The triplet radical pair is longer lived and can escape the solvent cage to give the substituted dibenzyl compound, **135**. The major arsenic containing product in the reaction is the arsenic oxide, **132**.

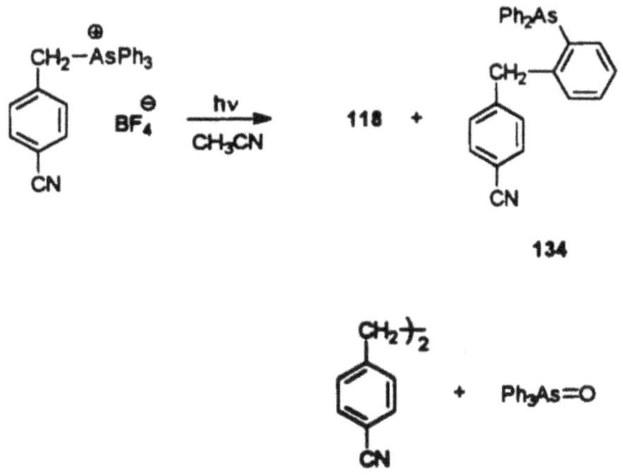

## K. Selenium

Ouchi and co-workers [153] reported the 248-nm excimer laser irradiation of 1,8-bis(phenylselenomethyl)naphthalene to give acenaphthalene via the diradical, 110. Their UV analysis suggests that the n,π* selenium absorption partially overlaps with the π,π* of the phenyl and naphthyl groups at 248 nm. The two-photon study found that the naphthylmethylselenium bond underwent radical cleavage with a higher efficiency than the sulfur and oxygen analogs at low fluence. The leaving group ability for this radical reaction was PhSe > PhS ≫ PhO. They rationalized that the photolysis was a result of selenium absorption, phenyl excitation, or excitation to the $S_2$ state of the naphthalene. The irradiations were carried out in degassed cyclohexane.

## V. CONCLUSIONS

Table 6 shows a summary of the results for the $ArCH_2$-X photolyses that have been reported in the literature and discussed in this review.

Ample evidence has been presented that the mechanism of photocleavage in benzylic substrates is a function, in a complex way, of the leaving group and the aromatic chromophore. There are cases where the cleavage is solely homolytic. It is less clear if any photocleavages occur entirely by the heterolytic pathway. On the basis of LFP transient studies, at least some cases seem to involve heterolytic

**Table 6**  Summary of $ArCH_2$—X Photocleavage Reactions[a]

| X | Radical-derived Pdts | Ion-derived Pdts | Intermediates Observed | Substituent Studies |
|---|---|---|---|---|
| —CN | | Y | Y | N |
| —C(OR)₂—R | | Y | Y | N |
| —C(OH)R₂ | Y | | | N |
| —C(CN)R₂ | | Y | | N |
| —cyclopropyl | | Y | | Y |
| —NR₂ | Y | | | Y |
| —NR₃⁺ | Y | Y | | Y |
| —OH | Y | Y | Y | Y |
| —OH₂⁺ | | Y | Y | Y |
| —OR | Y | Y | Y | Y |
| —O(C=O)R | Y | Y | Y | Y |
| —O(C=O)OR | Y | Y | | N |
| —O(C=O)NR₂ | Y | Y | | N |
| —OPO₃R₂ | | Y | | Y |
| —OP(OR)₂ | Y | N | Y | N |
| —O(SO₂)R | Y | | Y | N |
| —F | Y | Y | Y | Y |
| —PR₃⁺ | Y | Y | Y | Y |
| —SR | Y | | Y | Y |
| —SR₂⁺ | Y | | Y | Y |
| —SO₂R | Y | | | Y |
| —Cl, —Br, —I | Y | Y | Y | Y |
| —As | Y | Y | | N |
| —Se | Y | | | N |

[a] For radical- and ion-derived products, a blank space means that they were not reported or that the yields reported were less than 5%. For intermediates observed, a blank space means that LFP experiments were not attempted.

cleavage from the excited state as cations are already observable in times as short as 20 ps. Substituent studies have shown that there is a meta (and probably an ortho) effect on the homolytic cleavage rates and efficiencies. To the extent to which there is direct heterolytic cleavage, and consequently an increased polarity at the transition state, in the bond being broken, undoubtedly a much stronger sense of the substituents on the ring should be observed. The recent development of somewhat more routine LFP instrumentation (particularly at the picosecond time scale) should help to unravel some of these questions.

Several factors have been shown to clutter analysis of the cleavage mechanism as developed by product studies. Internal return has been observed in every case that has been designed to monitor it. Single-electron transfer, which gives ionic intermediates after initial homolytic cleavage, has been demonstrated in some cases and could, in principle, be occurring in each study where heterolytic cleavage is argued. The presence of oxygen, even at very low concentrations, has also been shown to have an impact on the pathway of cleavage in some cases; in particular, it can change the product ratios and the overall yields. Additionally, it may assist in singlet-to-triplet transitions.

Cleavage from the triplet excited state is clearly unlike the singlet cleavage reactions. The triplet excited state results in homolytic cleavage to a triplet radical pair and therefore the yields of out-of-cage radical coupled products are higher.

Finally, between the time of the review by Cristol and Bindel in 1983 [1] and the present, considerable progress has been made in understanding these seemingly simple photoreactions which involve cleavage of only one sigma bond. However, our current knowledge still does not allow us to make reliable predictions about the efficiency of reaction and product distribution for substrates that have not been previously examined. Surprises continue to occur and clearly there is considerable scope for new studies.

## REFERENCES

1.  Cristol, S.; Bindel, T.H. *Org. Photochem.* **1983**, *6*, 327–415.
2.  This scheme is a modification of one published recently in Ref. 3.
3.  McGowan, W.M.; Hilinski, E.F. *J. Am. Chem. Soc.* **1995**, *117*, 9019–9025.
4.  Zimmerman, H.E. *J. Am. Chem. Soc.* **1995**, *117*, 8988–8991.
5.  Pincock, J.A. *Acc. Chem. Res.* **1997**, *30*, 43–49.
6.  Deniz, A.A.; Li, B.; Peters, K.S. *J. Phys. Chem.* **1995**, *99*, 12209–12213.
7.  Lipson, M.; Deniz, A.A.; Peters, K.S. *J. Am. Chem. Soc.* **1996**, *118*, 2992–2997.
8.  Lipson, M.; Deniz, A.A.; Peters, K.S. *J. Phys. Chem.* **1996**, *100*, 3580–3586.
9.  McClelland, R.A. *Tetrahedron* **1996**, *52*, 6823–6858.
10. Das, P.K. *Chem. Rev.* **1993**, *93*, 119–144.
11. Binkley, R.W.; Flechtner, W. in *Synthetic Organic Photochemistry*, Horspool, W.M., Ed. Plenum Press, New York, 1984, p 375.

12. Falvey, D.E.; Banerjee, A. *J. Org. Chem.* **1997**, *18*, 6245–6251 and references therein.
13. Givens, R.S.; Weber, J.F.W.; Jung, A.H.; Park, C.-H. *Methods in Enzymology* **1998**, *291*, 1–29.
14. Fleming, S.A.; Rawlins, D.B.; Samano, V.; Robins, M.J. *J. Org. Chem.* **1992**, *57*, 5968–5976 and references therein.
15. Shirai, M.; Tsunooka, M. *Prog. Polym. Sci.* **1996**, *21*, 1–45.
16. Maslak, P.; Chapman, W.H.; Vallombroso, T.M.; Watson,B. A. *J. Am. Chem. Soc.* **1995**, *117*, 12380–12389.
17. Maslak, P. in *Topics in Current Chemistry*, Vol. *168*, Mattay, J., Ed. Springer-Verlag, Berlin, 1993, p 1.
18. Maslak, P.; Narvaez, J.N.; Vallombroso, T.M. *J. Am. Chem. Soc.* **1995**, *117*, 12373–12390.
19. Murov, S.L.; Carmichael, I.; Hug, G.L. *Handbook of Photochemistry*, 2nd Ed., Marcel Dekker, New York, 1993.
20. Silverstein, R.M.; Bassler, G.C.; Morrill, T.C. *Spectrometric Identification of Organic Compounds, 5th Ed.*, Wiley, New York, 1991, p 307.
21. Appleton, D.C.; Bull, D.C.; Givens, R.S.; Lillis, V.; McKenna, J.; McKenna, J.M.; Thackeray, S.; Walley, A.R. *J. Chem. Soc., Perkin II* **1980**, 77–82.
22. Foster, B; Gaillard, B.; Pincock, A.L.; Pincock, J.A.; Sehmbey, C. *Can. J. Chem.* **1987**, *65*, 1599–1607.
23. Bentley, T.W.; Llewellyn, G. in *Progress in Physical Organic Chemistry*, Vol *17*, Wiley, New York, 1990, pp 121–158.
24. Lissi, E.A.; Encinas, M.V. In *Handbook of Organic Photochemistry*, Scaiano, T., Ed.; CRC Press, Boca Raton, FL, Vol. II, 1989; pp 111–117.
25. Hilborn, J.W.; MacKnight, E.; Pincock, J.A.; Wedge, P.J. *J. Am. Chem. Soc.* **1994**, *116*, 3337–3346.
26. Stephenson, L.M.; Whitten, D.G.; Vesley, G.F.; Hammond, G.S. *J. Am. Chem. Soc.* **1966**, *88*, 3665.
27. Reference 19, p 95.
28. Slocum, G.H.; Schuster, G.B. *J. Org. Chem.* **1984**, *49*, 2177–2185.
29. Reference 19, p 85.
30. Reference 19, p 91.
31. Arnold, B.; Donald, L.; Jurgens, A.; Pincock, J.A. *Can. J. Chem.* **1985**, *63*, 3140–3146.
32. Scaiano, J.C. *J. Am. Chem. Soc.* **1980**, *102*, 7747–7753.
33. Manring, L.E.; Peters, K.S. *J. Phys. Chem.* **1984**, *88*, 3516–3520.
34. Bartl, J.; Steenken, S.; Mayr, H.; McClelland, R.A. *J. Am. Chem. Soc.* **1990**, *112*, 6918–6928.
35. McMillen, D.F.; Golden, D.M. *Annu. Rev. Phys. Chem.* **1982**, *33*, 493–532.
36. Sim, B.A.; Milne, P.H.; Griller, D.; Wayner, D.D.M. *J. Am. Chem. Soc.* **1990**, *112*, 6635–6646.
37. Wayner, D.D.M. in *Handbook of Organic Photochemistry*, Scaiano, T., Ed.; CRC Press, Boca Raton, FL, Vol. II, 1989; pp 363–367.
38. Fleming, S.A.; Jensen, A.W. *J. Org. Chem.* **1996**, *61*, 7040–7044.
39. Saeva, F.D. in *Topics in Current Chemistry*, Vol. *156*, Mattay, J., Ed. Springer-Verlag, Berlin, 1990, p 59–92.

40. Kawakami, J.; Iwamura, M. *J. Phys. Org. Chem.* **1994**, *7*, 31–42.
41. DeCosta, D.P.; Pincock, J.A. *Can. J. Chem.* **1992**, *70*, 1879–1885.
42. Palmer, I.J.; Ragazos, I.N.; Bernardi, F.; Olivucci, M.; Robb, M.A. *J. Am. Chem. Soc.* **1993**, *115*, 673–682.
43. Tokamura, K.; Ozaki, T.; Nosaka, H.; Saigusa (Ejiru), Y.; Itoh, M. *J. Am. Chem. Soc.* **1991**, *113*, 4974–4980.
44. Zimmerman, H.E.; Sandel, V.R. *J. Am. Chem. Soc.* **1963**, *85*, 915–922.
45. Mathivanan, N.; Cozens, F.; McClelland, R.A.; Steenken, S.J. *J. Am. Chem. Soc.* **1992**, *114*, 2198–2203.
46. Milne, P.H.; Wayner, D.D.M.; DeCosta, D.P.; Pincock, J.A. *Can. J. Chem.* **1992**, *70*, 121–127.
47. Michl, J.; Bonačić-Koutecký, V. *Electronic Aspects of Organic Photochemistry*, Wiley, New York, 1990.
48. The quotes reflect the fact that for $CH_3$—$NH_3^+$, the species that are formed by the two possible modes of heterolytic cleavage are not actually zwitterions. One is $CH_3^+$ $NH_3$, the other $CH_3^-$ $NH_3^{+2}$.
49. Reference 47, p 163.
50. Reference 47, p 296.
51. Wan, P.; Krogh, E.; Chak, B. *J. Am. Chem. Soc.* **1988**, *110*, 4073–4074.
52. Wan, P.; Budac, D.; Krogh, E. *J. Chem. Soc., Chem. Commun.* **1990**, 255–257.
53. Budac, D.; Wan, P. *J. Org. Chem.* **1992**, *57*, 887–894.
54. Shukla, D.; Wan, P. *J. Photochem. Photobiol. A: Chem.* **1998**, *113*, 53–64.
55. Wan, P.; Shukla, D. *Chem. Rev.* **1993**, *93*, 571–584.
56. Spears, K.G.; Gray, T.H.; Huang, D. *J. Phys. Chem.* **1986**, *90*, 779–790.
57. Cosa, J.J.; Gsponer, H.E. *J. Photochem. Photobiol. A: Chem.* **1989**, *48*, 303–311.
58. McClelland, R.A.; Kangasabapathy, V.M.; Banait, N.S.; Steenken, S. *J. Am. Chem. Soc.* **1989**, *111*, 3966–3972.
59. Pischel, U.; Abraham, W.; Schnabel, W.; Muller, U. *Chem. Commun.* **1997**, 1383–1384.
60. Budac, D.; Wan, P. *J. Photochem. Photobiol. A: Chem.* **1992**, *67*, 135–166.
61. Meiggs, T.O.; Miller, S.I. *J. Am. Chem. Soc.* **1972**, *94*, 1989–1996.
62. Krogh, E.; Wan, P. *J. Am. Chem. Soc.* **1992**, *114*, 705–712.
63. Bohne, C. in *Organic Photochemistry and Photobiology*, Horspool, W.M., Soon, P.-S., Ed. CRC Press, Boca Raton, 1994, pp 416–422, 423–429.
64. Wan, P.; Muralidharan, S. *J. Am. Chem. Soc.* **1988**, *110*, 4336–4345.
65. Steenken, S.; McClelland, R.A. *J. Am. Chem. Soc.* **1989**, *111*, 4967–4973.
66. Pienta, N.J.; Kessler, R.J.; Peters, K.S.; O'Driscoll, E.D.; Arnett, E.M.; Molter, K.E. *J. Am. Chem. Soc.* **1991**, *113*, 3773–3781.
67. Clifton, M.F.; Fenick, D.J.; Gasper, S.M.; Falvey, D.E.; Boyd, M.K. *J. Org. Chem.* **1994**, *59*, 8023–8029.
68. Hixson, S.S.; Franke, L.A.; Gere, J.A.; Xing, Y. *J. Am. Chem. Soc.* **1988**, *110*, 3601–3610.
69. Parman, T.; Pincock, J.A.; Wedge, P.J. *Can. J. Chem.* **1994**, *72*, 1254–1261.
70. Pincock, J.A.; Parman, T. unpublished results.
71. Er-Rhaimini, A.; Moshsinaly, N.; Mornet, R. *Tetrahedron Lett.* **1990**, *31*, 5757–5760.

72. Siskos, M.G.; Zarkadis, A.K.; Steenken, S.; Karakostas, N.; Garas, S.K. *J. Org. Chem.* **1998**, *63*, 3251–3259.
73. Holmes, C.P. *J. Org. Chem.* **1997**, *62*, 2370–2380.
74. Salerno, C.P.; Resat, M.; Magde, D.; Kraut, J. *J. Am. Chem. Soc.* **1997**, *119*, 3403–3404.
75. Ramesh, D.; Wieboldt, R.; Billington, A.P.; Carpenter, B.K.; Hess, G.P. *J. Org. Chem.* **1993**, *58*, 4599–4605 and references therein.
76. Rodebaugh, R.; Fraser-Reid, B.; Geysen, H.M. *Tetrahedron Lett.* **1997**, *38*, 7653–7656.
77. Mann, C.K. *Anal. Chem.* **1964**, *36*, 2424.
78. Eberson, L. *Electron Transfer Reactions in Organic Chemistry*; Springer-Verlag, New York, 1987.
79. Pincock, J.A.; Parman, T.; MacKnight, E. unpublished results.
80. Irie, M. *J. Am. Chem. Soc.* **1983**, *105*, 2078–2079.
81. Boyd, M.K. in *Organic Photochemistry, Molecular and Supramolecular Photochemistry*, Vol. *1*. Ramamurthy, V.; Schanze, K.S., Ed. Marcel Dekker, New York, 1997, pp 147–186.
82. Wan, P.; Yates, K.; Boyd, M.K. *J. Org. Chem.* **1985**, *50*, 2881–2886.
83. Minto, R.E.; Das, P.K. *J. Am. Chem. Soc.* **1989**, *111*, 8858–8866.
84. McClelland, R.A.; Banait, N.; Steenken, S. *J. Am. Chem. Soc.* **1989**, *111*, 2929–2935.
85. Johnston, L.J.; Lobaugh, J.; Wintgens, V. *J. Phys. Chem.* **1989**, *93*, 7370–7374.
86. Shukla, D.; Wan, P. *J. Photochem. Photobiol. A: Chem.* **1994**, *79*, 55–59.
87. Wan, P.; Krogh, E. *J. Am. Chem. Soc.* **1989**, *111*, 4887–4895.
88. Mecklenberg, S.L.; Hilinski, E.F. *J. Am. Chem. Soc.* **1989**, *111*, 5471–5472.
89. Turro, N.J.; Wan, P. *J. Photochem.* **1985**, *28*, 93–102.
90. Wan, P. *J. Org. Chem.* **1985**, *50*, 2583–2586.
91. Wan, P.; Chak, B. *J. Chem. Soc. Perkin II*, **1986**, 1751–1756.
92. Wan, P.; Chak, B.; Krogh, E. *J. Photochem. Photobiol. A: Chem.* **1989**, *46*, 49–57.
93. Hall, B.; Wan, P. *J. Photochem. Photobiol. A: Chem.* **1991**, *56*, 35–42.
94. Zimmerman, H.E.; Somasekhara, S. *J. Am. Chem. Soc.* **1963**, *85*, 922–927.
95. McClelland, R.A.; Kanagasabapathy, V.M.; Banait, N.; Steenken, S. *J. Am. Chem. Soc.* **1989**, *111*, 3966–3972.
96. Benn, R.; Dreeskamp, H.; Schuchmann, H.-P.; von Sonntag, C. *Z. Naturforsch.* **1979**, *34b*, 1002–1009.
97. Pitchumani, K.; Devanathan, S.; Ramamurthy, V. *J. Photochem. Photobiol. A: Chem.* **1992**, *69*, 201–208.
98. Pincock, J.A. in *Organic Photochemistry and Photobiology*, Horspool W.M., Soon, P.-S., Ed. CRC Press, Boca Raton, 1995, pp 393–407.
99. Hilborn, J.W.; Pincock, J.A. *J. Am. Chem. Soc.* **1991**, *113*, 2683–2686.
100. Hilborn, J.W.; Pincock, J.A. *Can. J. Chem.* **1992**, *70*, 992–999.
101. DeCosta, D.P.; Pincock, J.A. *J. Am. Chem. Soc.* **1993**, *115*, 2180–2190.
102. Pincock, J.A.; Wedge, P.J. *J. Org. Chem.* **1994**, *59*, 5587–5595.
103. Pincock, J.A.; Wedge, P.J. *J. Org. Chem.* **1994**, *60*, 4067–4076.
104. Kim, J.M.; Pincock, J.A. *Can. J. Chem.* **1995**, *73*, 885–895.
105. Nevill, S.M.; Pincock, J.A. *Can. J. Chem.* **1997**, *75*, 232–247.

106.  Iwamura, M.; Ishikawa, T.; Koyama, Y.; Sakuma, K.; Iwamura, H. *Tetrahedron Lett.* **1987**, *28*, 679–682.
107.  Cozens, F.L.; Pincock, A.L.; Pincock, J.A.; Smith, R. *J. Org. Chem.* **1998**, *63*, 434–435.
108.  Peach, J.M.; Pratt, A.J.; Snaith, J.S. *Tetrahedron* **1995**, *51*, 10013–10024.
109.  Rock, S.R.; Chan, S.I. *J. Org. Chem.* **1996**, *61*, 1526–1529.
110.  Shi, Y.; Corrie, J.E.T.; Wan, P. *J. Org. Chem.* **1997**, *62*, 8278–8279.
111.  Gee, K.R.; Niu, L.; Schaper, K.; Hess, G.P. *J. Org. Chem.* **1995**, *60*, 4260–4263.
112.  Sauers, C.K.; Jencks, W.P.; Groh, S. *J. Am. Chem. Soc.* **1975**, *97*, 5546.
113.  Chateauneuf, J.; Lusztyk, J.; Maillard, B.; Ingold, K.U. *J. Am. Chem. Soc.* **1988**, *110*, 6727.
114.  Cameron, J.F.; Fréchet, J.M.J. *J. Org. Chem.* **1990**, *55*, 5919–5922.
115.  Cameron, J.F.; Fréchet, J.M.J. *J. Am. Chem. Soc.* **1991**, *113*, 4303–4313.
116.  Beecher, J.E.; Cameron, J.F.; Fréchet, J.M.J. *J. Mater. Chem.* **1992**, *2*, 811–816.
117.  Burgess, K.; Jacutin, S.E.; Lim, D.; Shitangkoon, A. *J. Org. Chem.* **1997**, *62*, 5165–5168.
118.  Cameron, J.F.; Willson, C.G.; Fréchet, J.M.J. *J. Am. Chem. Soc.* **1996**, *118*, 12925–12937.
119.  Cameron, J.F.; Willson, C.G.; Fréchet, J.M.J. *J. Chem. Soc., Perkin Trans. 1*, **1997**, 2429–2442.
120.  Okada, S.; Yamashita, S.; Furuta, T.; Iwamura, M. *Photochem. Photobiol.* **1995**, *61*, 431–434.
121.  Givens, R.S.; Kueper, L.W. *Chem. Rev.* **1993**, *93*, 55–66.
122.  Givens, R.S.; Park, C.-H. *Tetrahedron Lett.* **1996**, *37*, 6259–6262.
123.  Givens, R.S.; Matuszewski, B. *J. Am. Chem. Soc.* **1984**, *106*, 6860–6861.
124.  Givens, R.S.; Matuszewski, B.; Athey, P.S.; Stoner, R.M. *J. Am. Chem. Soc.* **1990**, *112*, 6016–6021.
125.  Givens, R.S.; Singh, R. *Tetrahedron Lett.* **1991**, *48*, 7013–7016.
126.  Furuta, T.; Torigai, H.; Sugimoto, M.; Iwamura, M. *J. Org. Chem.* **1995**, *60*, 3953–3956.
127.  Furuta, T.; Torigai, H.; Osawa, T. Iwamura, M. *Chem. Lett.* **1993**, 1179–1182.
128.  Ordoukhanian, P.; Taylor, J.-S. *J. Am. Chem. Soc.* **1995**, *117*, 9570–9571.
129.  Pirrung, M.C.; Shuey, S.W. *J. Org. Chem.* **1994**, *59*, 3890–3897.
130.  Omelanzcuk, J.; Sopchik, A.E.; Lee, S.-G.; Akutagawa, K.; Cairns, M.; Bentrude, W.G. *J. Am. Chem. Soc.*, **1988**, *110*, 6908–6909.
131.  Koptyug, I.V.; Ghatlia, N.D.; Sluggett, G.W.; Turro, N.J.; Ganapathy, S.; Bentrude, W.G. *J. Am. Chem. Soc.* **1995**, *117*, 9486–9491.
132.  Cairns, S.W.; Bentrude, W.G. *Tetrahedron Lett.* **1989**, *30*, 1025–1028.
133.  Koptyug, I.V.; Sluggett, G.W.; Ghatlia, N.D.; Landis, M.S.; Turro, N.J. *J. Phys. Chem.* **1996**, *100*, 14581–14583.
134.  Fujio, M.; Toshihiro, T.; Akasaka, I.; Mishima, M.; Tsuno, Y. *Bull. Chem. Soc. Jpn.* **1990**, *63*, 1146–1153.
135.  Naitoh, K.; Yoneyama, K.; Yamaoka, T. *J. Phys. Chem.* **1992**, *96*, 238–244.
136.  Binkley, R.W.; Liu, X. *J. Carbohydrate Chem.* **1992**, *11*, 183–188.
137.  Seiler, P. von; Wirz, J. *Helv. Chim. Acta* **1972**, *53*, 2693–2711.
138.  Kosmrlj, B.; Krajl, B.; Sket, B. *Tetrahedron Lett.* **1995**, *36*, 7921–7924.

139. Steinmetz, M.G. *Chem. Rev.* **1995**, *95*, 1527–1588.
140. Fasani, E.; d'Alessandro, N.; Albini, A.; Mariano, P.S.; *J. Org. Chem.* **1994**, *59*, 829.
141. Kako, M.; Kakuma, S.; Hatakenaka, K.; Nakadaira, Y.; Yasui, M.; Iwasaki, F. *Tetrahedron Lett.* **1995**, *36*, 6293–6296.
142. Baciocchi, E.; DelGiacco, T.; Elisei, F.; Ioele, M. *J. Org. Chem.* **1995**, *60*, 7974–7983.
143. Griffin, C.E.; Kaufman, M.L. *Tetrahedron Lett.* **1965**, 773–776.
144. Oliveira, M.E.C.D.R.; Pereira, L.C.; Thomas, E.W.; Bisby, R.H.; Cundall, R.B. *J. Photochem.* **1985**, *31*, 373–379.
145. Breslin, D.T.; Saeva, F.D. *J. Org. Chem.* **1988**, *53*, 713–715.
146. Alonsono, E.O.; Johnston, L.J.; Scaiano, J.C.; Toscano, V.G. *Can. J. Chem.* **1992**, *70*, 1784–1794.
147. Imrie, C.; Modro, T.A.; Rohwer, E.R.; Wagener, C.C.P. *J. Org. Chem.* **1993**, *58*, 5643–5649.
148. Blakemore, D.C.; Gilbert, A. *J. Chem. Soc., Perkin Trans. 1* **1992**, *16*, 2265–2270.
149. Fleming, S.A.; Jensen, A.W. *J. Org. Chem.* **1993**, *58*, 7135–7137.
150. Fleming, S.A.; Rawlins, D.B.; Robins, M.J. *Tetrahedron Lett.* **1990**, *31*, 4995–4998.
151. Thyrion, F.C. *J. Phys. Chem.* **1973**, *77*, 1478–1482.
152. Sucholeiki, I. *Tetrahedron Lett.* **1994**, *35*, 7307–7310.
153. Ouchi, A.; Yabe, A.; Adam, W. *Tetrahedron Lett.* **1994**, *34*, 6309–6312.
154. Maycock, A.L.; Berchtold, G.A. *J. Org. Chem.* **1970**, *35*, 2532–2538.
155. Saeva, F.D.; Breslin, D.T.; Luss, H.R. *J. Am. Chem. Soc.* **1991**, *113*, 5333–5337.
156. Guo, Y.; Jenks, W.S. *J. Org. Chem.* **1995**, *60*, 5480–5486.
157. The benzyl group is not required for this photochemistry. Even dialkylsulfoxides can undergo cleavage upon irradiation at 248 nm. See Ref. 154.
158. Jenks, W.S.; Gregory, D.D.; Guo, Y.; Lee, W.; Tetzlaff, T. *Molecular and Supramolecular Photochemistry*, Vol. 1, Marcel Dekker, New York, 1997, pp 1–56.
159. Givens, R.S.; Hrinczenko, B.; Liu, J.H.-S.; Matuszewski, B.; Tholen-Collison, J. *J. Am. Chem. Soc.* **1984**, *106*, 1779–1789.
160. Givens, R.S.; Matuszewski, B. *Tetrahedron Lett.* **1978**, 861–864.
161. Amiri, A.S.; Mellor, J.M. *J. Photochem.* **1978**, *9*, 902–916.
162. Gould, I.R.; Tung, C.; Turro, N.J.; Givens, R.S.; Matuszewski, B. *J. Am. Chem. Soc.* **1984**, *106*, 1789–1793.
163. Reference 20, p 305.
164. Pitchumani, K.; Velusamy, P.; Banu, H.S.; Srinivasan, C. *Tetrahedron Lett.* **1995**, 1149–1152.
165. Cristol, S.J; Bindel, T.H. *J. Am. Chem. Soc.* **1981**, *103*, 7287–7293.
166. Gajewski, J.J.; Paul, G.C. *Tetrahedron Lett.* **1998**, *39*, 351–354.

# 6

# Photophysical Probes for Organized Assemblies

**Kankan Bhattacharyya**

Indian Association for the Cultivation of Science, Calcutta,
West Bengal, India

## I. INTRODUCTION

In nature, most chemical and biological processes occur in self-organized molecular assemblies. In these assemblies, the active chemical species is confined within a small region, a few nanometers in size. In such a nanoenvironment the "local" polarity, viscosity, and pH are often vastly different from those in a bulk medium. The proximity and the favorable disposition of the reactants and the substantially altered local properties in such an assembly exert profound influence on the structure, reactivity, and dynamics of the confined chemical species. Chemistry in organized media differs markedly from that in any homogeneous fluid medium and mimics the extremely efficient chemical processes in the biological systems [1–6]. Photophysical processes in organized assemblies are interesting particularly for two reasons. First, in many organized assemblies the rate of some photophysical processes changes dramatically by three to four orders of magnitude. Second, the remarkable sensitivity of the photophysical processes to the environment can be utilized to probe the local properties of the organized assemblies.

The organized assemblies include self-assembled molecular aggregates in polar liquids (e.g., micelles in water) or nonpolar liquids (e.g., reverse micelles or microemulsions in hydrocarbons), cage-like hosts soluble in many liquids (cyclodextrins or calixarenes), microporous solids (e.g., zeolites), semirigid materials (e.g., polymers, hydrogels, etc.), and so on. The most important feature of any organized assembly is the confinement of the photophysical probe, often along with many solvent molecules in a finite region in space. Such confinement results in significant changes in the local dielectric constant and viscosity, and imposes considerable constraint on the free movement of the probe and the confined solvent molecules. The ultimate goal of the photophysical studies in such organized assemblies is to understand the biological processes that occur in the hydrophobic pockets of various proteins or the membrane surfaces, etc. With this end in view, in Section II we will introduce some photophysical processes used as a probe for the organized assemblies. In Section III we will describe the architecture of several organized assemblies. Finally, in Section IV we will discuss the recent results on the dynamics of various photophysical processes in these assemblies and their implications.

In keeping with the fundamental importance of this subject and the intense recent activity in this area, several excellent reviews have already summarized different aspects of photophysical processes in organized assemblies. The development in this area up to 1991 was summarized in the volume edited by Ramamurthy [2–4]. Subsequently, several authors discussed the effect of cyclodextrins on many photophysical processes [5]. The binding of organic probes with cyclodextrins and other hosts in aqueous medium arises as a result of the hydrophobic effect. The different aspects of hydrophobic binding have also been reviewed extensively [1]. Many organized assemblies involve an interface between two drastically different media. The properties of molecules at various interfaces have been recently studied using a number of new experimental and theoretical techniques. This aspect is reviewed by Eisenthal [6a], Robinson et al. [6b], and others [6c–e].

## II.  PHOTOPHYSICAL PROCESSES

In this section we will discuss the salient features of several photophysical processes. Due to our interest in the dynamics of the photophysical processes we will emphasize factors that govern the dynamics of these processes.

### A.  Solvation Dynamics

Solvation, i.e., the interaction of the solvent molecules with a solute, plays a key role in most chemical processes that occur in fluid solutions. The dynamics of the solvation process can be followed by creating a dipole or an electron in a

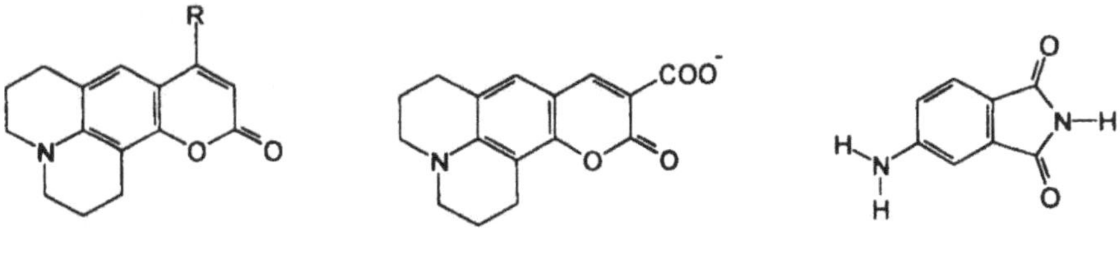

C 480 : R=CH$_3$
C 153 : R=CF$_3$

C 343

4 - AP

**Figure 1** Structure of a few fluorescence probes used in solvation dynamics studies.

polar liquid and then observing the gradual decrease in the energy of the system due to solvation. For this purpose, one uses a probe that is nonpolar or weakly polar in the ground state and is highly polar in the electronically excited state. A dipole can be created instantaneously by exciting such a probe with an ultra-short pico- or femtosecond light pulse. The structure of a few probes commonly used for the study of solvation dynamics is given in Fig. 1. When the probe solute is in the ground state, the polar solvent molecules remain randomly oriented around the nonpolar or weakly polar probe solute molecule (Fig. 2a). Immediately after creation of the dipole by an ultrashort pulse, the polar solvent molecules remain randomly oriented around the created dipole as the solvent relaxation is slower than the excitation process (Fig. 2b). After the creation of the dipole, the solvent molecules gradually reorient around the newly created electron or dipole. This process of reorientation of the solvent dipoles around an electron or a dipole is referred to as *solvation dynamics*. The system eventually reaches the fully solvated state (Fig. 2c). The solvation time, $\tau_s$, is defined as the time taken for the solvent molecules in going from the randomly oriented configuration (2b) to the fully solvated state (2c). As the solute dipole, in the excited state, is gradually

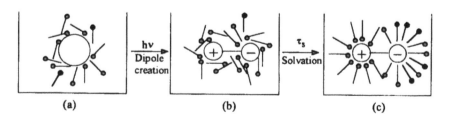

(a)  (b)  (c)

**Figure 2** Creation and solvation of a dipole in a polar solvent (match sticks denote solvent dipoles).

stabilized through solvation, the incompletely solvated species (2b) decays rapidly giving rise to a fast decay at the blue end of the emission spectrum. However, the solvated species (2c) grows with time, and this growth is observed at the red end of the emission spectrum. This results in the gradual decrease in the emission energy, i.e., red shift of the emission spectra with increase in time (Fig. 3). This phenomenon is known as the time-dependent Stokes shift (TDSS) [7–14]. The wavelength-dependent temporal decays of emission and TDSS are regarded as evidence of solvation dynamics. The TDSS technique is the most popular method of studying solvation dynamics. Recently, several other techniques have also been used to study solvation dynamics [7a,14,15]. The solvation dynamics is followed by the decay of the solvent response function, $C(t)$, which is defined as

$$C(t) = \frac{v(t) - v(\infty)}{v(0) - v(\infty)} \tag{1}$$

where $v(0)$, $v(t)$, and $v(\infty)$ are the emission frequencies at time zero, $t$, and infinity, respectively. The solvation time, $\tau_s$, is the time constant of the decay of the response function $C(t)$, so that $C(t) = \exp(-t/\tau_s)$. If the decay of $C(t)$ is multiexponential, e.g., $a_1 \exp(-t/\tau_1) + a_2 \exp(-t/\tau_2)$, one considers the average solvation time $\langle\tau_s\rangle = a_1\tau_1 + a_2\tau_2$. The solvation time is also known as the longitudinal relaxation time of the solvent. The frequency ($\omega$) dependence of the dielectric constant of a liquid is represented in the simplest case by the Debye dispersion relation:

$$\varepsilon(\omega) = \varepsilon_\infty + \frac{\varepsilon_0 - \varepsilon_\infty}{1 + i\omega\tau_D} \tag{2}$$

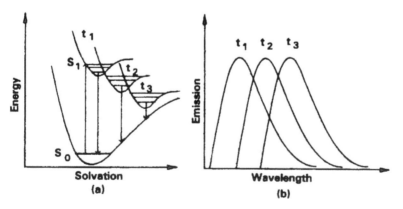

**Figure 3** (a) Change in energy of the solute due to solvation, with increase in time ($t_1 > t_2 > t_3$). (b) Time-resolved emission spectra displaying time-dependent Stokes shift.

where $\varepsilon(\omega)$ is the frequency-dependent dielectric constant, $\varepsilon_\infty$ is the limiting high-frequency dielectric constant, $\varepsilon_0$ is the static dielectric constant, and $\tau_D$ is the dielectric relaxation time of the medium. For such a liquid, according to the simple continuum theory, the solvation time is related to the dielectric relaxation time, $\tau_D$, as [8,9]

$$\tau_s = (\varepsilon_\infty/\varepsilon_0)\tau_D \qquad (3)$$

For most polar liquids, $(\varepsilon_\infty/\varepsilon_0) < 1$. As a result, the solvation time, $\tau_s$, is shorter than the dielectric relaxation time, $\tau_D$. Since water is by far the most important solvent for biological systems, in the next section we will discuss some recent results on the solvation dynamics and dielectric relaxation of water.

## 1. Solvation Dynamics and Dielectric Relaxation in Water

For water, $\varepsilon_\infty \approx 5$, $\varepsilon_0 \approx 80$, and the dielectric relaxation time ($\tau_D$) [16] is 10 ps. Thus, according to the simple continuum theory, the solvation time of water should be $\approx(5/80) \times 10$ or 0.6 ps. The first experimental study on the solvation dynamics of the water molecules around a dye molecule, coumarin 343, was performed by Barbara et al. [10a]. They reported solvation times of 0.16 and 1.2 ps with relative contributions of 1:2. Later, using a setup with a better time resolution, Fleming et al. [10b] detected an initial ultrafast Gaussian component $(a_g e^{-1/2\omega_g^2 t^2})$ with a frequency $\omega_g = 38.5$ ps$^{-1}$ and a slower biexponential decay with time constants of 126 and 880 fs, respectively. The ultrafast Gaussian component is ascribed to the intramolecular vibration and librational motions of the water molecules, and the slower component to their diffusive, reorientational motion. Using molecular hydrodynamic theory, Bagchi et al. discussed in detail the role of different vibrational modes on the solvation dynamics in water [12a]. They showed that the initial ultrafast part of the solvation dynamics is essentially controlled by the intramolecular vibration and the librational modes of water. This theory predicts a deuterium isotope effect at a long time. This has been supported by the 20% isotope effect calculated by Schwartz and Rossky [12b]. Very recently, the deuterium isotope effect has been detected in a time-resolved resonance heterodyne Kerr effect study of rhodamine 800 in water and $D_2O$ [15].

While the properties of the water molecules in bulk have been studied quite thoroughly for a very long time, much less is known about the behavior of water molecules in confined environments. The properties of water change drastically when the water molecules are confined in a small region (dimension less than 500 Å). Robinson et al. reported that the probes to solvent charge transfer processes are markedly affected when the solvent water molecules are confined in such a small region [6b]. Several groups have carried out extensive molecular dynamics simulations to explain the structure and dynamics of the confined water molecules at various other interfaces including the free water surface [6c-e,17]. These simulations indicate that the dielectric constant of water decreases by as much as a factor of 16, when the water molecules are confined in a region of dimension 27 Å [17].

The dynamics of the so-called biological water molecules in the immediate vicinity of a protein have been studied using dielectric relaxation [18], proton and [17]O NMR relaxation [19], reaction path calculation [20], and analytical statistical mechanical models [21]. While the dielectric relaxation time of ordinary water molecules is 10 ps [16], both the dielectric [18] and nuclear magnetic resonance (NMR) relaxation studies [19], indicate that near the protein surface the relaxation dynamics are bimodal with two components in the 10-ns and 10-ps time scale, respectively. The 10-ns relaxation time cannot be due to the motion of the peptide chains, which occurs in the 100-ns time scale. From the study of NMR relaxation times of [17]O at the protein surface, Halle et al. [19c,d] suggested dynamic exchange between the slowly rotating internal and the fast external water molecules.

To explain the bimodal dielectric relaxation in aqueous protein solutions, Nandi and Bagchi proposed a similar dynamic exchange between the "bound" and the free water molecules [21]. The bound water molecules are those that are attached to the biomolecule by a strong hydrogen bond. Their rotation is coupled with that of the biomolecule. The water molecules, beyond the solvation shell of the proteins, behave as free water molecules. The free water molecules rotate freely and contribute to the dielectric relaxation process, whereas the rotation of the doubly hydrogen-bonded 'bound' water molecules is coupled with that of the biomolecule and hence is much slower. The free and bound water molecules are in a process of constant dynamic exchange. The associated equilibrium constant, $K$, can be written as

$$K = \exp\left(-\Delta G^0/RT\right) \tag{4}$$

where $\Delta G^0$ is the difference in the hydrogen bond free energy, per mole of water, between a bond to a biopolymer and to a free water molecule. The frequency-dependent dielectric relaxation of the biological water is calculated using this model [21].

Using the conjugate peak refinement method, Fischer et al. [20a] calculated the reaction path of the motion of the biological water molecules. They also compared the computed transition state and activation energy to those in ice. Their calculation shows that the motion of the water molecules, buried in the proteins, involves exchange of two water hydrogen atoms and involves two successive rotations around orthogonal axes.

Relaxation of the water molecules near a protein surface depends strongly on the temperature and the charge of the protein side chain (i.e., pH of the medium) [18c,d]. In aqueous solution of a hydrophobic protein-based polymer, Urry et al. [18c] observed a temperature-dependent dielectric relaxation near 5 GHz. The water molecules, responsible for the fast relaxation, are known as the water of hydrophobic hydration. The number of such water molecules ($w_{hh}$) decreases to zero as the temperature increases beyond the transition temperature of hydrophobic folding. However, with increase in pH, as the glutamic acid side chain becomes charged, $w_{hh}$ decreases but does not go to zero. This indicates that even for the charged side chains there exists a firmly held layer of water of hydrophobic hydration.

In Section IV.A we will show that the substantially slower dielectric relaxation times of water in organized assemblies markedly slow down the solvation dynamics, in some cases by four orders of magnitude, compared to bulk water.

## B. Photoisomerization

The cis-trans isomerization about an olefinic double bond is forbidden in the ground state of a molecule because it involves a very large barrier, roughly equal to the $\pi$-bond energy. However, in the $\pi\pi^*$ excited state the $\pi$-bond order becomes zero and thus in the $\pi\pi^*$ excited state there is no restriction on the cis-trans isomerization. The interconversion of different rotational and geometrical isomers via the excited state is known as photoisomerization. The photoisomerization of an olefin (e.g., *trans*-stilbene) may be described in terms of the potential energy curves depicted in Fig. 4 [22]. Since in the excited state the isomerization process is almost barrierless, the system moves freely along the torsional coordinate. When it reaches the perpendicular geometry, it undergoes rapid transition from the excited electronic state to the nearly isoenergetic ground state. Once the system is at the peak of the barrier between the cis and the trans isomer, in the ground state it can go to either of them with equal probability, resulting in

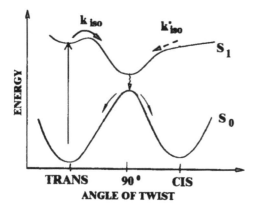

**Figure 4**  Potential energy curves depicting photoisomerization of *trans*-stilbene.

photoisomerization. Photoisomerization plays an important role in many chemical and biological processes, which include the vision process. The vision process
involves such a cis-trans photoisomerization about the $C_{11}-C_{12}$ double bond of
a retinyl polyene attached to the opsin part of a 7-helix membrane protein, rhodopsin [23–25].

The dynamics of the photoisomerization process depend on the friction
offered by the medium to the motion of the system along the torsional coordinate.
In Kramers' seminal work [26], the isomerization process is viewed as a one-
dimensional barrier crossing and the rate, $k_{iso}$, is given by

$$k_{iso} = \{[\zeta/2\omega_b I]^2 + 1\}^{1/2} - [\zeta/2\omega_b I] \tag{5}$$

where $\zeta$ denotes the friction, $\omega_b$ is the barrier frequency, and $I$ is the moment of
inertia of the reaction coordinate. If the friction is purely hydrodyamic in nature,

$$\zeta = 4\pi\eta dr^2 \tag{6}$$

where $\eta$ is the bulk viscosity, and $d$ and $r$ are molecular dimensions associated
with the isomerizing groups. At very high viscosity, the Kramers' relation reduces
to a simple relation where $k_{iso}$ becomes inversely proportional to the viscosity of
the medium. This is known as the Smoluchowski limit. The failure and success
of the Kramers' theory has been the subject of intense debate. The failure of
Kramers' theory is often ascribed to three factors. First, it is argued that the
friction is not described correctly by the hydrodynamic model. Several improved
models have been proposed, e.g., the time-dependent friction model where the
friction is assumed to be different at different parts of the potential barrier [27].
Second, for many probes (e.g., stilbene, diphenylbutadiene, etc.) the isomerization process is not strictly one-dimensional. Miller and Eisenthal [28a] demonstrated that the Kramers' theory is quite successful for the isomerization of 1,1'-
binaphthyl, which is strictly a one-dimensional process. Third, the microviscosity
around a probe molecule may not be same as the bulk viscosity. Several authors
tried orientational relaxation time as an empirical parameter for microviscosity
and found good correlation between the rate of isomerization and that of orientational relaxation [27b,c]. For many aromatic molecules, the translational diffusion
time is not directly proportional to the viscosity of the medium [28b]. To explain
the fractional viscosity dependence of the rate of isomerization, Bowman et al.
[28b] used a model proposed earlier by Zwanzig and Harrison [29]. According
to this model, the hydrodynamic radius of a molecule varies from solvent to
solvent due to the difference in the solute–solvent interactions. Sun and Saltiel
[30a], on the other hand, used the model proposed by Spernol and Wertz [30b]
to estimate the microviscosity of the medium.

For some probes, the activation barrier for the isomerization process and
hence the isomerization dynamics depend on the polarity of the media. For *trans-*

stilbene, Hicks et al. [31] observed that the slope of the isoviscous plots of $\ln(k_{iso})$ against $1/T$ decreases with increase in the viscosity. Since for alcohols, higher viscosity is associated with lower polarity, Hicks et al. [31] proposed that the barrier for the isomerization process decreases at higher viscosity and hence at lower polarity. However, for the cyanine dye 3,3'-diethyloxadicarbocyanine iodide (DODCI), Velsko and Fleming [32a] demonstrated that in alcoholic solvents the slopes of the isoviscous plots of $\ln(k_{iso})$ against $1/T$ do not vary much with increase in viscosity. This indicates that the photoisomerization of DODCI is more or less unaffected by the polarity of the medium [32]. Waldeck and co-workers [22a] studied this issue in considerable detail. They showed that a barrier for the isomerization process can be extracted only for solvents like nitriles where the solvent relaxation is much faster than the excited state isomerization process. In slower solvents like alcohols, the slow and incomplete solvation obscures observation of a well-defined barrier for the isomerization process.

Though Kramers' expression describes a somewhat complicated viscosity dependence and hence is difficult to apply to extract microviscosity, the relation becomes very simple at very high viscosity (Smoluchowski's limit). Since many organized assemblies possess very high microviscosity it is reasonable to assume that the Smoluchowski limit is valid for them. We will demonstrate that assuming the Smoluchowski limit, the microviscosities of some organized assemblies can be estimated quite accurately.

## C. Excited State Proton Transfer

For many molecules, due to extensive redistribution of electron densities, acid–base property in the excited state differs considerably from that in the ground state [33]. For instance, aromatic amines are weakly basic in the ground state. But many of them become acidic in the excited state and readily donate a proton to a proton acceptor to produce the anion in the excited state. Such a molecule, which behaves as an acid in the excited state, is called a photoacid; similarly, photobases are those that display basic properties in the excited state. In many cases, excited state proton transfer (ESPT) results in dual emission bands. One of these emission bands arises from the neutral excited state and bears mirror image relation with the absorption spectrum. The other emission band is due to the excited deprotonated (anion) or protonated species and exhibits a large Stokes shift.

## 1. Intermolecular Proton Transfer Processes

In an intermolecular proton transfer process, the proton is transferred from one molecule to another. The rate of deprotonation/reprotonation is obtained from the analysis of the temporal decay of the neutral and the anion emission [33c].

In this case, the basic issues are whether the acid–base equilibrium is attained within the excited state lifetime of the photoacid or the photobase. If the acid–base equilibrium is attained in the excited state, the $pK_a$ and $pK_b$ in the excited state may be determined by steady-state or time-resolved emission spectroscopy. The excited state $pK_a$ and $pK_b$ of many organic compounds have been summarized by Ireland and Wyatt [33b]. Evidently, the rate of deprotonation/reprotonation depends on the medium and particularly its pH. There have been several attempts to determine the exact number of solvent molecules needed to solvate and stabilize the ejected proton in an intermolecular proton transfer process. Detailed studies on proton transfer from naphthol to protic solvents in ultracold solvent clusters in supersonic jet have revealed that three ammonia and two piperidine molecules are needed to solvate the ejected proton, whereas for water clusters no proton transfer is observed in jets even for clusters containing 21 water molecules [34]. For liquid solutions, Robinson et al. suggested that $4 \pm 1$ water molecules are needed to solvate a proton to form a cluster $H^+(H_2O)_{4\pm1}$ [35]. These studies are relevant to an understanding of the excited state intermolecular proton transfer processes in organized assemblies. In an organized assembly the local pH is often very different from the bulk pH. Again, in an organized medium an adequate number of water molecules $(4 \pm 1)$ often is unavailable to solvate the ejected proton. As a result, the proton transfer processes in the organized assemblies differ considerably from those in the ordinary solutions.

Though, at first sight, lower accessibility of water or protic solvents to the ejected proton may imply a slower deprotonation rate, the actual situation depends on the probe used. The effect of different solvent mixtures on the ESPT process of various probe molecules has been the subject of several studies [35–43]. In a water-alcohol mixture, the deprotonation rate of protonated aminopyrene increases with alcohol concentration up to about 65–70%, whereas it decreases at higher alcohol concentrations [36]. However, for 1-naphthol, the deprotonation rate decreases monotonically as the alcohol content increases [38]. At high alcohol content the rise time of the anion of 1-naphthol is faster than the decay of the neutral form. This indicates that in alcohol-water mixtures, the anion (460 nm) and the neutral emission (360 nm) originate from different 1-naphthol molecules.

ESPT has been identified as the main nonradiative pathway in the excited state of ethidium bromide (EB, **I**), a popular DNA probe [46a]. In aqueous solution, on addition of DNA, EB intercalates in the double helix of DNA [44–46]. This causes a nearly 11-fold increase in the emission intensity and lifetime of EB. The emission quantum yield and lifetime of EB are very similar in methanol and glycerol, whose viscosities differ by a factor of 2000 [46a]. Thus, the fluorescence enhancement of EB on intercalation is not due to high local viscosity. Emission intensity of EB is low in highly polar, protic solvents, such as alcohol

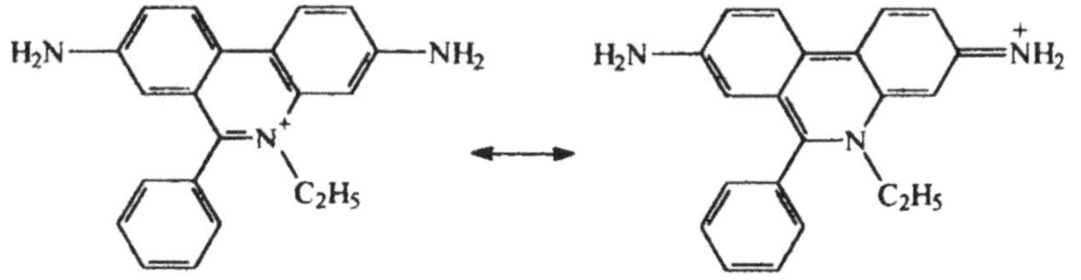

I II

and water, compared to polar, aprotic solvents, e.g., acetone or pyridine. Addition of water to acetone is found to quench fluorescence of EB. This suggests that water quenches emission of EB by abstraction of the amino proton [46a]. In the quinonoid structure of ethidium cation (**II**) the amino proton becomes quite acidic, even in the ground state, and its acidity is expected to increase further in the excited state, like other aromatic amines. If this conjecture is correct, the emission intensity of EB should depend on the hydrogen bond acceptor (HBA) basicity of the solvent, $\beta$, instead of the polarity. The HBA basicity, $\beta$, introduced by Kamlet et al. [47a], and other polarity scales of various solvents are elaborately discussed in many reviews [47]. The polarity of acetone [dielectric constant, $\varepsilon$ = 20.7 and $E_T(30)$ = 42] is less than that of another polar, aprotic solvent, acetonitrile [$\varepsilon$ = 37.5 and $E_T(30)$ = 46] [47a]. However, the HBA basicity, $\beta$, of acetone (0.48) is greater than that of acetonitrile (0.31) and thus acetone is a better proton acceptor than acetonitrile [47a]. Pal et al. [48] observed that in the more polar but weaker proton acceptor, acetonitrile, the fluorescence intensity and lifetime of EB are 1.25 ± 0.1 times those in acetone. This conclusively establishes that the high is HBA basicity of the solvent, the high is the nonradiative rates of EB, and hence the low is the emission intensity. Thus the nonradiative rates of EB are controlled by the HBA basicity of the solvent rather than the solvent polarity. This lends further support to the contention that ESPT is the main nonradiative pathway for EB. Pal et al. [48] further showed that in acetonitrile addition of water causes quenching of the EB emission, with a quenching constant, 1.7 ± 0.3 × $10^7$ $M^{-1}$ $s^{-1}$. In aqueous solution, hydroxyl ion quenches EB emission more dramatically with a quenching constant, 4.4 ± 0.4 × $10^{10}$ $M^{-1}$ $s^{-1}$. The more than 2000-fold quenching constant for the hydroxyl ion, compared to water, reinforces the proton abstraction model.

## 2. Excited State Intramolecular Proton Transfer

For some molecules, in the excited state, a hydrogen atom is transferred to a group within the same molecule. This is known as excited state intramolecular proton transfer (ESIPT) [33a,49–68]. The product of ESIPT is a phototautomer, i.e., a tautomer formed in the excited state. In this case, two emission bands are

III       IV       V

observed, a normal one having mirror image relation with the absorption spectrum and a Stokes-shifted tautomer band. This phenomenon is observed for molecules in which the proton donor and the acceptor group are in close proximity and intramolecularly hydrogen-bonded [49–51]. The ESIPT process was first proposed by Weller [33a] to explain the dual emission of methyl salicylate (MS) with emission maxima at 340 (UV) and 450 nm (blue). For MS, the UV emission at 340 nm is ascribed to the normal or open form (III) and the blue emission at 450 nm to the phototautomer (V). Klopffer and Naundorf first noticed the difference in the excitation spectra of the two forms and excitation wavelength dependence of the emission of MS [52]. This led them to propose the existence of two rotamers, "open" (III) and "closed" (IV), giving rise to the normal and the tautomer (V) emission. Helmbrook et al. [53a] and Felker et al. [53b] studied MS in supersonic jet. The origin of the excitation spectrum of MS exhibits a 99 $cm^{-1}$ blue shift on monodeuteration. From this Helmbrook et al. [53a] estimated that during ESIPT the hydrogen atom moves by a distance of 0.1–0.2 Å. Felker et al. [53b] showed that the 0-0 transitions for the open form, responsible for the normal emission, is at 309.6 nm and that of the closed form is at 332.7 nm. Using femtosecond depletion technique, Herek et al. [54] observed that in supersonic jet the ESIPT of MS occurs in 60 fs and exhibits no deuterium isotope effect. This time scale is longer than the time scale of the O-H stretching vibration (13 fs) and is roughly equal to the half period of the low-frequency vibrations [54]. Resonance Raman studies on 1-hydroxy-2-acetonaphthone (HAN) also suggests that the ESIPT process involves, instead of the O-H stretch, the low-frequency skeletal modes [55]. In liquid solutions, Schwartz et al. reported a proton transfer time of 240 fs for 3-hydroxyflavone [56a], whereas Sekikawa and Kobayashi [56b] and Mitra and Tamai [56c] reported a time constant of 160 fs for ESIPT in salicylideneamide and Nelson et al. [56d] reported a time constant of 320–500 fs for the ESIPT of dinitrobenzylpyridine. Tautomer emission band of 2,2′-hydroxyphenylbenzothiazole exhibits a rise time of 160 ± 20 fs while the rise time for its analog, deuterated at the hydroxyl group, is 140 ± 40 fs [50]. The long time of proton transfer and the absence of deuterium isotope effect suggest that the O-H (or O-D) stretching mode plays little or no role in the ESIPT process. The observed time constants match the periods of skeletal modes, between

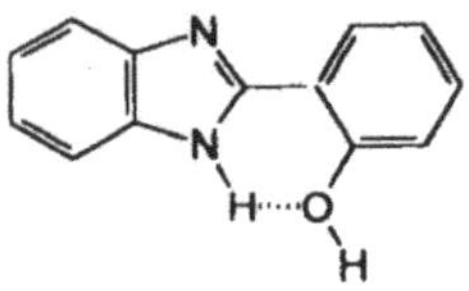

VI 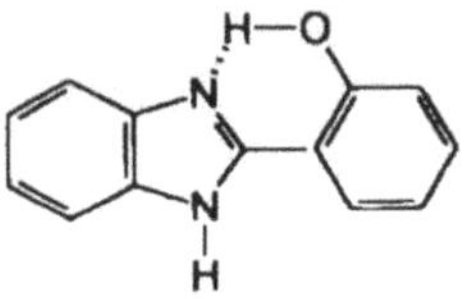                         VII 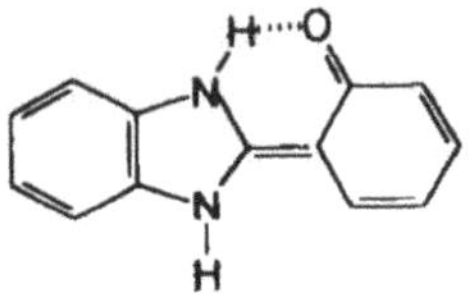                         VIII

100 cm$^{-1}$ and 200 cm$^{-1}$, which affect the mutual orientation of the donor and acceptor groups. These involve heavier groups and thus are not expected to be affected by the substitution of an H atom by a D atom.

The photophysics of the polycyclic quinone hypericin and its analogs has been studied quite exhaustively by Petrich and co-workers [57]. The interest in this compound stems from its proven antiviral activity, induced by light. Ultrafast time-resolved transient absorption studies indicate that ESIPT is the primary non-radiative process in hypericin.

The ESIPT process is seriously inhibited if the intramolecular hydrogen bond between the migrating proton and its terminus is disrupted. For instance, in 2,2'-hydroxyphenylbenzimidazole (HBI) the conformation VII undergoes ultrafast ESIPT to produce the tautomer VIII in the excited state from which the Stokes-shifted emission band at 460 nm originates. However, for HBI the conformation VI in which the migrating proton is not hydrogen-bonded to the terminus does not undergo ESIPT at all and gives predominantly the enol emission at 350 nm. For HBI, the difference in the temporal decay, excitation spectra, and temperature variation of the intensities of the 350-nm (enol) and the 450-nm (keto) band conclusively proves the existence of two distinctly different rotamers of HBI, responsible for the normal (enol) and the tautomer (keto) emission, respectively [58].

While MS undergoes ESIPT readily, its naphthalene analog does not exhibit ESIPT [59]. However, 1-hydroxyacetonaphthone and 1-hydroxynaphthaldehyde undergo ESIPT readily [59]. Obviously, ESIPT is feasible, i.e., exothermic, if in the excited state the keto form is stabler than the enol form. Various theoretical models have been employed to examine the potential energy surface and the energies of the enol and the keto forms. Nagaoka et al. proposed a nodal plane model to predict the exothermicity of the ESIPT process [60]. Das et al. [58a] applied semiempirical quantum chemical method (AM1) to construct the potential energy surface (PES) in the ground and excited states of HBI. Catalan et al. [61] computed PES of *o*-hydroxybenzoyl compounds by the B3LY6 method using 6-31G** basis set. Scheiner et al. [62] performed ab initio calculations using Gaussian 94 and 6-31G** basis set at MP2 and CIS level to explain the proton transfer process in the anionic analog of malonaldehyde. Guthrie [63] used a

multidimensional Marcus theory to show that the intrinsic barrier for proton transfer along a preformed hydrogen bond (1 kcal/mole) is lower than that for a bimolecular water-mediated proton transfer process (5 kcal/mole).

## 3.  Excited State Double Proton Transfer

Kasha et al. first proposed that the hydrogen-bonded dimer of 7-azaindole (7-
AI) undergoes excited state double-proton transfer (ESDPT) process (Scheme 1)
[69]. The main interest in this system originates from its relevance in the photoinduced mutation of hydrogen-bonded base pairs in DNA and the use of tryptophan
analog of 7-AI (7-azatryptophan) as a noninvasive probe for proteins [57b,70].
The first picosecond study on 7-AI in solution indicated that the ESDPT occurs
in less than 5 ps [71] while the early femtosecond studies in hexadecane yielded
time constants of 1.4 and 4.0 ps for the protonated and deuterated compound,
respectively [72]. More recently, the ESDPT in 7-AI is studied in much more
detail by femtosecond upconversion [73,74], mass spectrometry in supersonic jet
[75], and Coulomb explosion technique [76]. It is observed that the decay of
the dimer fluorescence of 7-AI in hydrocarbon solvents is biexponential with an
ultrafast component of 0.2 ps and a fast component of 1.1 ps [73]. The latter is
similar to the rise of the tautomer emission. Zewail et al. [75] attributed the
biexponential decay of the dimer in jet to a two-step proton transfer in which in
the first step an ion pair is produced which then undergoes another proton transfer
to produce the neutral tautomer. However, Sekiuchi and Tahara [73] argued that
if the ion pair were an intermediate it would have emitted at a wavelength very
different from that of the tautomer. From detailed time-resolved anisotropy studies they concluded that there are two excited states. The ultrafast decay is due
to the transition from the upper of this to the lower state and the fast component
is due to ESDPT, which occurs in the lowest state.

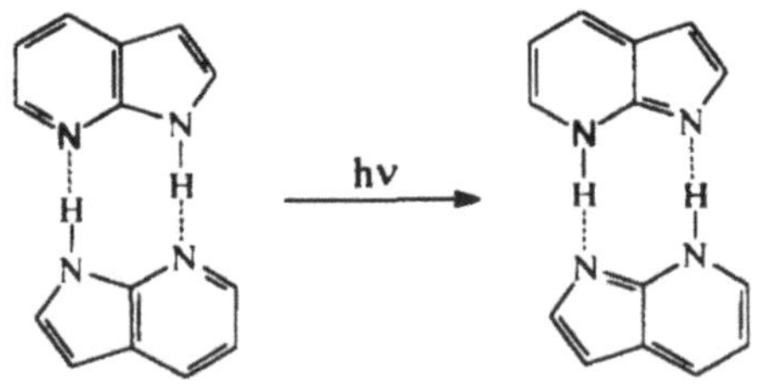

7-AI dimer                    Tautomer

**Scheme 1**  Excited state double-proton transfer (ESDPT) in 7-azaindole (7-AI) dimers.

### 4. Effect of Protic Solvents and Solvent-Mediated Proton Transfer

Though in ESIPT and ESDPT the proton does not come out of the molecules and hence solvation of proton is unimportant, the protic solvents exert considerable influence on the ESIPT and ESDPT process. For the ESIPT process in the protic solvents the differential solvation of the two conformers (**VI** and **VII** in the case of HBI) responsible for the normal (enol) and the tautomer (keto) emission affects the relative intensities of the two emissions. Furthermore, the protic solvents disrupt the intramolecular hydrogen bonds [58a]. The role of the mono- and poly-hydrates on the ESIPT and the ESDPT processes has been studied by several groups. Mcmorrow and Aartsma [77] proposed a two-step tautomerization model for the ESDPT process of 7-AI in protic solvents, such as alcohols, where in the first step the solute 7-AI undergoes solvation and in the second step the proton is transferred through a cyclic hydrogen-bonded species (Scheme 2a). It is interesting to note that while solvent-mediated ESDPT is very facile for 7-AI in liquid solutions, in ultracold supersonic jets solvent clusters like 7-AI $(H_2O)_n$, do not exhibit ESDPT [78]. This indicates that the geometry of the clusters in the liquid and gas phase is different, so that in the gas phase the cluster cannot surmount the barrier for the ESDPT process. Kasha et al. [79] and Petrich et al. [70] proposed that while cyclic monohydrates facilitate proton transfer by solvent mediation, in polyhydrates the ESIPT is disrupted. In the case of HBI a very dramatic increase of the enol emission is observed on microaddition of water to dioxane due to formation of polyhydrates (**IX**) [58a].

An interesting case of solvent-mediated proton transfer is proposed for 4-aminophthalimide (4-AP). In aprotic solvents like dioxane or acetonitrile (ACN), 4-AP exhibits intense emission ($\phi_f = 0.7$ in dioxane and 0.63 in ACN) with emission maximum at 435 nm in dioxane (458 nm in ACN) with very long lifetime (15 ns in dioxane and 14 ns in ACN) [81]. In protic solvents the emission maximum of 4-AP exhibits a very dramatic red shift to above 500 nm (520 nm

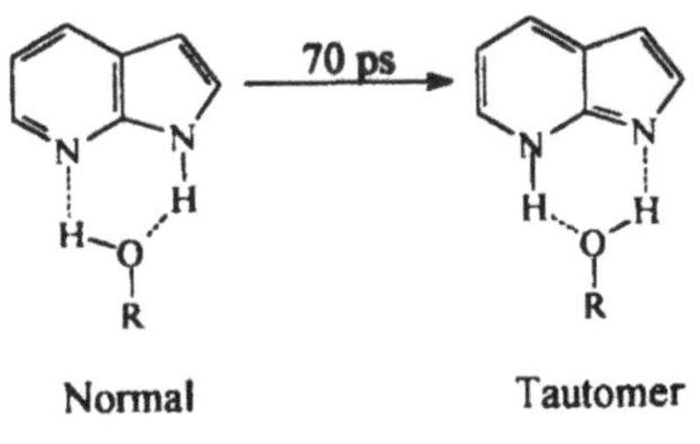

　　Normal　　　　　　Tautomer

**Scheme 2a**　Solvent-mediated proton transfer in 7-azaindole.

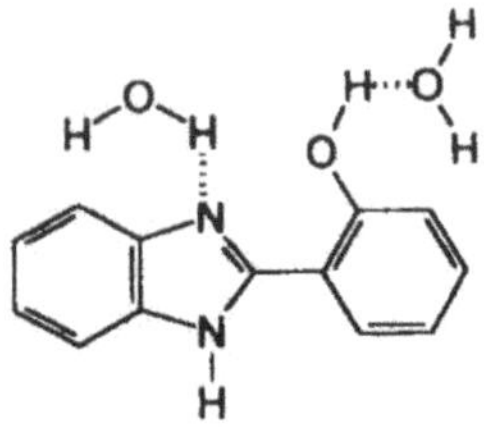

IX

in alcohols and 550 nm in water), and the emission quantum yield and lifetime (0.01 and 1.2 ns in water) also decrease markedly. In $t$-butanol [$E_T(30) = 43.6$], which is less polar than ACN [$E_T(30) = 46$] the emission maximum of 4-AP is 518 nm. This indicates that the remarkable emission sensitivity of 4-AP has nothing to do with solvent polarity. Harju et al. proposed that the remarkable sensitivity of 4-AP to protic solvents is due to a solvent-mediated proton transfer process occurring in the excited state of 4-AP (Scheme 2b) [82]. However, when the imide proton of 4-AP is replaced by an alkyl group, the emission properties remain as sensitive to protic solvents as 4-AP itself [81c]. This suggests the possibility of solvent-mediated proton transfer involving the amino protons of 4-AP through two solvent molecules (Scheme 2c). The strong interaction of 4-AP with water molecules has been studied in supersonic jets by Andrews et al. [83]. Appearance of a weak Stokes-shifted emission band on addition of methanol to solution of dipyridylcarbazole is ascribed to protic solvent-mediated double-proton transfer [84].

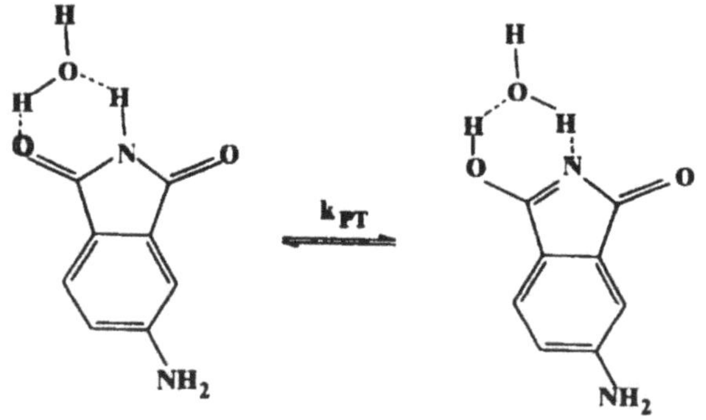

**Scheme 2b**  Excited state proton transfer in 4-aminophthalimide mediated by one solvent molecule.

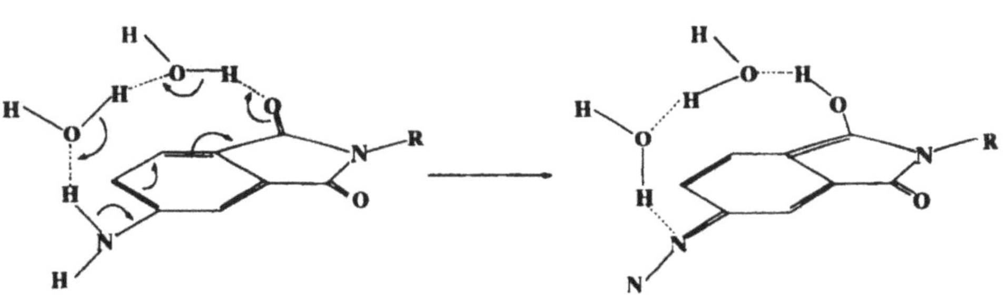

**Scheme 2c**  Excited state proton transfer in 4-aminophthalimide mediated by two solvent molecules.

## D.  Twisted Intramolecular Charge Transfer

When an electron donor and an acceptor are joined by a flexible single bond, on electronic excitation in the "nonpolar" excited state, often an electron is transferred from the donor to the acceptor intramolecularly and the donor and the acceptor undergo a torsional motion about the flexible bond joining them. This process is known as twisted intramolecular charge transfer (TICT) and the resulting highly polar state is known as a TICT state. This often results in dual emission from the nonpolar and the TICT state [85,86]. This phenomenon was first proposed to explain the dual emission of p-dimethylaminobenzonitrile (DMABN). Subsequently, this phenomenon was observed in many other systems [86]. The structures of some of the molecules that undergo TICT are given in Fig. 5. Since the dipole moment in the TICT state is much higher than that in the nonpolar or locally excited (LE) state, the relative energies of the TICT and LE state change markedly with solvent polarity. In a nonpolar solvent, the LE state remains lower in energy compared to the TICT state and hence a single emission band arising from the LE state is observed. However, in a polar solvent, due to larger solvation energy of the TICT state, it becomes stabler than the LE state and this causes dual emission from the LE and TICT states. As discussed in a previous review, several theoretical models have been applied to explain the solvent dependence of the TICT process [86]. More recently, Hynes et al. [87] and Sobolewski et al. [88] carried out detailed theoretical analysis to explain the polarity dependence of the TICT process. Sobolewski et al. [88] used CIS, CASSCF, and CASPT2 methods to calculate the reaction path for the TICT process in DMABN and its analogs. Unlike earlier calculations that considered a single internal coordinate, Sobolewski et al. took into account more than one internal coordinate [88]. Their calculations support the TICT hypothesis for DMABN. For DMABE, for which dual emission is not observed in a polar solvent like acetonitrile, they proposed strong quenching via the weakly fluorescent

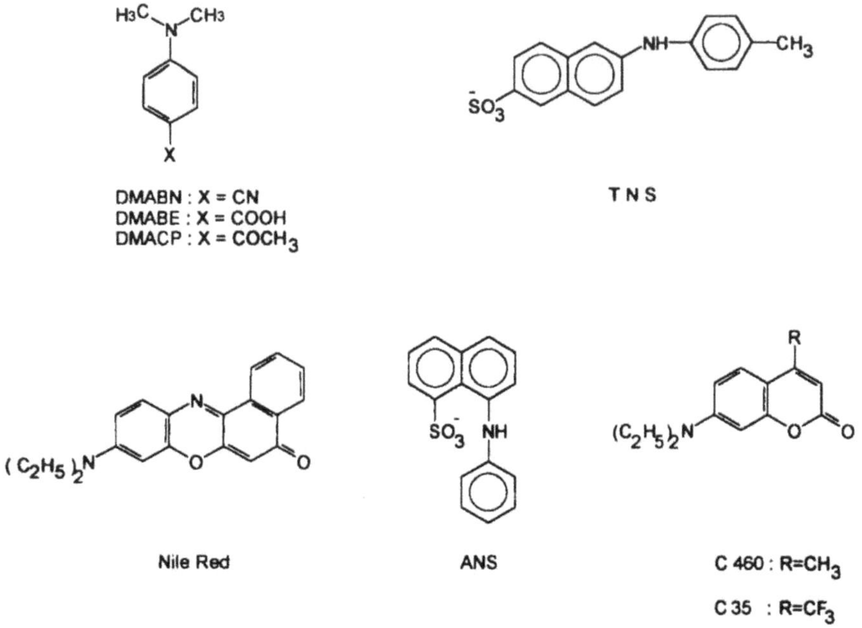

DMABN : X = CN
DMABE : X = COOH
DMACP : X = COCH$_3$

TNS

Nile Red

ANS

C 460 : R=CH$_3$

C 35  : R=CF$_3$

**Figure 5**   Structure of a few fluorescence probes that exhibit the TICT phenomenon.

rehybridized-ICT state. The effect of polar solvents on the spectra of molecules undergoing TICT in supersonic jet has been studied by many groups [86]. More recently, Ishida et al. [89] studied the dynamics of the intramolecular charge transfer process in ultracold clusters of 9,9'-bianthryl with water in supersonic jet using picosecond time-resolved spectroscopy. They observed that the LE state is converted to the unrelaxed CT state in a time scale of <20 ps. Following this, the unrelaxed CT state relaxes in 50 ps time scale to a new equilibrated state. The dynamics of the TICT process can be followed by the decay of the nonpolar emission and the rise in the TICT emission. The dynamics of the TICT process depends strongly on the polarity of the medium. Eisenthal et al. [90] first proposed that the energy barrier for the TICT process for DMABN decreases linearly with the polarity parameter $E_T(30)$ [47] of the medium as

$$E_B = E_B^0 - A[E_T(30) - 30] \tag{7}$$

where $E_B^0$ is the barrier in a hydrocarbon medium of $E_T(30) = 30$. The polarity-dependent barrier model is subsequently extended to other TICT probes [91,92].

In many cases the TICT state is nonemissive and for these the TICT process is simply a nonradiative decay process in the "nonpolar" excited state of these molecules. The nonradiative rate is obtained from the emission quantum yield ($\phi_f$) and lifetime ($\tau_f$), as $k_{nr} = (1 - \phi_f)/\tau_f$. The TICT rate is very sensitive to the polarity of the medium. For instance, in the case of TNS, $\phi_f$ changes 300 times from 0.3 in dioxane [$E_T(30) = 36$] to 0.001 in water [$E_T(30) = 63$] and $\tau_f$ changes over 100 times from 8 ns in dioxane to 0.06 ns in water. Thus the rate constant of TICT is a very sensitive indicator of the microscopic polarity of an organized media and has been used quite extensively for this purpose [86].

## III. ORGANIZED MEDIA

In this section we will discuss the architecture of a few organized media. Since structures of these organized media have already been discussed in a number of reviews, we will present only a brief overview rather than a comprehensive summary.

## A. Micelles

Amphiphilic surfactant molecules form spherical or nearly spherical aggregates called micelles, above a certain critical concentration, known as the critical micellar concentration (cmc) and above a critical temperature, called Kraft temperature [4,93]. The size of the micellar aggregates is usually 1–10 nm and the aggregation number, i.e., the number of surfactant molecules per micelle, ranges from 20 to 200. The structure of a typical cationic micelle is shown schematically in Fig. 6. The core of a micelle is essentially "dry" and consists of the hydrocarbon chains with the polar and charged headgroups projecting outward into the bulk

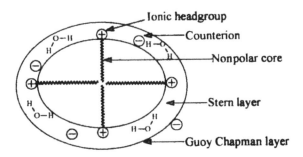

**Figure 6**  Structure of a cationic micelle.

water. The Stern layer, surrounding the core, comprises the ionic or polar head-groups, bound counterions, and water molecules. Between the Stern layer and the bulk water there is a diffuse layer, termed the Guoy-Chapman (GC) layer. The GC layer contains the free counterions and water molecules. In nonionic polyoxyethylated surfactants, the hydrocarbon core is surrounded by a palisade layer, which consists of the polyoxyethylene groups hydrogen-bonded to water molecules. The structure of a micelle is metastable and involves continuous exchange of monomers between the aggregates and those in bulk solution.

Recent small-angle x-ray and neutron scattering studies revealed detailed information on the structure of a variety of micelles [94,95]. According to these studies, the thickness of the Stern layer is 6–9 Å for cationic cetyltrimethylammonium bromide (CTAB) micelles and anionic sodium dodecyl sulfate (SDS) micelles, whereas the palisade layer is about 20 Å thick for neutral Triton X-100 (TX-100) micelles [94]. The radius of the dry, hydrophobic core of TX-100 is 25–27 Å and thus the overall radius of TX-100 micelle is about 51 Å The overall radius of CTAB and SDS micelles are about 50 Å and 30 Å respectively.

The structure of more complicated micelles and their phase transitions have recently been studied using small-angle neutron scattering (SANS) [96–99]. The CTAB micelles give rise to worm-like, very large micelles with unusual morphology and viscoelastic property on addition of sodium salicylate (NaSal) [96]. At low NaSal concentration, a slightly elongated micelle with low viscosity is observed. With increase in the concentration of NaSal, the viscosity of this system exhibits two maxima separated by a minimum. Aswal et al. [97] studied this phenomenon in considerable detail and showed that beyond the first maxima the structure of the micelle is rod-like and the micelles are highly polydisperse, i.e., the aggregation numbers are very different for different aggregates. De et al. studied the structure of a series of novel gemini or dimeric micelles of the form $(n\text{-}C_{16}H_{33})N^+Me_2\text{-}(CH_2O\,CH_2)_p\text{-}(n\text{-}C_{16}H_{33})$ [98], and $(n\text{-}C_{16}H_{33})N^+Me_2\text{-}(CH_2)_m\text{-}(n\text{-}C_{16}H_{33})$ [16-$m$-16] and their mixed micelles with CTAB [99]. They found that the shape of the gemini micelles depends strongly on the length and nature of the spacer unit, e.g., 16-$m$-16, $2Br^-$ form disk or cylindrical micelles for $m \leq 4$ and ellipsoidal or spherical micelles for $m \geq 5$.

Telgmann and Kaatze studied the structure and dynamics of micelles using ultrasonic absorption in the 100-KHz to 2-GHz frequency range [100]. They detected several relaxation times in the long ($\mu$s), intermediate (10 ns), and fast (0.1–0.3 ns) time scale. The longest relaxation time has been attributed to the exchange of monomer between bulk and the micelles, and the fastest to the rotation of the alkyl chains of the surfactants in the core of the micelle. The intermediate relaxation time has not been assigned to any particular motion. We will discuss later that the intermediate relaxation times in the 10-ns time scale may well be due to solvent relaxation in the Stern layer.

## B.  Reverse Micelles and Microemulsions

The reverse micelles refer to the aggregates of surfactants formed in nonpolar solvents, in which the polar head groups of the surfactants point inward while the hydrocarbon chains project outward into the nonpolar solvent (Fig. 7) [101–126]. Their cmc depends on the nonpolar solvent used. The cmc of aerosol-OT (sodium dioctyl sulfosuccinate, AOT) in a hydrocarbon solvent is about 0.1 mM [102]. The AOT reverse micelle is fairly monodisperse with aggregation number around 20 and is spherical with a hydrodynamic radius of 1.5 nm. No salt effect is observed for NaCl concentration up to 0.4 M. Apart from liquid hydrocarbons, recently several microemulsions are reported in supercritical fluids such as ethane, propane, and carbon dioxide [111–113].

The most important property of the reverse micelles is their ability to encapsulate a fairly large amount of water to form what is known as a "microemulsion." Up to 50 water molecules per molecule of the surfactant can be incorporated inside the AOT reverse micelles. Such a surfactant-coated nanometer-sized water droplet, dispersed in a nonpolar liquid, is called a "water pool" (Fig. 7). The radius ($r_w$) of the water pool varies linearly with the water-to-surfactant mole ratio, $w_0$. In $n$-heptane, $r_w$ (in Å) $\approx 2w_0$ [103b]. The structural information on the microemulsions, i.e., radius of the micellar aggregates and that of the water pool, has been obtained using dynamic light scattering [104,110], transient grating [105], SANS studies [113,120,121,125], small-angle x-ray scattering [114], ultrasound velocity measurements [108], Fourier transform infrared (FT-IR) [107,109], NMR, electron spin resonance (ESR), differential scanning calorimetry [106], and dielectric relaxation [126]. In contrast to AOT, which does not require any cosurfactant to form reverse micelles, cationic surfactants do not form reverse micelles in the absence of consurfactants [122a]. Several nonionic or neutral surfactants (e.g., Triton X-100, etc.) have recently been reported to form

**HYDROCARBON**

**Figure 7**  Structure of a microemulsion.

reverse micelles in pure and mixed hydrocarbon solvents [108,123]. Finally, apart from water, confinement of other polar solvents, such as acetonitrile, alcohol, and formamide, has been reported in such microemulsions [124].

The water molecules confined in the water pool of the microemulsions differ in a number of ways from ordinary water. In a microemulsion, the first two to four water molecules are held very tightly by the surfactant and all of the water molecules except the six most tightly held ones freeze at $-50°C$ [106]. The FT-IR [107] and the compressibility studies [108] indicate that the first three water molecules "lubricate" the dry surfactants. During this process the compressibility of the AOT microemulsion increases steeply. The next three water molecules solvate the counterion ($Na^+$ for Na-AOT) and starts the self-organization process. At this stage, the headgroups of AOT become linked by hydrogen bonds through the water molecules and the compressibility gradually decreases. For $w_0 > 6$, the water pool swells in size but the compressibility reaches a plateau. Around $w_0 = 13$, the first solvation shell of AOT becomes complete and up to this point the water structure remain severely perturbed inside the water pool. But even in the very large water pools, the compressibility of the microemulsions remains at least two times higher than that of ordinary water [108]. The behaviors of the microemulsions containing neutral surfactants are similar except that they exhibit only a monotonic increase of compressibility reaching a plateau and do not show the decrease of compressibility observed for the ionic surfactants arising from the solvation of the counterions [108]. In the water pool, there may be three kinds of water molecules: the "bound" ones near the polar headgroup of the surfactant and therefore held strongly, the "free" ones near the central region of the water pool, and the "trapped" ones between the surfactants. Using FT-IR spectroscopy, Jain et al. [109b] determined the relative amounts of the three kinds of water molecules. Obviously, the mobility of the free water molecules is expected to be faster than bound ones. In the solvation dynamics experiments to be discussed later, it will be seen that even the free water molecules in the water pools are significantly slower than ordinary water molecules. The number of free and bound water molecules in an AOT microemulsion are listed in Table 1.

## C. Lipid Vesicles

In a vesicle an aqueous volume (water pool) is entirely enclosed by a membrane that is basically a bilayer of lipid molecules [127–137]. In the case of the unilamellar dimyristoylphosphatidylcholine (DMPC) vesicles (radius ≈ 250 nm) there is only one such bilayer, whereas a multilamellar vesicle (radius ~ 1000 nm) consists of several concentric bilayers. Unilamellar vesicles can be produced from multilamellar vesicles by sonication. In such a system there are two kinds of

**Table 1**  Properties of AOT Microemulsion

| $r_w$ (Å) | $N$ | $w_0$ | $n$ | $n_w$ | $n_B$ |
|---|---|---|---|---|---|
| 10.0 | 5 | 4.1 | 34 | 140 | 109 |
| 16.5 | 8 | 8.2 | 77 | 628 | 372 |
| 29.8 | 15 | 16.4 | 225 | 3,698 | 1,338 |
| 55.4 | 28 | 32.9 | 722 | 23,765 | 4,339 |

$N$, number of zones within the water pool; $n$, aggregation number; $n_w$, total number of water molecules in a microemulsion; $n_B$, number of "bound" water molecules.
Data from Ref. 223.

water molecules present: those in the bulk and those entrapped within the water pool of the vesicles. The entrapped water pool of a small unilamellar DMPC vesicle is much bigger (radius ≈ 250 nm) than those of the water pool of the reverse micelles (radius < 10 nm). In recent years, several groups studied chain dynamics of lipids using SANS, conductivity and electron microscopy [127], ESR of spin-labeled lipids [128a,b], and fluorescence of pyrene-labeled lipids [128c]. Recent molecular dynamics (MD) simulations [129a,b], NMR [130a], crystal structure [131a,b], and other studies [131c] revealed detailed information on the structure of DMPC vesicles and the water molecules in its neighborhood. The most recent MD simulation indicates that above transition temperature (≈23°C) each DMPC molecule is hydrogen-bonded to about 4.5 water molecules that form an inner hydration shell of the polar headgroup of the lipids and about 70% of the DMPC molecules remain connected by the water bridges [129a].

The vesicles undergo phase transition at a well-defined temperature. Above the phase transition temperature, the viscosity of the lipid bilayer remains quite low and the permeability of the bilayer wall remains high so that small molecules can pass easily through the bilayer to enter the inner water pool. Below the transition temperature, viscosity of the lipid bilayer becomes high and also the permeability across the bilayer membrane decreases significantly. The change in the viscosity of the lipid bilayers with temperature is usually monitored by optical anistropy studies. The transport of small organic molecules across the lipid bilayer above transition temperature is most elegantly demonstrated in a recent surface second harmonic (SSH) generation study by Srivastava and Eisenthal [137]. The SSH signal is obtained as long as the molecule stays in an inhomogeneous region, i.e., at the lipid bilayer. Above the transition temperature of the lipids, the SSH signal decays in a time scale of 100 s, which denotes the residency time of the probe in the bilayer or the time taken by the probe to diffuse through the bilayer membrane from bulk water to the inner water pool. For lipids below the transition temperature no such time dependence of the SSH signal is observed, indicating that at a temperature below the transition temperature the bilayer membrane does not allow transport of molecules to the inner water pool.

## D.  Polymers and Hydrogels

Water-soluble polymers have generated considerable recent interest because of their versatile applications and compatibility with biological systems [138]. The properties of aqueous polymer networks and the water molecules attached to them have been studied by various techniques such as fluorescence [139], dielectric relaxation [140], computer simulation [141], and light scattering [142]. The microporous synthetic polymer hydrogels are inherently insoluble in water but can entrap a considerable amount of water within their polymer networks [143,144]. They are particularly interesting due to their diverse applications as biomaterials (e.g., contact lenses), as chromatographic packings, in devices for controlled release of drugs, and as electrophoresis gels. The polyacrylamide (PAA) hydrogel is obtained by polymerizing acrylamide in the presence of $N,N'$-methylene bisacrylamide as a crosslinker [145]. The pore size in such a gel can be varied by varying the concentration of the monomer (acrylamide) [143a,b]. Among the various types of hydrogels PAA is most suitable for photophysical studies as it is optically transparent over a wide range of concentrations of the monomer and the crosslinker. On absorption of water such a hydrogel swells in size. The swelling and other properties of this interesting semirigid material have recently been studied using light scattering [142], NMR, and calorimetry [146]. The bulk viscosity of any hydrogel is very high. However, since the hydrogels contain large pores even very large biological macromolecules like DNA pass through such hydrogels during gel electrophoresis. Several groups attempted to immobilize small probe molecules (e.g., Nile red) [147a] or proteins [147b,c] within the hydrogels. Using far-field fluorescence microscopy, Moerner et al. demonstrated that in PAA hydrogel, despite the fact that most Nile red molecules move freely, motion of a minute fraction (~2%) of them becomes severely restricted so that the Brownian motion of *individual* Nile red molecules may be recorded [147a].

## E.  Zeolites

Zeolites are open structures of silica in which some of the silicon atoms in the tetrahedral sites are replaced by aluminum ions [148–151]. Counterions like $Na^+$ and $K^+$ maintain the electroneutrality and reside freely at certain locations in the zeolite cages. Zeolites can be represented by the empirical formula $M_{2/n}Al_2O_3 \cdot xSiO_2 \cdot yH_2O$, where M is an alkali metal or an alkaline earth metal cation of valence $n$, $x > 2$, and $y$ varies from 0 to 10. Depending on the Si/Al ratio and the cations, zeolites can have various rigid and well-defined structures that can be classified into cage and channel types. For the ZSM-5 zeolite there are two intersecting channel systems. One system consists of straight channels with a free cross-section of $5.4 \times 5.6$ Å$^2$ and the other consists of sinusoidal channels with free diameter of $5.1 \times 5.5$ Å$^2$. Faujasite zeolites are made up of a nearly spherical supercage of diameter 13 Å, surrounded by sodalite cages of dimension 8 Å.

The structure and dynamics of zeolites have been studied by molecular dynamics (MD) simulation, Monte Carlo simulation [151,152], density functional theory [153a], and stochastic models [153b]. These studies indicate that the spatial locations are similar for different cations. The exact locations of the cations within a zeolite are determined by x-ray diffraction [155]. The mobile cations are responsible for the electrical conductivity of dehydrated zeolites [156a]. The mechanism of electrical conduction in a zeolite has been the subject of several studies. Conduction of dehydrated potassium zeolite L has been found to involve a thermally activated process [156b]. Dielectric properties of the zeolites depend on the degree of hydration [157]. The zeolites can act as a host for a large number of guest molecules. Neutron diffraction study suggests that in zeolite Y, cyclohexane stays in the 12-ring window site. This is in agreement with the MD simulations [158a]. Similar results are obtained for benzene in zeolites [158]. NMR line width and simulation studies indicate that in the faujasite zeolites the guest aromatic molecules hop from one cage to another in the nanosecond time scale [159].

Certain positions in the inner walls of the micropores of the zeolites serve as active sites, where catalytic conversions can take place. The size of the micropores and the location of cavities can be so adjusted that only one type of molecular species can reach the active sites. The regioselectivity in zeolites is demonstrated in the oxidation of alkenes to a single hydroperoxide [150c]. Zeolites contain many aciditic sites [154]. The number of different types of acidic sites in a zeolite depends on the method of activation of the zeolite. Using NMR, Cao et al. demonstrated that CaY zeolites activated by heating in air contain more Bronsted acid sites than those activated in vacuum and this explains the difference in reactivity of diarylethylenes in zeolite CaY [150d].

The photophysics and photochemistry of organic molecules change remarkably on encapsulation in zeolite [149]. The marked changes caused by the zeolites result partly from their rigid structure, which imposes considerable restriction on molecular motion within a zeolite. The presence of cations in close proximity with organic guest molecules exerts significant influences because of the strong local electric field produced by the cations and also the enhanced singlet–triplet transitions in the case of zeolites having heavy cations. The polarity and acid–base behavior of the zeolites also affect different photophysical processes.

## IV. PHOTOPHYSICAL PROCESSES IN ORGANIZED ASSEMBLIES

In this section we will discuss how the dynamics of various photophysical processes are affected inside organized assemblies and what information on the organized assemblies can be inferred from the photodynamic studies. It should be noted that due to molecular diffusion a probe molecule undergoes an excursion over a region of radius of a few nanometers within its excited state lifetime of several nanoseconds. Thus a fluorescent probe actually reports the property of a microenvironment of radius several nanometers [160,161]. For a homogeneous fluid, the slight uncertainty about the position of the fluorophore does not cause any serious problem. However, since an organized medium is a few nanometers in size, it is necessary to ensure that the fluorescent probe resides within the microenvironment under study. Fortunately, the position of the maximum, intensity, and temporal decay of the emission often changes drastically on incorporation in an organized medium and thus exhibits an unmistakable signature that the probe is really confined in the organized medium.

### A. Solvation Dynamics in Organized Assemblies

As discussed in Section II.A, the solvation dynamics indicates the mobility of the solvent molecules in a medium. Among all solvents, study of relaxation properties of water in organized media is most important because the water molecules present in the confined environments control the structure, dynamics, and reactivity of biological systems. The results of the dielectric relaxation experiments, discussed in Section II.A, suggest that the water molecules present in biological environments are substantially slower than those in ordinary water. We will now show that the solvation dynamics studies also indicate similar trends and reveal many other finer details.

### 1. Cyclodextrin

Fleming et al. [162a] first studied solvation dynamics of two laser dyes, coumarin 480 (C480) and coumarin 460 (C460), in γ-cyclodextrin (γ-CD) cavity using time-dependent Stokes shift (TDSS). The marked blue shift of the emission spectra and the increase in fluorescence lifetimes of the two probes on addition of γ-CD to their aqueous solutions indicates that the probes are located inside the γ-CD cavity [162b,c]. The initial component of solvation in γ-CD is found to be similar to that in bulk water (0.31 ps) [162a]. However, at longer times, the solvent response in γ-CD is at least three orders of magnitude slower than that in bulk water. In γ-CD, the slow relaxation reveals three components of 13 ps, 109 ps, and 1200 ps for C480. Molecular dynamics calculations indicate that in the γ-CD cavity there are 13 water molecules for C480/γ-CD complexes and 16 in the case of C460. These numbers resemble the number of water molecules present in the first solvent shell of the dyes in aqueous solutions. If the first solvent shell dominates the response at short times, the response at short times for the C480/γ-CD complex should have been different from that for C460 due to the presence of fewer water molecules for the former. However, since the initial Gaussian component is the same for the two dye molecules in γ-CD, it seems that it is incorrect to assume that the first solvent shell dominates the solvent response [162a].

Using molecular hydrodynamic theory (MHT), Nandi and Bagchi [163] showed that the slow solvation dynamics in γ-CD may be explained if one assumes complete freezing of the translational motions of the solvent molecules inside the γ-CD cavity. They further showed that the slow part of the response contributes about 10% to the total response. It is proposed that the collective response of the solvent molecules, rather than the contribution from the different solvent shells, dominates the inertial component of solvation.

## 2. Microemulsion

As noted in Section III.B, the surfactant-coated water droplets in water-in-oil microemulsions provide an elegant model for water molecules in confined environments. For a microemulsion the emission spectrum of the probe changes markedly when it is transferred from bulk hydrocarbon to the water pool. For example, absorption maxima of coumarin 480 (C480) in $n$-heptane and water are at 360 and 395 nm, respectively, whereas the emission maxima are at 410 and 490 nm, respectively [164]. In $n$-heptane solution of C480 on addition of AOT and subsequently water, a very prominent shoulder appears at 480 nm (Fig. 8) [165a]. The 480-nm band is assigned to the C480 molecules in the water pool of the microemulsion since the position and excitation peak (at 390 nm) of this band are very different from those of the C480 molecules in bulk $n$-heptane. The position of the 480-nm band indicates that polarity of the water pool resembles that of alcohol [165a]. Sarkar et al. studied the solvation dynamics of C480 in AOT/$n$-heptane/water microemulsions [165a]. They observed distinct rise time in the nanosecond time scale at the red end of the emission spectra. This indicates nanosecond solvation dynamics in the microemulsions. They observed that in a small water pool ($w_0 = 4$, $r_w = 8$ Å) the solvation time is 8 ns while for a very large water pool ($w_0 = 32$, $r_w = 64$ Å) the response is bimodal with a fast component of 1.7 ns and a slower component of 12 ns. Bright et al. studied the solvation dynamics of acrylodan-labeled human serum albumin (HSA) in AOT microemulsion by phase fluorimetry [165b]. They reported that the solvation time is about 8 ns for a small water pool ($w_0 = 2$) and 2 ns for a large water pool ($w_0 = 8$). For 4-aminophthalimide (4-AP) in a large water pool, the solvation dynamics is biexponential with an average solvation time of 1.9 ns [166]. In AOT microemul-

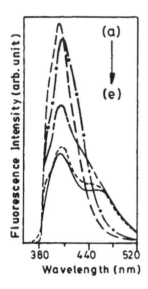

**Figure 8**  Emission spectra of $5 \times 10^{-5}$ M coumarin 480 in $n$-heptane. (a) 0 M AOT; (b–e) 0.09 M AOT; (b) $W_0 = 0$; (c) $W_0 = 4$; (d) $W_0 = 8$; (e) $W_0 = 32$. (Data from Ref. 165a.)

sions, the solvation time of 4-AP increases from 1.9 ns in $H_2O$ to 2.3 ns in $D_2O$, which displays a 20% deuterium isotope effect. The appearance of a nearly 2-ns component in the large water pools indicates that even in the large water pools of the microemulsions the water molecules are about 6000 times slower compared to bulk water (solvation time 0.31 ps [162a]). The wavelength-dependent temporal decays, time-resolved emission spectra, and decay of C(t) of 4-AP are depicted in Fig. 9.

A semiquantitative explanation of the 2-ns component may be as follows: The static polarity or the dielectric constant of the water pool of the AOT microemulsions can be obtained from the position of the emission maximum of the probes (C480 and 4-AP) [165,166]. For both probes, the water pool resembles an alcohol-like environment with an effective dielectric constant of 30–40. If one makes a reasonable assumption that the infinite frequency dielectric constant of water in the water pool of the microemulsions is the same as that of ordinary water, i.e., 5, and that the dielectric relaxation time is 10 ns as obtained for the biological systems [18b], then the solvent relaxation time should be about 1.67 ns, which is close to the observed solvation time in AOT microemulsions.

One might argue that the nanosecond dynamics observed in the water pool is not due to the slower water molecules but is due to the solvation by the $Na^+$ counterions present in the water pool for the AOT microemulsions. Nanosecond

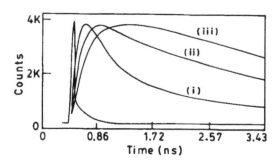

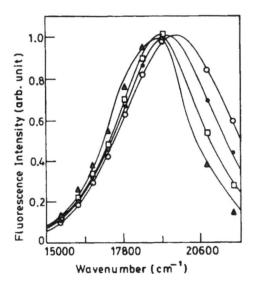

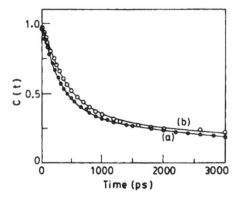

**Figure 9** Solvation dynamics of 4-aminophthalimide in AOT microemulsion. (i) Wavelength-dependent emission decays. (ii) Time-resolved emission spectra at 0 (O), 150 (●), 425 (□), and 1600 (▲) ps. (iii) Decay of solvent response function $C(t)$ in a microemulsion containing (a) water (●) and (b) $D_2O$ (O). (Data from Ref. 166.)

solvation dynamics due to ions, in solutions as well as molten salts, is well documented in the literature [167]. Datta et al. [168] studied the solvation dynamics of 4-AP in a microemulsion containing neutral surfactant Triton X-100 where no ions are present in the water pool. The Triton X-100 microemulsion also exhibits nanosecond solvation dynamics, which suggests that the ionic solvation dynamics has little or no role in the solvation dynamics observed in the water pool.

Levinger et al. studied the solvation dynamics of a charged dye coumarin 343 (C343) in lecithin [169] and AOT microemulsions [170,171] using femtosecond upconversion. For lecithin microemulsions [169], the solvent relaxation displays a very long component that does not become complete within 477 ps. This observation is similar to the nanosecond dynamics reported by Bright et al. [165b] and Sarkar et al. [165a]. For C343 in AOT, Levinger et al. observed that the decay characteristics of the emission intensity at different wavelengths display considerable differences for sodium and ammonium counterions [170]. However, they did not present a complete analysis of this result in terms of dynamic Stokes shift and the decay of the solvent response function $C(t)$. For Na-AOT, the solvation dynamics reported by Levinger et al. [171] for the charged probe C343 are faster than that reported by Bright et al. [165b] and Sarkar et al. [165a]. It is obvious that due to its inherent negative charge, AOT repels the negatively charged C343 probe from its vicinity. Thus the C343 anion is expected to reside in the central region of the water pool. Neutral probes like C480 and 4-AP may stay both in the central region of the pool and in the peripheral region close to the AOT molecules. The discrepancy in the results in the case of AOT microemulsions, reported by Levinger et al. [171], and those of Sarkar et al. [165a] and Bright et al. [165b], however, is too large to be explained in terms of different locations of the probes and merits further careful investigation.

Most recently, several groups studied solvation dynamics of nonaqueous solvents, such as formamide [172], acetonitrile, and methanol [173] in AOT microemulsions. Using a picosecond setup, Shirota and Horie [173] demonstrated that in the AOT microemulsions the solvation dynamics of acetonitrile and methanol is nonexponential and 1000 times slower than those in the pure solvents. They attributed the nonexponenial decay to the inherent inhomogeneous nature of the solvent pools. Evidently, the static polarity and relaxation properties of the entrapped polar solvents vary quite strongly as a function of the distance from the ionic headgroup of the AOT surfactants. As noted earlier, within its nanosecond excited lifetime, the probe passes through different layers of solvents of different relaxation properties within the pool. This quite reasonably may give rise to a nonexponential decay. However, it should be mentioned that nonexponential solvation dynamics is encountered even for homogeneous solutions. To explain the nonexponential behavior in homogeneous solutions, Castner et al. [174] earlier employed an "inhomogeneous continuum" model using a dielectric constant $\varepsilon_0(r)$, which depends on the distance $r$, from the probe. This model involves assumption of a particular mathematical form for $\varepsilon_0(r)$. Unfortunately, such a procedure is too complicated to apply in the case of the microemulsions and other organized assemblies.

## 3. Micelles

In the case of the micelles, there are three possible locations of the probe, namely, the bulk water, the "dry" micellar core, and the Stern layer. If the probe stays in the bulk water, obviously the solvation time will be in the subpicosecond time scale. Since the core of the micelle resembles aliphatic hydrocarbons, the probe is not expected to exhibit dynamic Stokes shift in the core. However, if the probe stays in the Stern layer, its solvation dynamics may be quite different from that in the bulk water because the mobility of the water molecules may be considerably constrained in the Stern layer. Solvation dynamics in micelles has been studied using C480 and 4-AP as probes [175,176]. Emission properties of the probes in the micelles are very different from those in water and in hydrocarbon [175,176]. This shows that the probes reside neither in bulk water nor in the core of the micelles, and hence are located in the Stern layer of the micelles. Sarkar et al. [175] and Datta et al. [176] studied solvation dynamics of C480 and 4-AP, respectively, in neutral (TX-100), cationic (CTAB), and anionic (SDS) micelles. It is observed that for SDS, CTAB, and TX-100, the average solvation times are 180 ps, 470 ps, and 1450 ps, respectively, for C480 [175] and 80, 270, and 720 ps for 4-AP [176]. The solvation times in micelles differ only by a factor of 2 for the two probes. This suggests that the solvation dynamics in the Stern layer of the micelles do not depend very strongly on the probe. It is readily seen that the solvation dynamics in the Stern layer of the micelles are three orders of magnitude slower than in bulk water (0.31 ps [162a]), about 10 times faster than in the water pool of the microemulsions [165,166], and slightly faster than the longest component of solvation dynamics in γ-CD [162a]. The main candidates causing solvation in the Stern layer of the micelles are the polar or ionic headgroups of the surfactants, the counterions (for SDS and CTAB), and the water molecules. Since the headgroups are tethered to the long alkyl chains their mobility is considerably restricted. The dynamics of such long alkyl chains occur in the 100-ns time scale [128] and hence are too slow to account for the subnanosecond solvation dynamics observed in the micelles. The role of ionic solvation by the counterions also appears to be minor because of the very similar time scale of the ionic (CTAB) and the neutral (TX-100) micelles.

For both the probes, it is observed that the solvation times are of the order TX > CTAB > SDS. Qualitatively, the difference in the solvation times in the three micelles may be ascribed to the difference in their structures [94]. As mentioned earlier, the thickness of the hydrated shell for TX-100 (25 Å) is much higher than that for SDS and CTAB (6–9 Å). Thus for SDS and CTAB, the probe remains close or partially exposed to the bulk water whereas for TX it remains well shielded from bulk water and hence exhibits slower solvation dynamics. The SANS studies indicate that CTAB micelles are drier than SDS micelles [94]. Thus the faster solvation dynamics in SDS, compared to CTAB, may be due to the more water-like environment in SDS. It is interesting to note that the time scale of solvation is similar to the intermediate range of dielectric relaxation times reported by Telgmann and Kaatze [100].

## 4.  Lipids

The state of solvation of a fluorescent probe, in the ground state, in the unilamellar and multilamellar vesicles is usually studied by red edge excitation spectroscopy (REES) [135]. The idea behind REES is that in such an inhomogeneous medium the probe molecules in different regions remain in different states of solvation and, as a result, exhibit different absorption and emission characteristics. This is reflected in the gradual shift in the emission maximum as the wavelength of excitation is varied. The REES gives information on the state of solvation in the ground state of the molecules and gives no dynamic information on the relaxation properties inside the vesicles. The dynamics of the surfactant chain in the lipid bilayer have been investigated using ESR [128]. However, the interesting issue of the dynamics of the water molecules inside the water pool of vesicles has been addressed only recently [177]. Datta et al. [177] studied C480 in sonicated unilamellar DMPC vesicles. The position of emission maximum of C480 in DMPC vesicles is once again different from that in bulk water and the hydrocarbon. This indicates that the probe stays in the inner water pool of the vesicle. Datta et al. [177] observed that the solvation dynamics of C480 in DMPC vesicles are highly nonexponential with two components of 0.6 ns (40%) and 11 ns (60%). This result is very similar to the solvation dynamics of the same probe in the large water pools of AOT microemulsions [165]. Thus the nanosecond solvation dynamics in lipids cannot be due to the chain dynamics of DMPC, which occur in the 100-ns time scale [128]. Since in the bulk water the solvation dynamics are much faster (0.31 ps [162a]), the results reported by Datta et al. [177] demonstrate restricted motion of the water molecules in the inner water pool of the vesicles.

## 5.  Polymer Hydrogels

The bulk viscosity of most polymers and particularly the semirigid hydrogels is very high. Thus at first sight one expects very slow relaxation of the water molecules in polymer matrices and polymer hydrogels. Contrary to this expectation, in the orthosilicate [178] and polyacrylamide [179] hydrogels, both solvation dynamics and rotational relaxation are found to occur in <50-ps time scale. The surprisingly fast solvation and rotational dynamics of small probe molecules in hydrogels may be attributed to the porous structure of the hydrogels, through which even large biomolecules pass easily. Datta et al. [179] demonstrated that the microenvironment of 4-AP in polyacrylamide (PAA) hydrogel is quite hetero-geneous. In the PAA hydrogel there are broadly two kinds of environments. One of them is water-like in which the 4-AP molecules exhibit emission maximum at 550 nm with lifetime 1.3 ns and the other is quite aprotic in which 4-AP molecules emit at 470 nm with a 7.2-ns lifetime. The recent steady-state anisot-ropy measurements by Claudia-Marchi et al. [180] demonstrate that for titania gels at the sol-gel transition point when the bulk viscosity increases abruptly, the emission anisotropy does not change perceptibly. Thus the microviscosity of the gel is very low in spite of the very high bulk viscosity. The NMR [181] and simulation [182] studies indicate that the diffusion coefficient of water molecules in polymer hydrogels is not appreciably slower than that in ordinary water and is smaller at most by a factor of 2 than that in ordinary water. Argaman and Huppert [183] studied solvation dynamics of coumarin 153 in polyethers and found very fast solvation times ranging from 50 fs to 100 ps. The fast solvation dynamics in polymer matrices are consistent with the dielectric relaxation studies [140a,b], which shows that except the highly water-soluble polymer, polyvinyl pyrrolidone, dielectric relaxation times of most aqueous polymer solutions are faster than the 100-ps time scale. In Table 2, we have summarized the solvation dynamics of water in different environments.

**Table 2**  Solvation Times of Water in Different Environments

| Medium | Average solvation time (ps) |
|---|---|
| Water | 0.31 |
| AOT reverse micelle, $r_w < 10$ Å | 8000 |
| AOT reverse micelle, $r_w > 20$ Å | 2000 |
| CTAB | 470,[a] 270[b] |
| Triton X-100 | 1450,[a] 720[b] |
| SDS | 180,[a] 80[b] |
| Lipid (DMPC) | 600 (40%), 11,000 (60%) |
| PAA hydrogel | <50 |

[a] For C480.
[b] For 4-AP.

## 6.  Liquid Crystal

It is observed that in the nematic phase of a liquid crystal, the solvation dynamics of coumarin 503 are biexponential [184a]. The slowest time constant decreases from 1670 ps at 311.5 K to 230 ps at 373 K. The solvation time is not affected by the nematic-isotropic phase transition. Thus, it appears that the local environment and not the long-range order controls the time-dependent Stokes shift. A theoretical model has been developed to explain the experimental findings. This model takes into account the reorientation of the probe as well as the fluctuation of the local solvent polarization. Similar results are also obtained for rhodamine 700 in the isotropic phase of octylcyanobiphenyl [184b].

## 7.  Zeolite and Nanoparticles

The solvation dynamics in microporous solids have been the subject of several recent studies. Sarkar et al. [185] showed that C480 exhibits wavelength-dependent decays and time-dependent Stokes shift in a solid host, faujasite zeolite 13X. They observed a highly nonexponential decay with an average solvation time of 8 ns. Interestingly, the 8-ns time constant is very close to the nanosecond solvation dynamics ($\langle \tau \rangle$ = 4.1 ns) observed in molten salts [167a]. In a faujasite zeolite the mobile components are the sodium ions and the probe dye molecule itself. Since in a faujasite zeolite the encapsulated guest molecules hop from one cage to another in the nanosecond time scale [159], the 8-ns relaxation time observed in zeolite may also arise as a result of the self-motion of the probe from one cage to another. The role of self-motion of solutes on the solvation dynamics has recently been discussed in detail by Biswas and Bagchi [186]. However, it is difficult to establish unequivocally whether the nanosecond dynamics observed in zeolite are due to ionic solvation or self-motion of the probe.

   Pant and Levinger [187] studied solvation dynamics of C343 in a suspension of nanodimensional zirconia particles of radius 2 nm in a water–acetone mixture (95:5, v/v). They observed two subpicosecond components similar to those in bulk but having different amplitudes resulting in a relaxation time faster than that in bulk solution. They also showed that the maximum Stokes shift is three times smaller for the dye molecules adsorbed on the zirconia particles compared to those in bulk solution.

## 8.  Proteins and DNA

One of the longstanding goals of biology is to unravel the dynamics occurring in complex biomolecules such as proteins and DNA [188,189]. Several groups reported that the solvation dynamics of protein-bound fluorophores are significantly slower compared to bulk water. Pierce and Boxer [189a] and Bashkin et al. [189b] reported that the solvation dynamics in the protein environments are nonexponential with a long component with time constant on the order of 10 ns.

It is interesting to note that this time scale is very close to the nanosecond component of dielectric relaxation earlier observed for the aqueous protein solutions [18,21]. More recently, Fleming et al. [190] and Beck et al. [191] used two third-order spectroscopic methods, namely, the three-photon echo shift (3PEPS) measurement and transient grating (TG) spectroscopy, to study the early events of solvation dynamics in proteins as well as in homogeneous solutions and polymer matrices. The 3PEPS technique permits separation of the inertial solvation response from the typically faster part of the dynamic Stokes shift, which involves displacement of the intrachromophoric vibrational modes. The 3PEPS results indicate the presence of an inertial solvation component arising from the protein–matrix surroundings of an embedded probe molecule. The TG signal appears to be more sensitive to the slower dynamical response compared to 3PEPS.

The static and dynamic properties of DNA have been studied by the temperature-dependent Stokes shift of the intercalated dye acridine orange [192] and by molecular dynamic simulation [193]. A large part of the Stokes shift of the intercalated dye in DNA is found to be frozen out at low temperature, as in the solution. Thus, the interior of DNA is found to have the diffusive and viscous dynamic characteristics of a fluid rather than the purely vibrational characteristics of a crystal. The results suggest that the probe dye molecule senses the movement of DNA and at high viscosity the rate of DNA motion is limited by the rate of solvent motion.

## B. Photoisomerization and Microviscosities of Organized Assemblies

The friction imparted by several organized assemblies and interfaces to the photoisomerization of organic molecules containing conjugated double bonds has been the subject of several recent studies [194–202]. The rate of photoisomerization of stilbene in various organized assemblies is found to be substantially slower than that in homogeneous medium. Retardation of the photoisomerization process is mainly due to the constraints imposed by these media on the rotation about the double bond. In aqueous methanol lifetime of *trans*-stilbene is 34 ps. However, in aqueous solution of α-cyclodextrin (α-CD) the lifetime of *trans*-stilbene is 137 ps. This demonstrates that the isomerization of stilbene inside the α-CD cavity is slower than that in aqueous methanol [194]. In β- or γ-CD, the fluorescence decays of *trans*-stilbene are biexponential with one component of 50 ps and another very slow component of several thousand picoseconds. The slow component corresponds to a very rigid microenvironment [194]. In a zeolite, isomerization of stilbene is highly retarded, so that its rate become comparable to the rate of intersystem crossing. This gives rise to a strong phosphorescence [195]. Holmes et al. observed a biexponential decay for stilbene in a lipid and assigned this to the presence of two sites [196]. Probe molecules containing a polar and

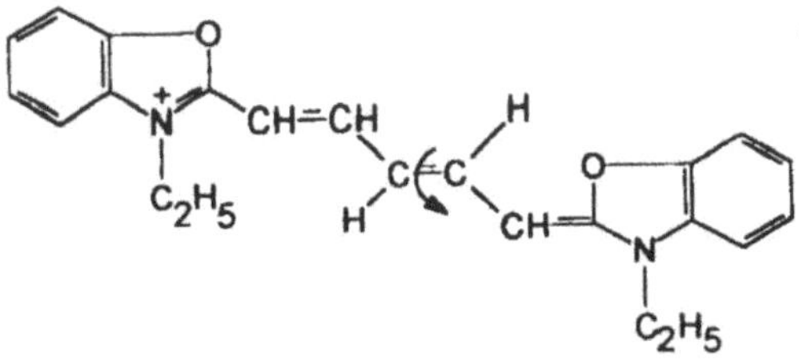

**Figure 10**   Structure of 3,3'-diethyloxadicarbocyanine iodide (DODCI).

a nonpolar portion prefer to stay at the interface between a polar and a nonpolar medium. The laser dye 3,3'-diethyloxadicarbocyanine iodide (DODCI, Fig. 10) and the triphenylmethane (TPM) dye malachite green (MG) (Fig. 11) both contain a charged portion and a large nonpolar part. The isomerization of DODCI and other cynaine dyes and MG has been studied at various interfaces, such as the air–water interface [197], microemulsions [198], micelles [199], DNA [200], and proteins [200]. As noted in Section II.B, photoisomerization is the main nonradiative pathway in the excited states of polyenes. Thus reduction in the rate of photoisomerization leads to an increase in emission lifetime and intensity. The recent time-resolved surface second harmonic generation (SSHG) experiments have demonstrated that for the air–water interface the friction against the photoisomerization is different for different probes [197]. Although for rod-shaped DODCI [197a] isomerization at the air–water interface is faster than that in bulk water, for nearly planar MG [197b] it is slower at the air–water interface. In the water pool of a microemulsion, photoisomerization of DODCI is nearly three times slower than in ordinary water [198]. The photoisomerization of DODCI is markedly slowed down at various micelle–water interfaces [199a]. Compared to aqueous solution, in CTAB, SDS, and TX-100 micelles, the rates of photoisomerization of DODCI are 20, 7, and 8, times slower, respectively [199a]. It may be recalled that at very high viscosity the rate of photoisomerization becomes inversely proportional to the viscosity of the medium (Smoluchowski limit). Assuming that the Smoluchowski limit and the same "slip/stick" boundary condi-

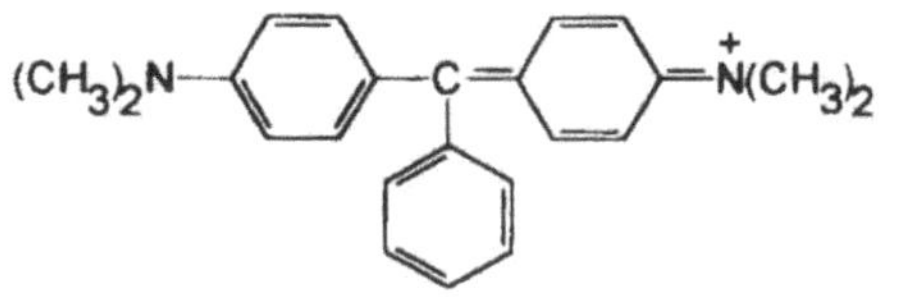

**Figure 11**   Structure of malachite green (MG).

tion hold for the highly viscous solvent $n$-decanol and the micelles, and then comparing the isomerization rates of DODCI in the three micelles with that in $n$-decanol, the microviscosities of CTAB, SDS, and TX-100 have been estimated to be $70.0 \pm 20$, $24.5 \pm 2$, and $26.0 \pm 2$ cP, respectively [199a]. Photoisomerization of DODCI is also studied in aqueous solution in the presence of salmon sperm DNA and bovine serum albumin (BSA) [200]. It is observed that in aqueous solutions DNA and BSA offer considerable friction to the isomerization process, like a highly viscous liquid. As a result, photoisomerization of DODCI is completely suppressed when DODCI binds to DNA and BSA.

For the TPM dyes, the main nonradiative decay pathway is the propellarlike torsional motion of the three phenyl rings. As a result, in medium of low viscosity the quantum yield and lifetime of emission are extremely low. Due to a marked increase in local viscosity, in the organized assemblies the torsional motion of the TPM dyes is significantly hindered resulting in a dramatic increase in the quantum yield and lifetime of emission of the TPM dyes. Baptista and Indig [201] reported a 1000-fold increase in the quantum yield and lifetime of emission of TPM dyes on binding to a protein BSA. This suggests that the local viscosity of the protein is several orders of magnitude higher than that of water. Tamai et al. [202] studied the microviscosities of the polyacryamide gels using TPM leuco dyes. On photoexcitation the cyano or hydroxyl group of the TPM leuco dyes is detached and the TPM cation is produced. Tamai et al. compared the emission properties of the TPM dyes in the hydrogel to those in homogeneous solutions. From this they estimated a microviscosity of 19–20 cP for the unswollen gels and about 10 cP for the swollen gels [202]. This once again demonstrates that while the bulk viscosity of the semirigid gels is many thousand times higher than that of water, the microviscosity is only 10–20 times higher. This is consistent with the very fast solvation and rotational relaxation time in gels reported earlier [178–182].

Since photoisomerization involves motion of one-half of the probe molecule against the other, it remains unaffected by the overall motions of the organized assemblies, such as the motion of different segments of DNA, or of the peptide chains of a protein or the surfactants in a micelle. In contrast, in optical anisotropy studies, the bending and twisting motion of the macromolecular chains of DNA, proteins, or micelles become superimposed on the orientational motion of the probe. As a result, photoisomerization seems to be a better probe for the microviscosity of complex biological assemblies.

## C. Proton Transfer Processes in Organized Assemblies

The p$K_a$ values of an acid and hydrogen ion concentration in an organized assembly often differ drastically from those in ordinary liquids [203–205]. The p$K_a$ values of proteins are very sensitive to the local dielectric constant. This in turn depends on the conformation of the protein, and polarization of the solvent and the protein [205]. Measurement of pH at a surface is one of the longstanding goals in chemistry. In general, the concentration of hydrogen ion at an interface, $C_s$, is related to the bulk value, $C_b$, as

$$C_s = C_b \exp (-e\Psi/kT) \tag{8}$$

where $\Psi$ is the surface potential and $e$ the electronic charge [203]. Similarly, the relation between $pK_a$ at a surface $pK_a(s)$ and that in the bulk $pK_a(b)$ is [203]:

$$pK_a(b) = pK_a(s) - \Psi/59.6 \tag{9}$$

Mukerjee and Banerjee [203a] used bromophenol indicators to estimate the pH and $pK_a$ at the micelle–water interface using absorption spectroscopy. Eisenthal et al. used SSHG to determine surface pH, $pK_a$, and potential ($\Psi$) at the air–water interface [204a,b]. The results depend strongly on the nature of charge of the surfactants. For instance, if the surfactant is negatively charged (e.g., $p$-alkyl phenolate) the hydrogen ion concentration at the surface is greater than that in the bulk, whereas the opposite is true for a cationic surfactant (e.g., $p$-alkyl anilinium) [204b]. More recently, An and Thomas [204c] used neutron scattering to estimate the $pK_a$ at the water surface. Apart from application of new experimental techniques to determine $pK_a$ in organized assemblies, various microscopic models have been developed to calculate theoretically the $pK_a$ of the ionizable groups of the proteins [205].

In AOT microemulsions, the presence of the negatively charged headgroup causes a sharp gradient in pH/pOH over the nanometer-sized water pool. Menger and Saito [206a] reported that the acid–base property of $p$-nitrophenol (PNP) gets substantially modified in AOT microemulsions. Although in bulk water, at pH 11.5, 95% of the PNP molecules remain in the anionic form, when an alkaline aqueous solution containing PNP is injected in the AOT microemulsion, no $p$-nitrophenolate anion is detected until the pH of the injected solution exceeds 11.5. On the basis of this, Menger and Saito concluded that the $pK_a$ of PNP in the AOT microemulsion is greater than that in bulk water (7.14) by more than four units. However, subsequent workers ascribed this phenomenon to the fact that the local hydroxyl ion concentration near the negatively charged AOT headgroup is substantially less than that in bulk water. Oldfield et al. [206b] showed that if a negatively charged group is attached to PNP, the probe remains in the water pool of the AOT microemulsions and its acid–base properties are similar to those in bulk water. Okazaki and Toriyama [207] studied the location of an organic acid at different pH values in AOT microemulsion using ESR spectroscopy. They observed that at low pH, when the molecule is in the neutral form, it stays close to the AOT–water interface, whereas at high pH the carboxylate anion is expelled from the AOT–water interface to the water pool. The pH within reverse micelles formed in supercritical $CO_2$ was recently measured by the fluorescence method. It is estimated to be 3.1–3.5 and is independent of the pressure of $CO_2$ and water loading [208].

The dynamics of the inter- and intramolecular excited state proton transfer processes in organized assemblies are often quite different from those in ordinary solutions [209]. The sharp local variation of pH in the water pool of the AOT microemulsions affects the intermolecular proton transfer process quite strongly. Fendler et al. [210] studied excited state deprotonation of the trinegatively charged probe hydroxypyrenetrisulfonate in AOT microemulsions. They observed that while in the large pools the proton transfer process is similar to that in bulk water, it is quite different in the small water pools ($w_0 < 7$). They concluded that in the large water pool, due to the electrostatic repulsion from the negatively charged AOT ions, the negatively charged probe remains in the large water pools, far from the AOT anion, and experiences an almost bulk water–like microenvironment. But in the small water pool, the very different local pH near the AOT anions renders the deprotonation/reprotonation behavior quite different from that in ordinary aqueous solutions. More interesting results are reported in the case of the proton transfer processes of a cationic probe in AOT microemulsions [48]. As discussed earlier, for the DNA probe ethidium bromide (EB), the main nonradiative pathway is excited state deprotonation. Hence, in ordinary aqueous solutions emission of EB is strongly quenched by the hydroxyl ions. However, in AOT microemulsion the hydroxyl ion does not quench the EB emission at all, even when a highly alkaline aqueous solution of EB (pH = 12.6) is injected into the reverse micelle. It is proposed that the anionic surfactant AOT strongly attracts the ethidium cation to the AOT–water interface but expels the hydroxyl anion from the AOT–water interface to the water pool. Hence, the hydroxyl anion cannot access the ethidium cation. Since the ethidium cation is forced to stay near the AOT anion, its emission properties are independent of $w_0$, i.e., size of the water pool [48].

Fleming et al. [211] reported that the deprotonation rate of 1-naphthol is retarded 20 times inside cyclodextrin cavities, whereas deprotonation of protonated aminopyrene occurs nearly three times faster. Since in water-alcohol mixtures the deprotonation rate for protonated aminopyrene increases with alcohol concentration up to about 65–70% and decreases at higher alcohol concentrations [36], the faster deprotonation of protonated aminopyrene inside the cyclodextrin cavity indicates that the polarity of the microenvironment is in-between pure water and 65% alcohol. For 1-naphthol, the deprotonation rate monotonically decreases as the alcohol content increases [38]. The slower deprotonation rate of 1-naphthol in cyclodextrin [211] is consistent with this. Mandal et al. [212] reported dramatic reduction in the rate of excited state deprotonation of 1-naphthol in micelles, from 35 ps in bulk water to the nanosecond time scale. The retardation of the rate of proton transfer of 1-naphthol is manifested in the dramatic enhancement of the neutral emission at 360 nm. Along with this there is a marked increase in the lifetime of the neutral emission at 360 nm and the rise time of the anion emission (460 nm) for CTAB, SDS, and TX-100R. For cationic CTAB, the rise time of the anion emission (600 ± 100 ps) is similar to the lifetime of decay at 360 nm. However, for TX-100R and SDS, the rise time of the anion emission (at 460 nm) is found to be faster than the decay of the neutral emission (at 360 nm). This indicates that in TX-100R and SDS there is no parental relation between the normal and the anion emission and they originate from the probe, 1-naphthol molecules, at distinctly different locations. This is consistent with the earlier observation [38] that in alcohol-water mixtures at high alcohol content the rise time of the anion emission is faster than the decay time of the neutral form. For TX-100R and SDS, the rise times of the 460-nm band are 1.8 ± 0.1 ns and 600 ± 100 ps, respectively. The corresponding decay times at 360 nm are 2.5 ± 0.1 ns and 1.8 ± 0.1 ns, respectively [212]. The dramatic reduction in the rate of deprotonation of 1-naphthol in micelles is attributed to the nonavailability of adequate number of protons to solvate the proton in the micellar environment. The ESPT process of 1-naphthol is also affected by lipids [213] and found to report faithfully the transition temperature of the lipids [213a].

The excited state intramolecular proton transfer (ESIPT), and hence the dual emission of the molecules exhibiting ESIPT, is also found to be affected by various organized assemblies. Zewail et al. reported appreciable reduction in the ESIPT rate of 2-(2'-hydroxyphenyl)-4-methyloxazole from 300 fs in bulk aprotic solvents to pico- and nanosecond time scales when encapsulated in β-cyclodextrin [214]. The relative polarities of the binding sites HSA and apomyoglobin have been estimated from the study of the emission properties of 4-hydroxy-5-azaphenanthrene in these proteins [215]. 10-Hydroxybenzo-[h]-quinoline (HBQ) has been found to exhibit slight enhancement of the tautomer emission in micelles [216a] as well as in α-, β-, and γ-cyclodextrins [216b]. The enhanced tautomer emission has been ascribed to a reduction in nonradiative rates in restricted environments. Dual emission of 2-(2'-hydroxyphenyl)benzimidazole (HPBI) has been used to monitor the microenvironments of micelles [217a,b] and microemulsions [217c]. ESIPT process of 3-hydroxyflavone (3-HF) has been studied in micelles [218a], microemulsions [218b], and phospholipid vesicles [218c]. The tautomer emission of 3-HF has been shown to be a sensitive probe for the estimation of cmc of surfactants [218a]. 3-HF senses two kinds of environments in the reverse micelles of AOT [218b]. With an increase in the water content of the microemulsion, the tautomer emission of 3-HF increases. However for 7-azaindole (7-AI) in AOT microemulsions, the tautomer emission is almost absent whereas the intensity of the normal emission gradually decreases on addition of water [218d]. A fluorescence anisotropy study of the tautomer emission has been reported to yield the correct value of the transition temperatures of the synthetic lipids dipalmitoylphosphatidylcholine (DPPC) and dimyristoylphosphatidylcholine (DMPC) [218c]. Such behavior, along with the position of the emission band, suggests that 3-HF molecules reside in the bilayer matrix. Fuji et al. studied the keto-enol tautomerization of 9-anthrol in sol-gel glasses with different Si/Al ratios [219]. They identified four emitting species and detected proton transfer from a surface Bronsted acid site of the -O-Si-O-Al-O- network to the anthrone.

## D.  TICT and Micropolarity of the Organized Assemblies

The dielectric constant or the local polarity of an organized assembly, or at the interface between two media of drastically different polarities, is often determined using solvatochromic dyes whose absorption or emission maxima depend on the polarity of the medium [47]. From the position of the charge transfer absorption band of dodecylpyridinium iodide, Mukerjee and Ray [220] estimated the dielectric constant of the micelle–water interface for dodecyltrimethylammonium iodide to be $36 \pm 2$. More recently, using surface second harmonic generation, Eisenthal et al. [221a,b] recorded the absorption spectrum of a solvatochromic probe at the air–water interface and at a liquid–liquid interface. From the position of the absorption maximum of the probe they inferred that the polarity $(\pi^*)$ of the air–water interface is 0.22, which is close to that of $CCl_4$ (0.26) and butyl ether (0.21). Perera et al. [221c] used visible attenuated total internal reflection absorption spectroscopy to measure the polarity of heptane–water interface and reported that the $E_T(30)$ for this interface is 47.7, which indicates a polarity close to that of $n$-octanol. Benjamin investigated the effect of different interfaces on the spectra of solvatochromic probes at an interface using an electrostatic continuum theory [222].

Since the dynamics of the twisted intramolecular charge transfer (TICT) process is very sensitive to the polarity of the medium, the local polarity of an organized medium may also be determined from the rate of the TICT process. For TNS, which is nearly nonfluorescent in water ($\phi_f = 10^{-3}$ and $\tau_f = 60$ ps), the emission quantum yield and lifetime increases nearly 50 times on binding to cyclodextrins and more than 500 times on binding to a neutral micelle, TX –100 [86]. Such a dramatic increase in the emission intensity and lifetime arises because of the marked reduction of the nonradiative TICT process inside the less polar microenvironment of the cyclodextrins and the micelle. Determination of the micropolarities of various organized assemblies using TICT probes has been surveyed quite extensively in several recent reviews [5b–d,f,86]. Therefore, in this chapter we will focus only on some selected works not covered in the earlier reviews.

Inside the water pool of the microemulsions the TICT process of several probes (ANS [223], TNS [224], Nile red [225], etc.) is observed to be significantly retarded compared to bulk water and the lifetime of the probes increase from the picosecond time scale in water to several nanoseconds in the water pool.

The retardation of the TICT process inside the water pool of the microemulsion compared to ordinary water is ascribed to the lower static polarity of the water pool compared to bulk water. The polarity or $E_T(30)$ of the pool is obtained from the observed rate of TICT and by comparing them with the values obtained in homogeneous solutions. Karukstis et al. analyzed the emission spectra of 6-propionyl-2-(dimethyl-amino)-naphthalene (PRODAN) in microemulsions and showed that the PRODAN molecules at different sites within the microemulsions exhibit different emission spectra [226]. More recently, using picosecond total internal reflection, Bessho et al. [227] determined the emission lifetime of ANS at the water–heptane interface. They observed a nanosecond component in the decay of ANS at the water–heptane interface, indicating that the interface is considerably less polar than bulk water.

The polarity of the supercages of solid faujasite zeolites has been estimated using TICT probes. For this purpose, Ramamurthy et al. [228] used DMABN and Sarkar et al. [229] used Nile red as the probe. Both studies indicate that the polarity of the faujasite zeolite resembles that of 1:1 methanol–water mixture. Kim et al. [230] studied TICT of dimethylaminobenzoic acid (DMABE) in faujasite Y zeolites. In Y zeolites the lifetime of the TICT emission and the TICT/LE emission intensity ratio of DMABE are greater than those in polar solvents. This is attributed to hydrogen bonding between DMABE and the zeolites. Kim et al. [231] studied the effect of neutral $SiO_2$ colloids of diameters 300–400 Å in acetonitrile on the dual emission of DMABE. For 10 μM DMABE, they found that up to a concentration of 0.3 μM $SiO_2$ the intensity of the TICT band increases whereas that of the normal band (LE) decreases. The intensity of the TICT emission decreases at $SiO_2$ concentrations above 0.3 μM. Kim et al. [231] attributed this to the formation of hydrogen bonds between the $SiO_2$ particles and the DMABE molecule. At high $SiO_2$ concentrations due to submonolayer coverage of the silica particles the TICT emission decreases.

Fendler et al. studied the TICT emission of DMABN in lipids [232]. They reported that in the lipids the TICT band of DMABN exhibits substantial blue shifts from 504 nm at a temperature above the gel transition temperature to 489 nm below the transition temperature. They further showed that the fusion of the lipids caused by addition of $CaCl_2$ can be conveniently studied from the shifts in the emission maximum of the TICT band of DMABN.

The internal cavity of the cyclic polysaccharide cyclodextrin (CD) is highly nonpolar and hydrophobic, and it is known to retard the TICT process markedly when the probe gets encapsulated inside the cavity [5b–d,86]. More recently, a linear polysaccharide dextrin has also been reported to provide a hydrophobic surface to which the probe TNS binds due to hydrophobic effect and, as a result of lower local polarity, the TICT rate is retarded nearly 50 times [233]. The effect of salting-in (urea, $LiClO_4$, etc.) and salting-out (LiCl) agents on the hydrophobic binding of TNS with dextrin has been studied in detail [233]. Interestingly another linear polysaccharide, dextran, affects the emission properties of TNS very slightly. This is attributed to the difference in the stereochemistry of dextrin and dextran. For dextrin, one surface is highly polar with all of the hydroxyl groups of the sugar pointing outward from it, whereas the other surface is completely devoid of hydroxyl groups and hence is nonpolar and hydrophobic. For dextran both surfaces contain hydroxyl groups and hence are polar [233]. The sensitivity of the TICT probes to different kinds of sugar surfaces has recently been utilized to develop new biosensors. Hillardi et al. [234] showed that a maltose-binding protein labeled with the TICT probe acrylodan exhibits a 60–180% fluorescence enhancement on addition of maltose, compared to the ligand free form.

Kim et al. studied effect of CD on the emission properties of DMABE in aqueous solutions and reported that α-CD enhances the nonpolar (LE) emission and β-CD enhances the TICT emission [235]. In aqueous solutions of DMABN, addition of α-CD is reported to cause an enhancement of the TICT emission while the LE band remains unaffected [236]. However, Al-Hassan et al. reported that when an aqueous solution of DMABN containing α-CD is kept for a long time (3 months) the LE emission is enhanced [237]. Unpurified DMABN samples contain a well-known impurity that absorbs at wavelengths longer than 340 nm and gives rise to an emission around 360 nm, i.e., in the region very similar to that of the nonpolar emission of DMABN [238]. Al-Hassan et al. did not report the absorption spectra of the 3-months-old DMABN solutions; further more, the excitation spectrum of the old solutions used by them displays significant intensity at wavelengths above 340 nm. Thus, one cannot rule out the possibility of the presence of an impurity in the 3-months-old solution. Matsushita and Hikida [239] studied the effect of α-CD on the emission properties of p-dimethylaminoacetophenone (DMACP) in aqueous solutions. DMACP does not give TICT emission in aqueous solutions. However, on addition of α-CD both the LE and TICT emission of DMACP is enhanced. To explain the larger enhancement of the LE emission they proposed a 1:2 complex in which the DMACP molecule is completely enclosed by two α-CD molecules [239].

The extraordinary sensitivity of the TICT probe TNS has recently been exploited to probe the interaction of CD and surfactants [240]. Study of such interactions is important due to the potential application of CD in targeted drug delivery and particularly for understanding how CD affects cell membrane surfactants. The important issues are whether CD preserves the structure of the membranes and releases the drug encapsulated in its cavity. The cmc of several ionic surfactants (alkyl sulfates, sulfonates, and tetraalkylammonium halides) [241,242] as well as neutral surfactants (Triton X-100, Igepal, etc.) [243] have been reported to increase on addition of CD, while their aggregation numbers remain more or less unchanged. The emission quantum yield of TNS in TX-100 micelles is nearly 10 times that of TNS bound to CD whereas the lifetime of the micelle-bound TNS is about 5 times higher. On addition of TX-100 to an aqueous solution containing CD, the emission intensity of TNS initially exhibits a slight decrease, and beyond a particular concentration of TX-100 the emission intensity and lifetime of TNS increase abruptly. The point of abrupt increase of emission intensity and lifetime of TNS gives the apparent cmc of TX-100 in the presence of CD. The apparent cmc of TX-100 increases significantly on addition of β-CD. In 10 mM β-CD, the apparent cmc of TX-100 is 7.25 ± 0.25 mM. This is 28 ± 1 times the cmc of TX-100 in water [240]. However, in the presence of α-CD at similar concentrations, the cmc of TX-100 remains more or less unaffected [240]. This indicates that TX-100 molecules bind very strongly with the large β-CD and very weakly with the small α-CD. This is in sharp contrast to the behavior of the linear surfactants for which both α-CD and β-CD cause an increase in cmc. The dramatic difference in the interaction of TX-100 with α- and β-CD may be attributed to the difference in size between α- and β-CD. Due to its large size the TX-100 molecule cannot be inserted in the small α-CD cavity and thus addition of α-CD leaves the cmc of TX-100 unchanged. However, TX-100 is easily accommodated in the bigger β-CD cavity and thus is rendered unavailable for the formation of the micelles causing an increase in the cmc. The initial decrease in the emission intensity of TNS at TX-100 concentrations below the apparent cmc is ascribed to the competitive binding of TNS and TX-100 with β-CD. This causes displacement of TNS from the CD cavity by the TX-100 surfactant molecules. The binding constant of TX-100 with β-CD is estimated to be 9400 ± 1300 L $M^{-1}$.

Very recently, Drickamer et al. [244] studied the effect of high pressure on two benzylidene malononitriles, dimethylamino benzylidene malononitrile (DMABMN) and julolidene malononitrile (JDMN) (Fig. 12), in a polymer matrix (Polymethyl Methacrylate). In the excited state, the flexible analog, DMABMN, can twist about the olefinic double bond and the single bond, connecting the dimethylamino group with the aromatic ring. For the rigid analog JDMN, the latter is prevented and the molecule can twist only about the double bond. Dra-

(a)                                              (b)

**Figure 12**  Structure of (a) DMABNMN and (b) JDMN.

matic differences are observed in the pressure dependence of emission intensity of the two derivatives. For JDMN, up to about 20 kbar the emission intensity increases about 6 times and above 20 kbar the emission intensity increases very slightly. However, the flexible analog, DMABN, exhibits a nearly 50 times emission enhancement at high pressure. This shows that torsion about the N-ring bond plays a more effective role in the nonradiative decay of these molecules.

## V. CONCLUSION

The dynamics of photophysical processes are strongly influenced by a wide range of organized assemblies. As a result, the dynamics of the photophysical processes are sensitive probes for these organized assemblies. The new results demonstrate that the water molecules in different organized and biological assemblies are significantly slower than ordinary water molecules. This result is likely to provide new impetus to study the behavior of water molecules in confined and organized environments. It will be important to identify the vibrational modes responsible for the slow solvation dynamics observed in many organized assemblies. It will be most interesting if new collective modes are detected in the organized assemblies. It is also demonstrated that the local polarity, pH, and viscosity in an organized media are markedly different from those in the ordinary solutions and often vary quite drastically over a small distance. The marked difference in the chemical dynamics in organized environments from those in ordinary solutions and their implications in different natural and biological phenomena, highlighted in this chapter, suggest a promising future of this subject. Ultrafast lasers and the various sophisticated theoretical models that revolutionized chemical dynamics in ordinary liquids in the 1980s have just begun to be applied to study organized assemblies and still the field is largely unexplored. Unfortunately, due to the inherent complexity of the organized assemblies, many of the results discussed in this chapter are explained so far in terms of somewhat qualitative or semiquantitative models. In summary, recent results have enabled us to formulate several interesting questions. However, quantitative answers to most of these questions are still lacking. One naturally expects a lot of activity in this area in the coming years. The new developments in this area, apart from enhancing our understanding of many complex phenomena, will bring us closer to nature and life. The enhanced understanding about the structure and function of organized assemblies should enable us to mimic the natural processes more closely employing various organized assemblies.

## ACKNOWLEDGMENTS

It is a pleasure to thank my students N. Sarkar, K. Das, S. Das, A. Datta, D. Mandal, and S. K. Pal. Special thanks are due to Professor B. Bagchi for many stimulating and helpful discussions. Thanks are also due to Department of Science and Technology and Council of Scientific and Research, Government of India for generous research grants.

## REFERENCES

1.  (a) Breslow, R. *Chem. Rev.* **1998**, *98*, 1997. (b) Breslow, R. *Acc. Chem. Res.* **1991**, *24*, 159. (c) Blokzijl, W.; Engberts, J.B.F.N. *Angew. Chem. Int. Ed. Engl.* **1993**, *32*, 1545. (d) Pindur, U.; Lutz, G.; Oho, C. *Chem. Rev.* **1993**, *93*, 741.
2.  Ramamurthy, V., Ed. *Photochemistry in Organized and Constrained Media*, VCH, New York, **1991**.
3.  Turro, N.J.; Garcia-Garibay, M. in ref. 2, p.1.
4.  (a) Bohne, C.; Redmond, R.W.; Scianao J.C. in ref. 3, p. 79. (b) Kalyansundaram, K. in ref. 3, p. 39.
5.  (a) Special issue, eds. D'Souza, V.T.; Lipkowitcz, K.B. *Chem. Rev.* **1998**, *98*, 1741–2076. (b) Bortulos, P.; Monti, S. *Adv. Photochem.* **1995**, *21*, 1. (c) Konnors, K.A. *Chem. Rev.* **1997**, *97*, 1325. (d) De Silva, A.P. *Chem. Rev.* **1997**, *97*, 1515. (e) Wenz, G. *Angew. Chem. Int. Ed. Engl.* **1994**, *33*, 741. (f) Warner, I.M. *Anal. Chem.* **1998**, *70*, 477R.
6.  (a) Eisenthal, K.B. *Chem. Rev.* **1996**, *96*, 1343. (b) Zhu, S.-B.; Singh, S.; Robinson, G.W. *Adv. Chem. Phys.* **1994**, *85*, 627. (c) Bittner, E.R.; Rossky, P.J. *J. Chem. Phys.* **1995**, *103*, 8130. (d) Benjamin, I. *Acc. Chem. Res.* **1995**, *28*, 233; *Annu. Rev. Phys. Chem.* **1997**, *48*, 401. (e) Sokhan, V.P.; Tildesky, P.J. *Mol. Phys.* **1997**, *92*, 625.
7.  (a) Fleming, G.R.; Cho, M. *Annu. Rev. Phys. Chem.* **1996**, *47*, 109. (b) Stratt, R.M. *Acc. Chem. Res.* **1995**, *28*, 201. (c) Rossky, P.J.; Simon, J.D. *Nature* **1994**, *369*, 471. (d) Hynes, J.T. In *Ultrafast Spectroscopy*; Simon, J.D., Ed. Kluwer, Dordrecht, **1994**.
8.  (a) Maroncelli, M. *J. Mol. Liq.* **1993**, *57*, 1. (b) Horng, M.-L.; Gardecki J.A.; Maroncelli, M. *J. Phys. Chem. A* **1997**, *101*, 1030.
9.  (a) Bagchi, B.; Biswas, R. *Acc. Chem. Res.* **1998**, *31*, 181. (b) Bagchi, B.; Chandra, A. *Adv. Chem. Phys.* **1991**, *80*, 1.
10. (a) Jarjeba, W.; Walker, G.C.; Johnson, A.E.; Kahlow, N.A.; Barbara P.F. *J. Phys. Chem.* **1988**, *92*, 7039. (b) Jimenez, R.; Fleming, G.R.; Kumar, P.V.; Maroncelli, M. *Nature* **1994**, *370*, 263. (c) Silva, C.; Walhout, P.K.; Yokoyama, K.; Barbara, P.F. *Phys. Rev. Lett.* **1998**, *80*, 1086.
11. (a) Shirota, H.; Pal, H.; Tominaga, K.; Yoshihara, K. *J. Phys. Chem.* **1996**, *100*, 14575. (b) Pal, H.; Nagasawa, Y.; Tominaga, K.; Yoshihara, K. *J. Phys. Chem.* **1996**, *100*, 11964.
12. (a) Nandi, N.; Roy, S.; Bagchi, B. *J. Chem. Phys.* **1995**, *102*, 1390. (b) Schwartz, B.J.; Rossky, P.J. *J. Chem. Phys.* **1996**, *105*, 6997.
13. (a) Benigno, A.J.; Ahmed, E.; Berg, M. *J. Chem. Phys.* **1996**, *104*, 7382. (b) Fourkas, J.T.; Benigno, A.J.; Berg, M. *J. Chem. Phys.* **1993**, *99*, 8552.
14. (a) Castner, E.W., Jr., Chang, Y.J.; Chu, Y.C.; Walraten, G.E. *J. Chem. Phys.* **1995**, *102*, 653. (c) Neelakanandan, M.; Pant, D.; Quitevis, E.L. *Chem. Phys. Lett.* **1997**, *265*, 283.

15. Zolotov, B.; Gan, A.; Fainberg, B.D.; Huppert, D. *Chem. Phys. Lett.* **1997**, *265*, 418.

16. Kaatze, U. *Chem. Phys. Lett.* **1993**, *203*, 1.

17. (a) Zhang, L.; Davies, H.T.; Kroll, D.M.; White, H.S. *J. Phys. Chem.* **1995**, *99*, 2878. (b) Liu, Q.; Brady, J.W. *J. Phys. Chem. B* **1997**, *101*, 1317.

18. (a) Pethig, R. In *Protein–Solvent Interactions*; Gregory, R.B., ed. Marcel Dekker, New York, **1995**, p.265. (b) Mashimo, S.; Kuwabara, S.; Yagihara, S.; Higasi, K. *J. Phys. Chem.* **1987**, *91*, 6337. (c) Urry, D.W.; Peng, S.; Xu, J.; McPherson, D.T. *J. Am. Chem. Soc.* **1997**, *119*, 1161. (d) Urry, D.W. *Angew. Chem. Int. Ed. Engl.* **1993**, *32*, 814.

19. (a) Fukuzaki, M.; Miura, N.; Sinyashiki, N.; Kunita, D.; Shioya, S.; Haida, M.; Mashimo, S. *J. Phys. Chem.* **1995**, *99*, 431. (b) Belton, P.S. *J. Phys. Chem.* **1995**, *99*, 17061. (c) Denisov, V.P.; Peters, J.; Horlein, H.D.; Halle, B. *Nature Struct. Biol.* **1996**, *3*, 505. (d) Denisov, V.P.; Halle, B. *Faraday Disc.* **1996**, *103*, 227.

20. (a) Fischer, S.; Verma, C.S.; Hubbard, R. *J. Phys. Chem.* **1998**, *102 B*, 1797. (b) Fischer, S.; Karplus, M. *Chem. Phys. Lett.* **1992**, *194*, 252.

21. (a) Nandi, N.; Bagchi, B. *J. Phys. Chem. B* **1997**, *101*, 10954. (b) Nandi, N.; Bagchi, B. *J. Phys. Chem. A* **1998**, *102*, 8217.

22. (a) Waldeck, D.H. *Chem. Rev.* **1991**, *91*, 415. (b) Saltiel, J. *J. Am. Chem. Soc.* **1969**, *89*, 1036.

23. (a) Lin, S.H.; Groesbeck, M.; van der Hoef, I.; Verdegen, P.; Lugtenberg, J.; Mathies, R.A. *J. Phys. Chem. B* **1998**, *102*, 2787. (b) La Penna, G.; Buda., F.; Bifone, A.; de Groot, H.J.M. *Chem. Phys. Lett.* **1998**, *294*, 447. (c) Schoenlein, R.W.; Peteanu, L.A.; Mathies, R.A.; Shank, C.V. *Science*, **1991**, *254*, 412.

24. (a) Takeuchi, S.; Tahara, T. *J. Phys. Chem. A* **1997**, *101*, 3052. (b) Yamaguchi, S.; Hamaguchi, H. *Chem. Phys. Lett.* **1998**, *287*, 694.

25. (a) Todd, D.; Fleming, G.R.; Lean, J.M. *J. Chem. Phys.* **1992** *97*, 8915. (b) Yamaguchi, S.; Hamaguchi, H. *J. Chem. Phys.* **1998**, *109*, 1397.

26. (a) Kramers, H.A. *Physica (The Hague)*, **1940**, *7*, 284. (b) Hanggi, P.; Talkner, T.; Borkovec, M. *Rev. Mod. Phys.* **1990**, *62*, 251.

27. (a) Bagchi, B.; Oxtoby, D.W. *J. Chem. Phys.* **1983**, *78*, 2735. (b) Rothenberger, G.; Negus, D.K.; Hochstrasser, R.M. *J. Chem. Phys.* **1983**, *79*, 215. (c) Courtney, S.H.; Kim, S.K.; Caninica, S.; Fleming, G.R. *J. Chem. Soc. Faraday Trans. II* **1986**, *82*, 2065. (d) Flom, S.R.; Nagarajan, V.; Barbara, P.F. *J. Phys. Chem.* **1986**, *90*, 2085. (e) Hynes, J.T. *J. Stat. Phys.* **1986**, *42*, 149.

28. (a) Miller, D.P.; Eisenthal, K.B. *J. Chem. Phys.* **1985**, *83*, 5076. (b) Bowman, R.M.; Miller, D.P.; Eisenthal, K.B. *Chem. Phys. Lett.* **1989**, *155*, 99.

29. Zwanzig, R.; Harrison, A.K. *J. Chem. Phys.* **1985**, *83*, 5861.

30. (a) Sun, Y.-P.; Saltiel, J. *J. Phys. Chem.* **1989**, *93*, 8310. (b) Spernol, A.; Wirtz, K. Z. *Naturjarsch.* **1953**, *8*, 522.

31. Hicks, J.M.; Vandersall, M.T.; Sitzman, E.V; Eisenthal, K.B. *Chem. Phys. Lett.* **1987**, *135*, 413.

32. (a) Velsko, S.P.; Fleming, G.R. *Chem. Phys.* **1982**, *65*, 59. (b) Jaraudis, J. *J. Photochem.* **1980**, *13*, 35.

33. (a) Weller, A. *Prog. React Kinetics* **1961**, *1*, 188. (b) Ireland, J.F.; Wyatt, P.A. *Prog. Phys. Org. Chem.* **1976**, *12*, 131. (c) Shizuka, H. *Acc. Chem. Res.* **1985**, *18*, 141.

34. (a) Douhal, A.; Lahmani, F.; Zewail, A.H. *Chem. Phys.* **1996**, *207*, 477. (b) Kim, S.K.; Li, S.; Bernstein, E.R. *J. Chem. Phys.* **1991**, *81*, 3119.
35. Robinson, G.W.; Thistlewaite, P.S.; Lee, J. *J. Phys. Chem.* **1986**, *90*, 4224.
36. Pines, E.; Fleming, G.R. *J. Phys. Chem.* **1991**, *95*, 10448.
37. Webb, S.P.; Yeh, S.W.; Philips, L.A.; Tolbert, M.A.; Clark, J.H. *J. Am. Chem. Soc.* **1984**, *106*, 7286.
38. Lee, J.; Robinson, G.W.; Webb, S.P.; Philips, L.A.; Clark, J.H. *J. Am. Chem. Soc.* **1986**, *108*, 6538.
39. Lee, J.; Robinson, G.W.; Webb, S.P; Philips, L.A.; Clark, J.H. *J. Am. Chem. Soc.* **1990**, *112*, 1353.
40. Shizuka, H.; Ogiwara, T.; Narita, A.; Sumitani, M.; Yoshihara, K. *J. Phys. Chem.* **1986**, *90*, 670.
41. Krishnan, R.; Lee, J.; Robinson, G.W. *J. Phys. Chem.* **1990**, *94*, 63.
42. Pines, E.; Huppert, D.; Agmon *J. Chem. Phys.* **1988**, *88*, 5620.
43. Htun, M.T.; Suwaiyan, A.; Baig, A.; Klein, U.K.A. *J. Phys. Chem. A* **1998**, *102*, 8230.
44. (a) Robinson, B.H.; Drobny, G.P. *Annu. Rev. Biophys. Biomol. Struct.* **1995**, *24*, 523. (b) Kumar, C.V. In *Photochemistry in Organized and Constrained Media*; V. Ramamurthy, Ed. VCH, New York, **1991**, p. 783.
45. (a) Millar, D.P.; Robbins, R.J.; Zewail, A.H. *J. Chem. Phys.* **1982**, *76*, 2080. (b) Barklay, M.D.; Zimm, B.H. *J. Chem. Phys.* **1979**, *70*, 2991. (c) Bhattacharyya, S.; Mandal, S. *Biochim. Biophys. Acta* **1997**, *1323*, 29.
46. (a) Ohmstead III, J.; Kearns D.R. *Biochemistry* **1977**, *16*, 3647. (b) Angerer, L.M.; Georghiou, S.; Moudrianakis, E.N. *Biochemistry* **1974**, *13*, 1075. (c) Le Pecq, J.B.; Paoletti, C. *J. Mol. Biol.* **1971**, *59*, 43.
47. (a) Kamlet, M.J.; Abbouid, J.L.M.; Taft, R.W. *Prog. Phys. Org. Chem.* **1981**, *13*, 485. (b) Buncel, E.; Rajagopal, S. *Acc. Chem. Res.* **1990**, *23*, 226. (c) Reichardt, C. *Chem. Rev.* **1994**, *94*, 2319.
48. Pal, S.K.; Mandal, D.; Bhattacharyya, K. *J. Phys. Chem. B* **1998**, *102*, 11017.
49. (a) Barbara, P.F.; Walsh, P.K.; Brus, L.E. *J. Phys. Chem.* **1989**, *93*, 29. (b) Sengupta, P.K.; Kasha, M. *Chem. Phys. Lett.* **1979**, *68*, 382.
50. Elsaesser, T. In *Femtosecond Chemistry.*; Manz, J.; Woste, L.; Eds. VCH, Weinheim, **1994**, p. 563.
51. (a) Douhal, A. *Science* **1997**, *276*, 221. (b) Benderskii, V.A.; Goldanskii, V.I.; Makarov, D.E. *Phys. Rep.* **1993**, *233*, 195. (c) Ormson, S.W.; Brown, R.G. *Prog. React. Kinet.* **1994**, *18*, 45. (d) Formosinho, S.J.; Arnaut, L. *J. Photochem. Photobiol. A* **1993**, *75*, 21.
52. Klopffer, W.; Naundorf, G. *J. Lumin.* **1974**, *8*, 457.
53. (a) Helmbrook, L.; Kinny, J.F.; Kohler, B.E.; Scott, G.W. *J. Phys. Chem.* **1983**, *87*, 280. (b) Felker, P.M.; Lambert, W.R.; Zewail, A.H. *J. Chem. Phys.* **1982**, *87*, 1603.
54. Herek, J.L.; Pedersen, S.; Banares, L.; Zewail, A.H. *J. Chem. Phys.* **1992**, *97*, 9046.
55. Peteanu, L.A.; Mathies, R.A. *J. Phys. Chem.* **1992**, *96*, 6910.
56. (a) Schwartz, B.J.; Peteanu, L.A.; Harris, C.B. *J. Phys. Chem.* **1992**, *96*, 3591. (b) Sekikawa, T.; Kobayashi, T. *J. Phys. Chem. B* **1997**, *101*, 10645. (c) Mitra, S.; Tamai, N. *Chem. Phys. Lett.* **1998**, *282*, 391. (d) Takeda, J.; Chung, D.D.; Zhou,

J.; Nelson, K.A. *Chem. Phys. Lett.* **1998**, *290*, 341. (e) Pfeiffer, M.;Lau, A.;Lenz, K.; Elsaesser, T. *Chem. Phys. Lett.* **1997**, *268*, 258.

57. (a) Das, K.; Dertz, E.; Paterson, J.; Zhang, W.; Kraus, G.A.; Petrich, J.W. *J. Phys. Chem. A* **1998**, *102*, 1479. (b) Kraus, G.A.; Jhang, W.; Fehr, M.J.; Petrich, J.W. *Chem. Rev.* **1996**, *96*, 523.

58. (a) Das, K.; Sarkar, N.; Ghosh, A.K.; Majumdar, D.; Nath. D.N.; Bhattacharyya, K. *J. Phys. Chem.* **1994**, *98*, 9126. (b) Das, K.; Sarkar, N.; Nath. D.N.; Bhattacharyya, K. *Chem. Phys. Lett.* **1992**, *198*, 443. (c) Sinha H.K.; Dogra, S.K. *Chem. Phys.* **1986**, *102*, 337.

59. Tobita, S.; Yamamato, M.; Kurahayashi, N.; Tsukagoshi, R.; Nakamura, Y.; Shizuka, H. *J. Phys. Chem. A* **1998**, *102*, 5206.

60. Nagaoka, S.; Shinde, Y.; Mukai, K.; Nagashima, U. *J. Phys. Chem. A* **1997**, *101*, 293.

61. Catalan, J.; Palomer, J.; de Paz, J.L.G. *J. Phys. Chem. A* **1997**, *101*, 7915.

62. Scheiner, S.; Kar, T.; Cuma, M. *J. Phys. Chem. A* **1997**, *101*, 5901.

63. Guthrie, J.P. *J. Am. Chem. Soc.* **1996**, *118*, 12886.

64. McGurry, P.F.; Jockusdi, S.; Fujiwara, Y.; Kaprinidas, N.A.; Turro, N.J. *J. Phys. Chem. A* **1997**, *101*, 764.

65. Mosquera, M.; Penedo, Rodriguez, M.C.R.; Rodriguez-Prieto, F. *J. Phys. Chem.* **1996**, *100*, 5398.

66. (a) Suzuki, T.; Okeryama, U.; Ichiura, T. *J. Phys. Chem. A* **1997**, *101*, 7047. (b) Ilich, P. *J. Mol. Struct.* **1995**, *354*, 37.

67. Rios, M.A.; Rios, M.C. *J. Phys. Chem. A* **1998**, *102*, 1560.

68. (a) Swiney, T.C.; Kelley, D.F. *J. Chem. Phys.* **1993**, *99*, 211. (b) Hineman, M.F.; Brucker, G.A.; Kelley, D.F.; Bunstein, R. *J. Chem. Phys.* **1992**, *95*, 3341.

69. Taylor, C.A.; El-Bayomi, M.A.; Kasha, M. *Proc. Natl. Acad. Sci USA*, **1969**, *63*, 253.

70. (a) Chen, Y.; Gai, F.; Petrich, J.W. *J. Phys. Chem.* **1994**, *98*, 2203. (b) Chen, Y.; Rich, R.L.; Gai, F.; Petrich, J.W. *J. Phys. Chem.* **1993**, *97*, 1770.

71. Hetherington, W.M.; Micheels, R.H.; Eisenthal, K.B. *Chem. Phys. Lett.* **1979**, *66*, 230.

72. Share, P.; Pereira, M.;Sarisky, M.; Repinec, S.; Hochstrasser, R.M. *J. Lumin.* **1991**, *48/49*, 204.

73. Takeuchi, S.; Tahara, T. *J. Phys. Chem. A* **1998**, *102*, 7740.

74. Chachisvilis, M.; Fiebig, T.; Douhal, A.; Zewail, A.H. *J. Phys. Chem. A* **1998**, *102*, 669.

75. Douhal, A.; Kim, S.K; Zewail, A.H. *Nature* **1995**, *378*, 260.

76. Folmer, D.E.; Poth, L.; Winiewski, E.S.; Castleman, A.W., Jr. *Chem. Phys. Lett.* **1998**, *287*, 1.

77. Mcmorrow, D.; Aartsma, T.J. *Chem. Phys. Lett.* **1986**, *125*, 581.45.

78. Nakajima, A. Hirano, M.; Hasumi, R.;Kaya, K.; Watanabe, H.; Carter, C.C.; Williamson, J.M.; Miller T.A. *J. Phys. Chem. A* **1997**, *101*, 392.

79. Chou, P.T.; Martinez, M.L.; Cooper, W.C.; McMorrow, D.; Collins, S.T.; Kasha, M. *J. Phys. Chem.* **1992**, *96*, 5203.

80. (a) Gordon, M.S. *J. Phys. Chem.* **1996**, *100*, 3974. (b) Moog, R.S.; Maroncelli, M. *J. Phys. Chem.* **1991**, *95*, 10359.

81. (a) Soujanya, T.; Fessenden, R.W.; Samanta, A. *J. Phys. Chem.* **1996**, *100*, 3507.
    (b) Soujanya, T.; Krishna, T.S.R; Samanta, A.*J. Phys. Chem.* **1992**, *96*,8544. (c)
    Yuan, D.; Brown, R.G. *J. Phys. Chem. A* **1997**, *101*, 3461.
82. Harju, T.O.; Huizer, A.H.; Varma, C.A.G.O. *Chem. Phys.* **1995**, *200*, 215.
83. Andrews, P.M.; Beyer, M.B.; Troxler, T.; Topp, M.R. *Chem. Phys. Lett.* **1997**, *271*,
    19.
84. Herbich, J.; Dobkowski, J.; Thummel, R.P.; Hegde, V.; Waluk, J. *J. Phys. Chem.
    A* **1997**, *101*, 5839.
85. Grabowski, Z.R. *Pure Appl. Chem.* **1993**, *65*, 1751.
86. Bhattacharyya, K.; Chowdhury, M. *Chem. Rev.*, **1993**, *93*, 507.
87. (a) Kim, H.-J.; Hynes, J.T. *J. Photochem. Photobiol. A* **1997**, *105*, 337. (b) Fonesca,
    T.; Kim, H.-J.; Hynes, J.T. *J. Photochem. Photobiol. A* **1994**, *82*, 67.
88. Sobolewski, A.L.; Sudholt, W.; Domcke, W. *J. Phys. Chem. A* **1998**, *102*, 2716.
89. Ishida, T.;Fujimura, Y.; Fujiwara, T.; Kajimoto, O. *Chem. Phys. Lett.*, **1998**, *288*,
    433.
90. Hicks, J.M.; Vandersall, M.T.; Babarogic, Z.; Eisenthal, K.B. *Chem. Phys. Lett.*,
    **1985**, *116*, 18.
91. (a) Chang, T.L.; Cheung, H.C. *Chem. Phys. Lett.*, **1990**, *173*, 343. (b) Das, K.;
    Sarkar, N.; Nath, D.; Bhattacharyya, K. *Spectrochim. Acta A* **1992**, *47*, 1701.
92. Chang, Y.J.; Cong, P.; Simon, J.D. *J. Phys. Chem.* **1995**, *99*, 7382.
93. (a) Attwood, D.; Florence, A.T. *Surfactant Systems*. Chapman and Hall, London,
    **1983**. (b) Zana, R. *Surfactant Solution: New Methods of Investigation*. Marcel Dek-
    ker, New York, **1987**. (c) Almgren, M. *Adv. Coll. Int. Sci.* **1992**, *41*, 9. (d) Gehlan,
    M.; DeSchryver, F.C. *Chem. Rev.* **1993**, *93*, 199. (e) Kalyansundaram, K. *Microhet-
    erogeneous Systems*. Academic Press, New York, **1987**.
94. (a) Paradies, H.H. *J. Phys. Chem.* **1980**, *84*, 599. (b) Berr, S.S. *J. Phys. Chem.*
    **1987**, *91*, 4760. (c) Berr, S.S.; Coleman, M.J.; Jones, R.R.M.; Johnson, J.S. *J. Phys.
    Chem.* **1986**, *90*, 6492. (d) Berr, S.S.; Caponetti, E.; Jones, R.R.M.; Johnson, J.S.;
    Magid, L.J. *J. Phys. Chem.* **1986**, *90*, 5766.
95. (a) Phillies, G.D.J.; Yambert, J.E. *Langmuir* **1996**, *12*, 3431. (b) Phillies, G.D.J.;
    Hunt, R.H.; Strang, K.; Sushkin, N. *Langmuir* **1995**, *11*, 3408.
96. Magid, L.J. *J. Phys. Chem. B* **1998**, *102*, 4064.
97. Aswal, V.K.; Goyal, P.S.; Thiyagarajan, P.*J. Phys. Chem. B* **1998**, *102*, 246
98. De, S.; Aswal, V.K.; Goyal, P.S.; Bhattacharya, S. *J. Phys. Chem. B* **1998**, *102*,
    6152.
99. De, S.; Aswal, V.K.; Goyal, P.S.; Bhattacharya, S. *J. Phys. Chem.* **1996**, *100*,
    11664.
100. Telgmann, T.; Kaatze, U. *J. Phys. Chem. A* **1997**, *101*, 7758 and 7766.
101. (a) Cazabat, A.M. In *Physics of Amphiphiles: Micelles, Vesicles and Microemul-
    sions*, Degiorgio, V.; Corti, M., Eds. North-Holland, Amsterdam, **1985**, p. 7. (b)
    Bourrel, M.; Schechter, R.S. *Microemulsions and Solid Systems: Formation, Sol-
    vency and Physical Properties*. Surfactant Science Series, Vol. 30. Marcel Dekker,
    New York, *1988*.
102. Jean, Y.C.; Ache, J.H. *J. Am. Chem. Soc.* **1978**, *100*, 984, 6320.
103. (a) Eastoe, J. *Langmuir* **1992**, *8*, 1503. (b) Eastoe, J.; Young, W.K.; Robinson,
    B.H.J. *Chem. Soc. Faraday Trans.* **1990**, *86*, 2883.

104. Zulauf, M.; Eicke, H.-F. *J. Phys. Chem.* **1979**, *83*, 480.

105. Cao, Y.N; Chen, H.X.; Diebold, G.J.; Sun, T.; Zimmt, M.B. *J. Phys. Chem. B* **1997**, *101*, 3005.

106. Hausier, H.; Haering, G.; Pande, A.; Luisi, P.L. *J. Phys. Chem.* **1989**, *93*, 7869.

107. Christopher, D.; Yarwood, J.; Belton, P.S.; Hills, B.P. *J. Colloid. Int. Sci.* **1992**, *152*, 465.

108. Amararene, A.; Gindre, M.; Le Huerou, J.-Y.; Nicot, C.; Urbach, W.; Waks, M. *J. Phys. Chem. B* **1997**, *101*, 10751.

109. (a) De, T.; Maitra, A. *Adv. Coll. Int. Sci.* **1995**, *59*, 95. (b) Jain, T.K.; Varshney, M.; Maitra, A. *J. Phys. Chem.* **1989**, *93*, 7409.

110. (a) Moulik, S.P.; Pal, B.K. *Adv. Coll. Int. Sci.* **1998**, *78*, 99. (b) Mukherjee, K.; Mukherjee, D.C.; Moulik, S.P. *Langmuir* **1993**, *9*, 1727.

111. (a) Niemeyer, E.D.; Bright, F.V. *J. Phys. Chem. B* **1998**, *102*, 1474. (b) Heitz, M.P.; Carlier, C.; deGrazia, J.; Harrison, K.L.; Johnston, K.P.; Randolph, T.W.; Bright, F.V. *J. Phys. Chem. B* **1997**, *101*, 6707.

112. Johnston, K.P.; Harrison, K.L.; Clarke, M.J.; Howdla, S.; Heitz, M.P.; Bright, F.V.; Carlier, C.; Randolph, T.N. *Science* **1996**, *271*, 624.

113. Zielinsky, R.G.; Kline, S.R.; Kalev, E.W.; Rosov, N. *Langmuir* **1997**, *13*, 3934.

114. Fulton, J.L.; Pfund, D.M.; McClain, J.B.; Romack, T.J.; Maury, E.E.; Combes, J.R.; Samulski, E.T.; De Simone, J.M.; Capel, M. *Langmuir* **1995**, *11*, 4241.

115. (a) Cassin, G.; Duda, Y.; Holovko, M.; Badiali, J.P.; Pileni, M.P. *J. Chem. Phys.* **1997**, *107*, 2683. (b) Kotlarchyk, M.; Huang, J.S.; Chen, H. *J. Phys. Chem.* **1985**, *89*, 4382.

116. Kalev, E.W.; Billman, J.F.; Fulton, J.L. Smith, R.D. *J. Phys. Chem.* **1991**, *95*, 458.

117. Lipgens, S.; Schubel, D.; Schlicht, L.; Spilgies, J.-H.; Ilgenfritz, G. *Langmuir* **1998**, *14*, 1041.

118. Cassini, G.; Badiali, J.P.; Pileni, M.P. *J. Phys. Chem.* **1995**, *99*, 12941.

119. Quist, P.O.; Halle, B. *J. Chem. Soc. Faraday Trans. I* **1988**, *84*, 1033.

120. Bardez, E.; Vy, N.C.; Zemb, T. *Langmuir* **1995**, *11*, 3374.

121. (a) Capuzzi, S.; Pini, F.; Gambi, C.M.L.; Monduzzi, M.; Baglioni, P. *Langmuir* **1997**, *13*, 6927. (b) Khan, A.; Fonhill, K.; Lindman, B. *J. Coll. Int. Sci.* **1984**, *101*, 193.

122. (a) Lang, J.; Lalem, N.; Zana, R. *J. Phys. Chem.* **1991**, *95*, 9533. (b) Verbeck, A.; Voortmans, G.; Jachers, C.; De Schryver, F.C. *Langmuir* **1989**, *5*, 766.

123. (a) Zhu, D.-M.; Wu, X.; Schelly, Z.A. *Langmuir* **1992**, *8*, 1538, 48. (b) Zhu, D.-M.; Wu, W.; Schelly, Z.A. *J. Phys. Chem.* **1992**, *96*, 2382, 712. (c) Caldaru, H.; Caragheorgheopal, A.; Dimonie, M.; Donescu, D.; Dragutan, I.; Marinescu, N. *J. Phys. Chem.* **1992**, *96*, 7109.

124. (a) Samii, A.A.-Z.; de Savignac, A.; Rico, I.; Lattes, A. *Tetrahedron*, **1985**, *41*, 3683. (b) Riter, R.E.; Kimmel, J.R.; Undiks, E.P.; Levinger, N. *J. Phys. Chem. B* **1997**, *101*, 8292.

125. (a) Perez-Casa, S.; Castillo, R.; Costas, M. *J. Phys. Chem. B* **1997**, *101*, 7043. (b) Compere, A.L.; Griffith, W.L.; Johnson, J.J.; Caponetti, E.; Chillura-Martino, D.; Triolo, R. *J. Phys. Chem. B* **1997**, *101*, 7139.

126. (a) Mittleman, D.M.; Nuss, M.C.; Colvin, V.L. *Chem. Phys. Lett.* **1997**, *275*, 332.

(b) D'Angelo, M.; Fioretto, D.; Onori, G.; Palmieri, L.; Santucci, A. *Phys. Rev. E* **1996**, *54*, 993 and 4620.

127. Bergmeir, M.; Gradzielski, M.; Hoffman, H.; Mortensen, K. *J. Phys. Chem. B* **1998**, *102*, 2837.

128. (a) Cassol, R.; Ge, M.-T.; Ferrarini, A.; Freed, J.H. *J. Phys. Chem. B* **1997**, *101*, 8782. (b) Sung-Suh, M.M.; Kevan, L. *J. Phys. Chem. A* **1997**, *101*, 1414. (c) Jutila, A.; Kinnunen, P.K.J. *J. Phys. Chem. B*, **1997**, *101*, 7635.

129. (a) Pasenkiewicz-Gierula, M.; Takaoka, V.; Miyagawa, H.; Kitamura, K.; Kusumi, A. *J. Phys. Chem. A* **1997**, *101*, 3677. (b) Lopez Cascales, J.; Berendsen, H.J.C.; Dela Tome, J.G. *J. Phys. Chem.* **1996**, *100*, 862. (c) Alper, H.E.; Bassolino-Klimar, D.; Stouch, T.R. *J. Chem. Phys.*, **1993**, *99*, 5547.

130. (a) Nagle, J.F. *Biophys. J.*, **1993**, *64*, 1476. (b) Vanderkooi, G. *Biochemistry*, **1991**, *30*, 10760.

131. (a) Pearson, R.M.; Pascher, I. *Nature*, **1979**, *81*, 499. (b) Borle, F.; Seelig, J. *Biochim. Biophys. Acta*, **1983**, *735*, 131.

132. See for example, New, R.R.C., Ed. *Liposomes: A Practical Approach*. Oxford University Press, Oxford, 1990.

133. (a) Dhami, S.; Philips, D. *J. Photochem. Photobiol. A*, **1996**, *100*, 77. (b) Song, X.; Geiger, C.; Vadas, S.; Perlstein, J.; Whitten, D.G. *J. Photochem. Photobiol. A* **1996**, *102*, 39.

134. Lakowicz, J.R.; Keating-Nakamoto, S. *Biochemistry* **1984**, *23*, 3013.

135. (a) Demchenko, A.P.; Ladokhin, A.S. *Eur. Biophys. J.* **1988**, *15*, 569. (b) Chattopadhyay, A.; Mukherjee, S. *Biochemistry* **1993**, *32*, 3804.

136. de Haas, K.H.; Blom, C.; van den Ende, D.; Devits, M.H.G.; Haveman, B.; Mellema, J. *Langmuir* **1997**, *13*, 6658.

137. Srivastava, A.; Eisenthal, K.B. *Chem. Phys. Lett.* **1998**, *292*, 345.

138. (a) Shalaby, S.; McCormick, C.; Butler, G. Eds. *Water Soluble Polymers. ACS Symposium Series* 467, American Chemical Society, Washington, DC, 1991. (b) Jenkins, R.D.; Delong, L.; Bassett, D.R. In *Hydrophilic Polymer*, Glass, J.E. Eds. *ACS Advances in Chemistry* 248, American Chemical Society, Washington DC, 1996, p. 425.

139. (a) Kumacheva, E.; Rharbi, Y.; Mitchel, R.; Winnik, M.A.; Guo, L.; Tam, K.C. and Jenkins, R.D. *Langmuir* **1997**, *13*, 182. (b) Eckert, A.R.; Martin, T.J.; Webber, S.E. *J. Phys. Chem. A* **1997**, *101*, 1646. (c) Yekta, A.; Xu, B.; Duhamel, J.; Brochard, P.; Adiwidjaja, H.; Winnik, M.A. *Macromolecules* **1995**, *28*, 956.

140. (a) Shiyanishiki, N.; Yagihara, S.; Arita, I.; Mashimo, S.; *J. Phys. Chem. B* **1998**, *102*, 3249. (b) Shiyanishiki, N.; Asaka, N.; Mashimo, S.; Yagihara, S. *J. Chem. Phys.* **1990**, *93*, 760. (c) Menzel, K.; Rupprecht, A.; Kaatze, U. *J. Phys. Chem. B* **1997**, *101*, 1255.

141. Olender, R.; Nitzan, A. *J. Chem. Phys.* **1995**, *102*, 7180.

142. (a) Bromberg, L.; Grossberg, A. Yu.; Suzuki, Y. and Tanaka, T. *J. Chem. Phys.* **1997**, *106*, 2906. (b) Ohmine, I. and Tanaka, T. *J. Chem. Phys.* **1982**, *77*, 8725. (c) Ohmine, I.; Tanaka, T. *Phys. Rev. Lett.* **1978**, *40*, 820. (d) Ohmine, I.; Tanaka, T. *Phys. Rev.* **1978**, *A17*, 8725. (e) Zhang, Y.B.; Zhang, Y.X. *Macromolecules* **1996**, *29*, 2494.

143. (a) Rodbard, D. In *Methods of Protein Separation*. N. Catsimpoolas, Ed. Plenum

Press, New York 1976 Vol. 2, p. 145.; (b) Chrambach, A; Rodbard, D. *Science* **1971**, *172*, 440; (c) Raymond S.; Nakamichi, M. *Anal. Biochem.* **1962**, *3*, 23; (d) White, M.L. *J. Phys. Chem.* **1960**, *664*, *1563*.

144.   Mathur, A.M.; Moorjani, S.K.; Scranton, A.B. *J. Macromol. Sci. Rev. Macromol. Chem. Phys.* **1996**, *C36*, 405.

145.   Janas, V.F.; Rodriguez, F.; Cohen, C. *Macromolecules* **1980**, *13*, 977.

146.   (a) Walderhans, H.; Nystrom, B. *J. Phys. Chem. B* **1997**, *101*, 1524. (b) Uemura, Y.;McNulty, J.; Macdonald P. *Macromolecules* **1995**, *28*, 4150. (c) Inomata, H.; Goto, S.; Ohtake, K.; Saito, S. *Langmuir* **1993**, *8*, 687. (d) Park, T.G.; Hoffman, A.S. *Macromolecules* **1993**, *26*, 5045. (e) Lele, A.K.; Hirve, M.M.; Badiger, M.V.; Mashelkar, R.A. *Macromolecules* **1997**, *30*, 157.

147.   (a) Dickson, R.M.; Norris, D.J.; Tzeng, Y.-L.; Moerner, W.E. *Science* **1996**, *274*, 966. (b) Dickson, R.M.; Cubitt, A.M.; Tsien, R.Y.; Moerner, W.E. *Nature* **1997**, *388*, 355. (c) Ellerby, L.M.; Nishida, C.R.; Nishida, F.; Yamanaka, S.A.; Dunn, B.; Valentine, J.S.; Zink, I. *Science* **1992**, *255*, 1113.

148.   van Beklum H.; Flanigen E.M.; Jansen, J.C., Eds. *Introduction to Zeolite Science and Practice.* Elsevier, Amsterdam, **1991**.

149.   (a) Ramamurthy, V.; Eaton, D.F.; Casper, J.V. *Acc. Chem. Res.* **1992**, *25*, 299. (b) Ramamurthy, V. in ref. 2, p. 303.

150.   (a) Robbins, R.; Ramamurthy, V. *J. Chem. Soc. Chem. Commun.* **1997**, 1071. (b) Leibovitch, M.; Olovsson, G.; Sundrababu, G.; Ramamurthy, V.; Scheffer, J.R.; Trottler, J. *J. Am. Chem. Soc.* **1996**, *118*, 1219. (c) Branalean, L.; Brousmiche, D.; Jayatirtha Rao, V.; Johnston, L.J.; Ramamurthy, V. *J. Am. Chem. Soc.* **1998**, *120*, 4926. (d) Kao, H.-M.; Grey, C.P.; Pitchumani, K.; Lashminarasimhan, P.H.; Ramamurthy, V. *J. Phys. Chem. A* **1998**, *102*, 5627.

151.   (a) Lachet, V.; Boutin, A.; Tavitian, B.; Fuchs, A.H. *J. Phys. Chem. B* **1998**, *102*, 9224. (b) Demontis, P.; Suffriti, G.B. *Chem. Rev.* **1997**, *97*, 2845. (c) Higgins, F.M.; Watson, G.W.; Parker, S.C. *J. Phys. Chem. B* **1997**, *101*, 9964.

152.   (a) Smit, B.; Maesen, T.L.M. *Nature* **1995**, *374*, 42. (b) van Tassel, R.P.; Davis, H.T.; McCornick, A.V. *J. Phys. Chem.* **1993**, *98*, 8919.

153.   (a) Campana, L.; Selloni, A.; Goursot, A. *J. Phys. Chem. B* **1997**, *101*, 9932. (b) Tsekov, R.; Smirniotis, Y.S. *J. Phys. Chem. B* **1998**, *102*, 9305.

154.   Farneth, W.E.; Gorte, R.J. *Chem. Rev.* **1995**, *95*, 615.

155.   Jang, S.B.; Kim, U.S.; Kim, Y.; Seff, K. *J. Phys. Chem.* **1994**, *98*, 3796.

156.   (a) Mongensen, N.H.; Shore, E. *Solid State Ionics*, **1995**, *77*, 51. (b) Anderson, P.A.; Armstrong, A.R.; Porch, A.; Edwards, P.P.; Woodball, L.J. *J. Phys. Chem. B* **1997**, *101*, 9892.

157.   Abdoulaye, A.; Chabanis, G.; Giuntini, J.C.; Vanderschcuren, J.; Zanchetta, J.V.; Di Ranzo, F. *J. Phys. Chem. B* **1997**, *101*, 1831.

158.   (a) Vitale, G.; Mellot, C.F.; Bull, M.; Cheetham, A.K. *J. Phys. Chem. B* **1997**, *101*, 4559. (b) Bull, L.M.; Cheetham, A.K.; Powell, B.M.; Ripmeester, J.A.; Ratcliffe, C.I. *J. Am. Chem. Soc.* **1995**, *117*, 4328.

159.   Schrimp, G.; Tavitian, B.; Espinat, D. *J. Phys. Chem.* **1995**, *99*, 10932.

160.   Toptygin, D.; Svodva, J.; Konopasek, I.; Brand L. *J. Chem. Phys.* **1992**, *96*, 7919.

161.   (a) Ware, W.R. in ref. 2, p. 563. (b) James, D.R.; Ware, W.R. *Chem. Phys. Lett.* **1985**, *120*, 485.

162. (a) Vajda, S.; Jimenez, R.; Rosenthal, S.J.; Fidler, V.; Fleming, G.R.; Castner, E.W., Jr. *J. Chem. Soc. Faraday Trans.* **1995**, *91*, 867. (b) Nag, A.; Chakrabarty, T.; Bhattacharyya, K. *J. Phys. Chem.* **1990**, *94*, 4203. (c) Bergmark, W.R.; Davies, A.; York, C.; Jones, G. II *J. Phys. Chem.* **1990**, *94*, 5020.

163. Nandi, N.; Bagchi, B. *J. Phys. Chem.* **1996**, *100*, 13914.

164. Jones, G. II; Jackson, W.P.; Choi, C.Y. *J. Phys. Chem.* **1985**, *89*, 294.

165. (a) Sarkar, N.; Das, K.; Das, S.; Datta, A.; Bhattacharyya, K. *J. Phys. Chem.* **1996**, *100*, 10523. (b) Lundgren, J.S.; Heitz, M.P.; Bright, F.V. *Anal. Chem.* **1995**, *67*, 3775.

166. Das, S.; Datta A.; Bhattacharyya, K. *J. Phys. Chem. A* **1997**, *101*, 3299.

167. (a) Bart, E.; Melstein, A.; Huppert, D. *J. Phys. Chem.* **1995**, *99*, 9253. (b) Neria, E.; Nitzan, A. *J. Chem. Phys.*, **1994**, *100*, 3855. (c) Chandra, A.; Wei, D.; Pattey, G.N. *J. Chem. Phys.* **1993**, *98*, 4959.

168. Mandal, D.; Datta, A.; Pal, S.K.; Bhattacharyya, K. *J. Phys. Chem. B* **1998**, *102*, 9070.

169. Willard, D.M.; Riter, R.E.; Levinger, N.E. *J. Am. Chem. Soc.* **1998**, *120*, 4151.

170. Riter, R.E.; Undiks, E.P.; Levinger, N.E. *J. Am. Chem. Soc.* **1998**, *120*, 6062.

171. Riter, R.E.; Willard, D.M.; Levinger, N.E. *J. Phys. Chem. B* **1998**, *102*, 2705.

172. Riter, R.E.; Undiks, E.P.; Kimmel, J.R.; Pant, D.D.; Levinger, *J. Phys. Chem. B* **1998**, *102*, 7931.

173. Shirota, H.; Horie, K. *J. Phys. Chem. B* **1999**, *103*, 1437.

174. Castner, E.W. Jr.; Fleming, G.R.; Bagchi, B.; Maroncelli, M. *J. Chem. Phys*, **1988**, *89*, 3519.

175. Sarkar, N.; Datta, A.; Das, S.; Bhattacharyya, K. *J. Phys. Chem.* **1996**, *100*, 15483.

176. Datta, A.; Mandal, D.; Pal, S.K.; Bhattacharyya, K. *J. Mol. Liq.* **1998**, *77*, 121.

177. Datta, A.; Pal, S.K.; Mandal, D.; Bhattacharyya, K. *J. Phys. Chem. B* **1998**, *102*, 6114.

178. Jordan, J.D.; Dunbar, R.A.; Bright, F.V. *Anal. Chem.* **1995**, *67*, 2436.

179. Datta, A.; Das, S.; Mandal, D.; Pal, S.K.; Bhatacharyya, K. *Langmuir* **1997**, *13*, 6922.

180. Claudia-Marchi, M.; Bilmes, S.A.; Negri, R. *Langmuir* **1997**, *13*, 3655.

181. Hsu, T.-P.; Ma, D.S.; Cohen, C. *Polymer* **1983**, *24*, 1273.

182. Netz, P.A.; Dorfmuller, T. *J. Phys. Chem. B* **1998**, *102*, 4875.

183. Argaman, R.; Huppert, D. *J. Phys. Chem. A* **1998**, *102*, 6215.

184. (a) Saielli, G.; Polomeno, A.; Nordio, P.L.; Bertolini, P.; Ricci, M.; Righini, R. *J. Chem. Soc. Faraday* **1998**, *94*, 121. (b) Ferrante, C. Rau, J.; Deeg, F.W.; Brauchle, C. *J. Lumin.* **1998**, *76* & *77*, 64.

185. Das, K.; Sarkar, N.; Das, S.; Datta, A.; Bhattacharyya, K. *Chem. Phys. Lett.* **1996** *249*, 323.

186. Biswas, R.; Bagchi, B. *J. Phys. Chem.* **1996**, *100*, 4261.

187. Pant, D.; Levinger, N.E. *Chem. Phys. Lett.* **1998**, *292*, 200.

188. (a) McGregor, R.B.; Weber, G. *Nature*, **1986**, *319*, 70. (b) DeToma, R.P.; Easter, T.H.; Brand, L. *J. Am. Chem. Soc.* **1976**, *98*, 5001. (c) Lakowicz, R. In *Principles of Fluorescence Spectroscopy*. Plenum Press, New York, **1983**, Chap. 8.

189. (a) Pierce, D.W.; Boxer, S.G. *J. Phys. Chem.* **1992**, *96*, 5560. (b) Baskhin, J.S.; McLendon, G.; Mukamel, S.; Marohn, J. *J. Phys. Chem.* **1990**, *94*, 4757.

190. Joo, T.; Jai, Y.; Yu, J.-Y.; Jonas, D.M.; Fleming, G.R. *J. Phys. Chem.* **1996**, *100*, 2399.
191. Homoelle, B.J.; Edington, M.D.; Diffey, W.M.; Beck, W.F. *J. Phys. Chem. B* **1998**, *102*, 3044.
192. Brauns, E.B.; Murphy, C.J.; Berg, M. *J. Am. Chem. Soc.*, **1998**, *120*, 2449.
193. (a) Young, M.A.; Jayaram, B.; Beveridge, D.L. *J. Phys. Chem. B* **1998**, *102*, 7666. (b) Mazur, A.K. *J. Am. Chem. Soc.* **1998**, *120*, 10298.
194. Duveneck, G.L.; Sitzmann, E.V.; Eisenthal, K.B.; Turro, N.J. *J. Phys. Chem.* **1989**, *93*, 7166.
195. Ramamurthy, V.; Casper, J.V.; Corbin, D.R. *Tetrahedron Lett.* **1990**, *31*, 1097.
196. Holmes, A.S.; Birch, D.J.S.; Sanderson, A.; Aloisi, G.G. *Chem. Phys. Lett.* **1997**, *266*, 309.
197. (a) Sitzmann, E.V.; Eisenthal, K.B. *J. Phys. Chem.* **1988**, *92*, 4579. (b) Shi, X.; Borguet, E.; Tarnovsky, A.N.; Eisenthal, K.B. *Chem. Phys.* **1996**, *205*, 167.
198. Datta, A.; Pal, S.K.; Mandal, D.; Bhattacharyya, K. *Chem. Phys. Lett.* **1997**, *278*, 77.
199. (a) Pal, S.K.; Datta, A.; Mandal, D.; Bhattacharyya, K. *Chem. Phys. Lett.* **1998**, *288*, 793. (b) Greiser, F.; Lay, M.; Thistlewaite, P.J. *J. Phys. Chem.* **1985**, *89*, 2065.
200. Mandal, D.; Pal, S.K.; Datta, A.; Bhattacharyya, K. *J. Res. Chem. Intermed* (in press).
201. Baptista, M.S.; Indig, J.L. *J. Phys. Chem. B* **1998**, *102*, 4678.
202. Tamai, N.; Ishikawa, M.; Kitamura, N.; Masuhara, H. *Chem. Phys. Lett.* **1991**, *184*, 398.
203. (a) Mukerjee, P.; Banerjee, K. *J. Phys. Chem.* **1964**, *68*, 3567. (b) Hartley, G.S.; Roe, J.W. *Trans. Faraday Soc.* **1940**, *36*, 101.
204. (a) Zhao, X.; Ong, S.; Eisenthal, K.B. *Chem. Phys. Lett.* **1993**, *202*, 513. (b) Castro, A.; Bhattacharyya, K.; Eisenthal, K.B. *J. Chem. Phys.* **1991**, *95*, 1310. (c) An, S.W.; Thomas, R.K. *Langmuir*, **1997**, *13*, 6881.
205. (a) Bashford, D.; Karplus, M. *Biochemistry*, **1990**, *29*, 10219. (b) Sham, Y.Y.; Chu, Z.T.; Warshel, A. *J. Phys. Chem. B* **1997**, *101*, 4458.
206. (a) Menger, F.M.; Saito, G. *J. Am. Chem. Soc.* **1978**, *100*, 4376. (b) Oldfield, C.; Robinson, B.H.; Freedman, R.B. *J. Chem. Soc. Faraday Trans.* **1990**, *86*, 833.
207. Okazaki, M.; Toriyama, K. *J. Phys. Chem.* **1989**, *93*, 5027.
208. Niemeyer, E.D.; Bright, F.V. *J. Phys. Chem. B* **1998**, *102*, 1474.
209. Ramamurthy, V.; Eaton, D.F. *Acc. Chem. Res.* **1988**, *21*, 300.
210. (a) Politi, M.J.; Brandt, O.; Fendler, J.H. *J. Phys. Chem.* **1985**, *89*, 2345. (b) Escabi-Perez J.R.; Fendler, J.H. *J. Am. Chem. Soc.* **1978**, *100*, 2234.
211. Hansen, J.E.; Pines, E.; Fleming, G.R. *J. Phys. Chem.* **1992**, *96*, 6904.
212. Mandal, D.; Pal, S.K.; Bhattacharyya, K. *J. Phys. Chem. A* **1998**, *102*, 9710.
213. (a) Sujatha, J.; Mishra, A.K. *Langmuir*, **1998**, *14*, 2256. (b) Il'ichev, Yu. V.; Demy-ashkevich, A.B.; Kuzmin, M.G.; Lemmetiyen, H. *J. Photochem. Photobiol. A* **1993**, *74*, 51.
214. Chachisvilis, M.; Fiebig, T.; Zewail, A.H. *J. Phys. Chem. A* **1998**, *102*, 1657.
215. (a) Stynik, A.; Kasha, M. *Proc. Natl. Acad. Sci. USA* **1994**, *91*, 8627. (b) Stynik, A.; DelValle, J.C. *J. Phys. Chem.* **1995**, *99*, 13028.

216. (a) Roberts, E.L.; Dey, J.; Warner, I.M. *J. Phys. Chem. A* **1997**, *101*, 5296. (b) Roberts, E.L.; Chou, P.-T.; Alexander, T.A.; Agbaria, R.A.; Warner, I.M. *J. Phys. Chem.* **1995**, *99*, 5431.

217. (a) Das, K.; Sarkar, N.; Das, S.; Datta, A.; Bhattacharyya, K. *J. Phys. Chem.* **1995**, *99*, 17711. (b) Das, S.K.; Dogra, S.K. *J. Chem. Soc. Faraday Trans.* **1998**, *94*, 139. (c) Das, K.; Sarkar, N.; Das, S.; Datta, A.; Bhattacharyya, K. *J. Photochem. Photobiol.* **1998**, *99*, 17711.

218. (a) Sarkar, M.; Sengupta, P.K. *Chem. Phys. Lett.* **1991**, *179*, 68. (b) Guha Ray, J.; Sengupta, P.K. *Chem. Phys. Lett.* **1994**, *230*, 75. (b) Guha Ray, J.; Sengupta, P.K. *Spectrochim. Acta* **1997**, *53A*, 905. (d) Guha Ray, J.; Sengupta, P.K. *Chem. Phys. Lett.* **1994**, *230*, 75.

219. Fuji, T.; Kodaira, K.; Kawauchi, O.; Tanaka, N.; Yamashita, H.; Aupo, M. *J. Phys. Chem. B* **1997**, *101*, 10631.

220. Mukerjee, P.; Ray, A. *J. Phys. Chem.* **1966**, *70*, 2144.

221. (a) Wang, H.; Borguet, E.; Eisenthal, K.B. *J. Phys. Chem. A* **1997**, *101*, 713. (b) Wang, H.; Borguet, E.; Eisenthal, K.B. *J. Phys. Chem. B* **1998**, *102*, 4927. (c) Perora, J.M.; Stevens, G.W.; Greiser, F. *Colloids Surf. A* **1995**, *95*, 185.

222. Benjamin, I. *J. Phys. Chem. A* **1998**, *102*, 9500.

223. Cho, C.H.; Chung, M.; Lee, J.; Nguyen, T.; Singh, S.; Vedamuthu, M; Yao, S.; Zhu, S.-B.; Robinson, G.W. *J. Phys. Chem.* **1995**, *99*, 7806.

224. Mandal, D.; Pal, S.K.; Datta, A.; Bhattacharyya, K. *Anal. Sci.* **1998**, *14*, 199.

225. Datta, A.; Mandal, D.; Pal, S.K.; Bhattacharyya, K. *J. Phys. Chem. B* **1997**, *101*, 10221.

226. Karukstis, K.K.; Frazier, A.; Loftus, C.L.; Tuan, A.S. *J. Phys. Chem. B* **1998**, *102*, 8163.

227. Bessho, K.; Uchida, T.; Yamauchi, A.; Shiyoa, T.; Teramae, N. *Chem. Phys. Lett.* **1997**, *264*, 381.

228. Ramamurthy, V.; Sanderson, D.R.; Eaton, D.F. *Photochem. Photobiol.* **1992**, *56*, 251.

229. Sarkar, N.; Das, K.; Nath, D.N.; Bhattacharyya, K. *Langmuir* **1994**, *10*, 326.

230. Kim, Y.; Lee, B.I.; Yoon, M. *Chem. Phys. Lett.* **1998**, *286*, 466.

231. Kim, Y.; Cheon, H.W.; Yoon, M.; Song, N.W.; Kim, D. *Chem. Phys. Lett.* **1997**, *264*, 673.

232. Guillaume, B.C.R.; Yosen, D.; Fendler, J.H. *J. Phys. Chem.* **1991**, *95*, 7489.

233. Das, K.; Sarkar, N.; Das, S.; Bhattacharyya, K.; Balasubramanian, D. *Langmuir* **1995**, *11*, 2410.

234. Hillardi, G.; Zhou, L.Q.; Hibbert, L.; Cass, A.E.G. *Anal. Chem.* **1994**, *66*, 3840.

235. Kim, Y.H.; Cho, D.W.; Yoon, M.; Kim, D. *J. Phys. Chem.* **1996**, *100*, 15670.

236. (a) Nag, A.; Bhattacharyya, K. *Chem. Phys. Lett.* **1988**, *151*, 474. (b) Cox, G.S.; Hauptman, P.; Turro N. *Photochem. Photobiol.* **1984**, *39*, 597.

237. (a) Al-Hassan, K.A.; Klein, U.K.A.; Suwaiyan, A. *Chem. Phys. Lett.* **1993**, *212*, 581. (b) Al-Hassan, K.A., private communication.

238. Nakashima, N.; Mataga, N. *Bull. Chem. Soc. Jpn.* **1973**, *46*, 3016.

239. Matsushita, Y.; Hikida, T. *Chem. Phys. Lett.* **1998**, *290*, 349.

240. Pal, S.K.; Datta, A.; Mandal, D.; Das, S.; Bhattacharyya, K. *J. Chem. Soc. Faraday* **1998**, *94*, 3471.

241. Gaitano, G.-G.; Crespo, A.; Compostizo A; Tardajos, G. *J. Phys. Chem. B* **1997**, *101*, 4413.
242. Jobe, D.J.; Reinsborough, V.C.; Wetmore, S.D. *Langmuir*, **1995**, *11*, 2476.
243. (a) Nelson, G.; Warner, I.M. *Carbohydrate Res.*, **1989**, *192*, 305. (b) Smith, V.K.; Ndou, T.T.; Munoz de la Pena, A.; Warner, I.M. *J. Incl. Phenomena Mol. Recog.* **1991**, *10*, 471.
244. Dreger, Z.A.; White, J.O.; Drickamer, H.G. *Chem. Phys. Lett.* **1998**, *290*, 399.

# Index